D1538446

POLLUTION CRITERIA FOR ESTUARIES

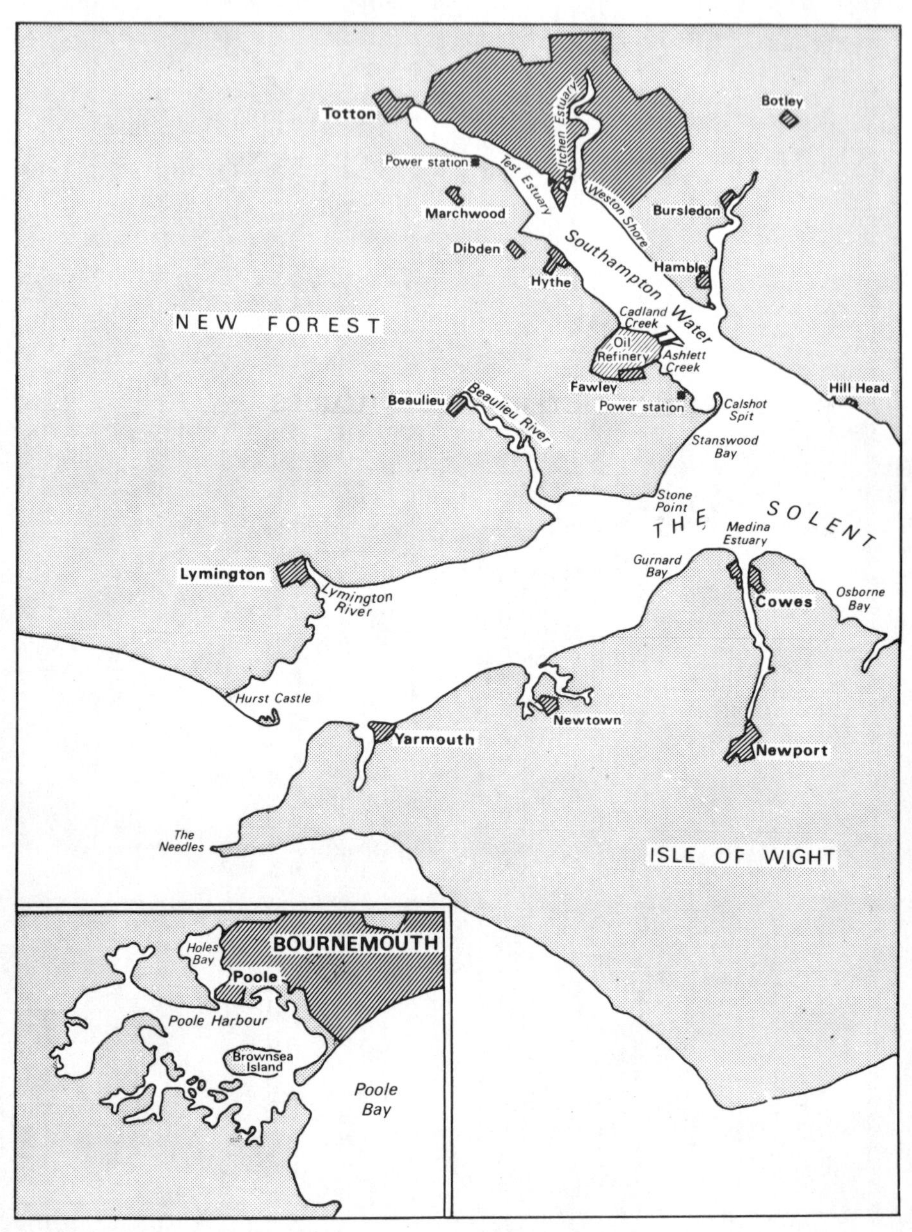

Totton
Itchen Estuary
Botley
Power station
Test Estuary
Weston Shore
Marchwood
Bursledon
Dibden
Southampton Water
Hamble
Hythe
NEW FOREST
Cadland Creek
Oil Refinery
Ashlett Creek
Fawley
Hill Head
Beaulieu
Beaulieu River
Power station
Calshot Spit
Stanswood Bay
Stone Point
THE SOLENT
Medina Estuary
Gurnard Bay
Lymington
Lymington River
Cowes
Osborne Bay
Hurst Castle
Newtown
Yarmouth
Newport
The Needles
ISLE OF WIGHT
Holes Bay
BOURNEMOUTH
Poole
Poole Harbour
Brownsea Island
Poole Bay

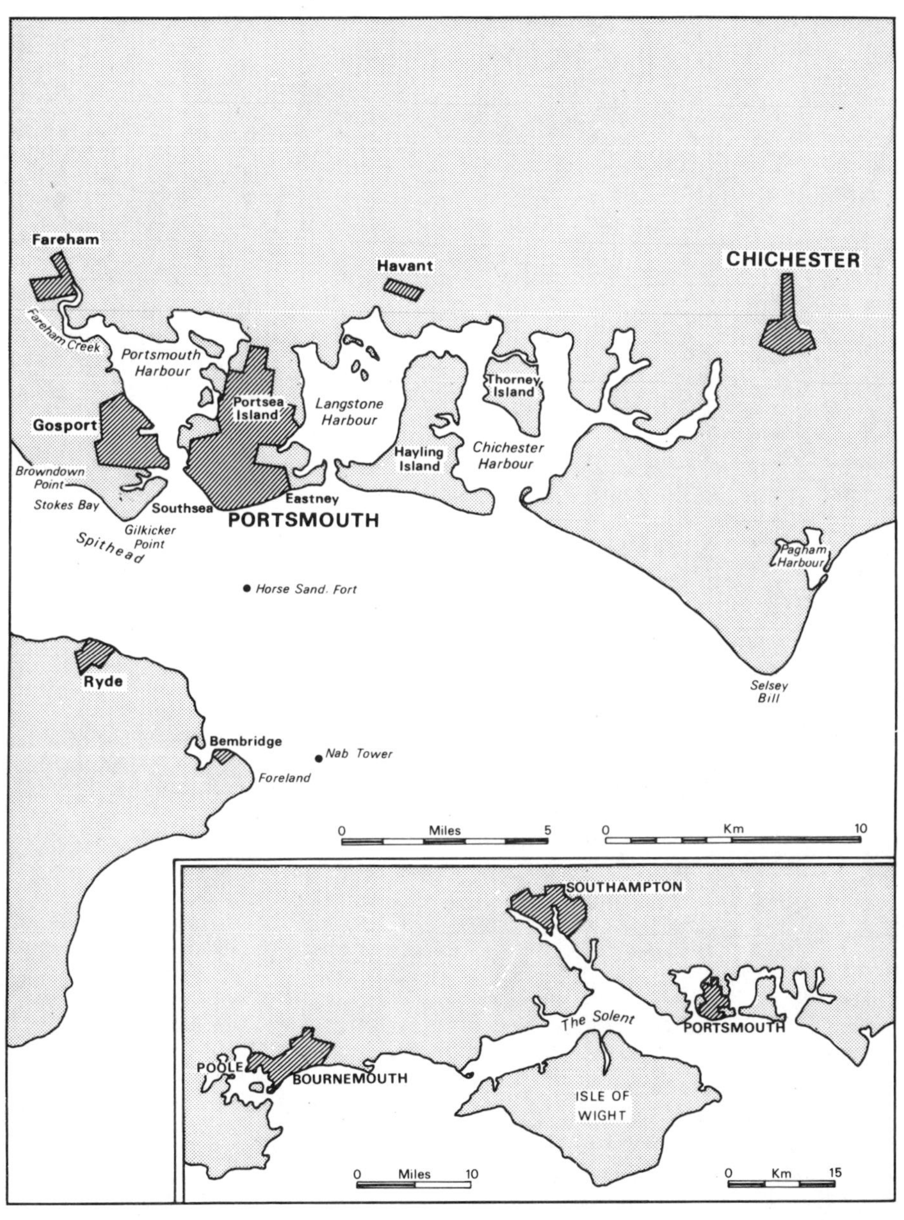

Fareham
Havant
CHICHESTER
Fareham Creek
Portsmouth Harbour
Thorney Island
Portsea Island
Langstone Harbour
Gosport
Hayling Island
Chichester Harbour
Browndown Point
Stokes Bay
Southsea
Eastney
PORTSMOUTH
Gilkicker Point
Spithead
Pagham Harbour
Horse Sand Fort
Ryde
Selsey Bill
Bembridge
Nab Tower
Foreland
0 Miles 5
0 Km 10
SOUTHAMPTON
The Solent
PORTSMOUTH
POOLE
BOURNEMOUTH
ISLE OF WIGHT
0 Miles 10
0 Km 15

POLLUTION CRITERIA FOR ESTUARIES

Proceedings of the Conference held at the University of Southampton, July 1973

edited by:

P. R. Helliwell
Department of Civil Engineering, University of Southampton

J. Bossanyi
Department of Extra-Mural Studies, University of Southampton

A Halsted Press Book

JOHN WILEY & SONS
New York

First Published 1975
Published in the U.S.A.
by Halsted Press, a Division
of John Wiley & Sons, Inc.
New York

Library of Congress Cataloging in Publication Data
Main entry under title:

Pollution criteria for estuaries.

"A Halsted Press book."
Includes index.
1. Estuarine pollution—Congresses. I. Helliwell, P. R., ed. II. Bossanyi, J., ed. III. Southampton, Eng. University. Dept. of Civil Engineering. IV. Southampton, Eng. University. Dept. of Extra-mural Studies.
GC97. P64 628.1'686'168 74-26695
ISBN 0-470-36920-5

PRINTED BY Unwin Brothers Limited
THE GRESHAM PRESS OLD WOKING SURREY ENGLAND

Produced by 'Uneoprint'

A member of the Staples Printing Group

CONFERENCE STATEMENT

PREAMBLE

1. These recommendations have been formulated with the problems of the estuaries and 'Harbours' of the coast of central southern England in mind. Many of them will be equally applicable in other estuarine waters.

2. The pressure of population on the estuaries and tidal inlets is resulting in proliferation of factories, marinas, pollution and traffic congestion both on water and land. These make heavy demands on and are detrimental to the natural environment and may adversely affect the quality of life.

OBJECTIVES AND RESEARCH NEEDS FOR ESTUARY MANAGEMENT

3. We endorse the recommendation of the Royal Commission on Environmental Pollution (Third Report, 1972) that management should aim at the establishment, maintenance and improvement of the diversity of the biota and habitats both in the body of the water and along the shores.

4. There is a pressing need to study and control conservative toxic materials.

5. Floating matter such as identifiable sewage solids, litter, oil, etc. must be prevented from finding its way into estuarine waters.

6. The causes and effects of beach pollution must be studied so that criteria for their quantification can be established.

MONITORING REQUIREMENTS AND APPLICATION OF MATHEMATICAL MODELS

7. Water quality and quantity criteria to achieve the objectives set out in paragraphs 3 to 6 of this statement should be based upon relationships which require to be established between water quality and quantity, and appropriate biological effects.

8. Measurements should be made at selected points of physical, chemical and biological parameters in the main water body, along the margins and in the intertidal zone. Human activity should also be studied since usage of the estuary can be employed as an indication of quality.

9. Where feasible and appropriate, continuous measurement techniques should be used. Otherwise observations should be made over selected tidal cycles.

10. The data collected in the monitoring programme are to provide information upon which standards can be set; to give warning of changes in the estuarine system; to use as a basis for making policy decisions for action to control pollution; and to justify action and expenditure.

11. The form of the monitoring programme may well be shaped *inter alia* by the data requirements of mathematical models of the estuary. Such models may be of two basic types: predictive or quality control.

12. Predictive models require precise definition of objectives. They are of value to controlling authorities by reason of their ability to

(i) forecast the behaviour of polluting loads
(ii) test sensitivity of the system to changes in polluting loads or other conditions
(iii) accommodate economic and social factors
(iv) provide a means of keeping pace with the changing requirement of the controlling authority whilst taking account of the accumulating data

13. The preparation of a Manual of Methods for Study of the Estuarine Environment should be arranged.

RECOMMENDATION ON DIRECT DISCHARGES TO ESTUARIES

14. No discharge of any kind or from any source should be allowed to an estuary without at least efficient maceration of solid particles. This includes discharge from boats. All storm sewage discharges not conveyed by the main foul sewage outfall ought to receive either primary sedimentation or efficient fine screening and should be made below the level of low water of ordinary spring tides wherever possible.

15. Storm sewage overflows require further special study, both of pollution effects and engineering problems.

16. All trade effluents must be effectively controlled at source.

17. Special consideration should be given to cases where there are shell fisheries and to cases where bathing is practised or likely to develop.

LEGAL FRAMEWORK AND ORGANISATION

18. An adequate and flexible system is required for setting and reviewing controlling standards of quality for all discharges to tidal waters.

19. The new water authorities must be encouraged to make immediate arrangements for surveys. These will provide data for use in design of programmes for improving all estuaries. Joint committees for estuaries should be established, as provided by the Water Act 1973. The Water Research Centre should be involved in these programmes and committees.

20. The new authorities will need from their inception an adequate and properly qualified staff for the estuary monitoring work.

21. It is necessary that The Government implement their promised legislation to revise the Rivers (Prevention of Pollution) Acts as soon as possible. This will raise the level of control of discharges to coasts and estuaries to that already existing for inland waters. A provision should be included to establish an independent watch over the monitoring activities of the Regional Water Authorities, in order to ensure that statutory bodies are liable to the same control as private concerns.

22. Regional Water Authorities should accept a responsibility for education of the public in the understanding and protection of the environment.

CONFERENCE OPENING

Mark Woodnutt, M.P. for the Isle of Wight

Eldon Griffiths, the Under Secretary of State for the Department of the Environment, in a written message says, 'The Agenda for the conference certainly promises some interesting sessions and should provide an opportunity for valuable discussion on this important subject. I should be glad if you would convey to the Chairman and to the delegates my good wishes for the success of the conference and I look forward to receiving the results of its deliberations.'

We are meeting in the knowledge that the Government is playing its full part in improving the conditions of our estuaries. It has already announced its intention to introduce legislation which will help to achieve this and it is the intention to give the new Regional Water Authorities full control of discharges to tidal rivers, estuaries and the sea, and more control over the discharges of sewage from boats. The Jeger working party on sewage disposal, in their report *Taken for Granted,* (1) and the Royal Commission on Environmental Pollution in their Third Report(2) have both considered the problems of pollution in estuaries, and the Government as part of the river pollution survey exercise are currently collecting information about the Solent, The Humber, the Wash, controlled parts of the Bristol Channel, parts of Morcambe Bay and the Solway, for inclusion in a report for publication in early 1974. The report of this conference will be published at a very appropriate time.

There is a conflict of opinion between the Hampshire River Authority, the Southampton Port Health Officer, the Southern Sea Fishery Committee and the Solent Protection Society about the condition of the Solent waters. Dr. Southgate undertook a study in 1970 of the existing drainage situation in South Hampshire Plan area.(3) He gave his independent views of the whole question of discharges of sewage into the sea from the plan area and one of his conclusions was, 'Judged by chemical evidence Southampton Water and especially the Solent would be regarded as substantially unpolluted and no quantity of evidence of recent damage to fisheries was submitted though the sewage discharged to the Solent includes a dry weather flow of eighteen million gallons per day from Portsmouth and Gosport alone'. In spite of that view expressed way back in 1970 situations and circumstances never remain the same but I have little doubt that the Solent is one of the cleanest stretches of coastal water in the Country.

This conference must look to the future and be very mindful of the fact that the problem of waste disposal in all its forms is increasing at an alarming rate annually. I have said that the Government is of the opinion that control of discharges should be extended to tidal waters, estuaries and the sea. The Government feels that without more comprehensive control it is not possible to tackle adequately the gross pollution in many major estuaries, to avoid conditions on beaches or in inland or coastal waters which might cause a hazard to public health or be detrimental to amenity, to protect fish and sea fishing or any risk to public health from the consumption of sea food or to fulfil its international agreements on protection of the sea.

The alternative possibilities are that the new control should be exercised by the Sea Fisheries Committees, by the Regional Water Authorities, or by some combination of these authorities. The Government has decided that local control by Regional Water Authorities rather than by the Sea Fisheries Committee is preferable. The Regional Water Authorities will already have an organisation for the control of discharges to rivers and estuaries, and for the Sea Fisheries Committee to duplicate this organisation would be a waste of resources. Moreover, as the sea is believed to receive most of its pollution from rivers and estuaries it would be irrational to attempt to control direct pollution of coastal waters independently of the control of estuaries and rivers. It is therefore proposed that the control of discharges to the sea should be administered by the Regional Water Authorities and it follows that the existing powers of Sea Fisheries Committees to control pollution will be ended when this legislation is introduced. As the Regional Water Authorities will cover larger areas than the existing River Authorities they will be better able to deal with the regional character of the problem of sea pollution.

The Government intends to present a report in early 1974 and then it intends to follow it up with legislation in the next session of Parliament. When you go to New Town Creek to the oyster beds, I hope you will reflect on the fact that it was at one time the Central Electricity Generating Board's intention to build there an atomic power station and may be you will feel as we did in the Isle of Wight that that would probably have been a worse form of pollution than the pollution we are discussing today. I congratulate the organisers on their initiative in holding this conference. I am delighted it is so well supported and I wish you all the greatest success in your deliberations.

REFERENCES

1 Ministry of Housing and Local Government. 'Taken for Granted', Report of the Working Party on Sewage Disposal (The Jeger Report), H.M.S.O., London, 1970.

2 Department of the Environment, Royal Commission on Environmental Pollution. Third Report. 'Pollution in some British Estuaries' and Coastal Waters, H.M.S.O., London, 1972.

3 Southgate, B. A. 'Main Drainage and Sewage Disposal'. Unpubl. report to the South Hampshire Plan Advisory Committee, Guildhall, Portsmouth, August, 1970.

EDITORIAL NOTES AND ACKNOWLEDGEMENTS

The Conference brought together 160 delegates with a wide range of expertise and interests relevant to the subject matter, which was dealt with both in broad outline of the whole field, and in detail in some specific areas. Delegates came from central government agencies and research establishments, river authorities, local authorities, from industry and from universities. Biological, chemical, physical, planning, engineering and amenity interests were represented.

The inlets along the coastline of West Sussex, Hampshire and the eastern part of Dorset—from Pagham to Poole—are free from widespread gross pollution. Resident and holiday populations along the shores are however increasing very rapidly together with a considerable growth of industry and commerce. By the standards of the 'industrial estuaries', such as the Thames, Mersey, Tyne, or Clyde, the local waters are not polluted. Yet there are constant comments on declining amenity and claims that the waters are in fact increasingly contaminated.

The Conference discussed the research and monitoring necessary to define the pollutional status. At the Final Session a Statement was drawn up by consensus to show subjects on which delegates were broadly in agreement, and to form a basis for more detailed consideration of monitoring schemes.

The interest of Mr. Mark Woodnutt in coming to open the Conference is gratefully acknowledged. The five session chairmen ensured the smooth running of the discussions. They were Mr. H. Speight, Director of Technical Services, Hampshire River Authority; Mr. H. R. Oakley, Senior Partner, J. D. and D. M. Watson, Consulting Engineers; Dr. A. L. Downing; Director, Water Pollution Research Laboratory; Dr. Jenifer Baker, Head of the Oil Pollution Research Unit of the Field of Studies Council; and Mr. J. A. Wakefield, Chairman, Coastal Anti-Pollution League Ltd. All of the Chairmen contributed to the discussion in addition to their other duties. Mr. Martin Lovett aided the work of the final session with his accurate and witty summary of the proceedings.

It goes without saying that the work of the fourteen authors is fully acknowledged. All gave of their skill, knowledge and experience in aspects of this difficult field of endeavour.

The work behind the scenes by the Administrative Secretary, Mr. M. A. McSweeney, by the audio-visual aids team, and by the stewards was essential to the Conference. The clerical staff of the Department of Civil Engineering have been involved in much typing of papers and correspondence.

Finally, the assistance of a number of people and organisations involved with the Conference tours must be mentioned. These tours, covering the whole area of interest of the Conference, were designed to illustrate problems raised in the technical sessions.

The list includes;

Mr. W. J. Thompson and Mr. S. Wray, of Hamworthy Engineering, Poole, who showed a sewage disposal unit designed for operation on ships.

Mr. P. Miller and Mr. C. L. Bennett, of Poole Corporation, who described the sewage treatment and land reclamation work around Holes Bay, which is part of Poole Harbour.

Mr. B. Silk of Poole Corporation, and Miss M. E. Dennis, of the Nature Conservancy, Furzebrook, who conducted a water tour in Poole Harbour and showed the many and contrasting activities in the Harbour.

The Station Superintendent and his staff at Poole Power Station, C.E.G.B., for hospitality and a talk on problems caused by sea weed at the cooling water intakes.

Mr. P. D. V. Savage and the staff at the Marine Laboratory, C.E.G.B., Fawley, for hospitality and a tour of the laboratory.

Mr. David May, owner of the Lymington Marina, for showing delegates the marina and talking about it, and to Mr. Maldwin Drummond, who described some other developments in the Lymington Estuary.

Mr. P. M. Chandler, of Lemon and Blizard, Consulting Engineers, for a description of the West Wight Sewage Disposal Scheme, and also the Cowes and Newport Sewage Disposal Scheme.

Mr. C. C. Lucas, Director, Newtown Oyster Fishery, for describing the work of his plant, and showing it to delegates.

Mr. G. T. Jewry, Engineer to Newport Borough, for a description of the Cowes and Newport Sewage Disposal Scheme.

Mr. S. L. Wright, of Hampshire River Authority, for a description of the pressure on the Hamble River, and for his comments on Fareham Creek. Also to the Engineer and Surveyor of Fareham U.D.C. and the Consultants to the Council, for arranging for representatives to address the delegates on site.

The Director of Technical Services, Havant and Waterlooville U.D.C., for allowing members of his staff to describe the refuse disposal and reclamation, and sewage disposal operation at the head of Langstone Harbour.

Mr. D. R. Houghton, Officer-in-charge of the Exposure Trials Station at Eastney, Portsmouth, for arranging a water tour of Langstone Harbour and to Miss J. N. Dunn for a description of her work in the Harbour.

The Engineer and staff of Portsmouth City Council for a visit to the Eastney Outfall Works.

CONTENTS

Conference Statement

Conference Opening

Editorial Notes and Acknowledgements

The Solent as a Recreational Resource. Maldwin Drummond. 1.1

Discussion 1.26

Legal Controls and Legislation. A.S.Wisdom. 2.1

Discussion 2.11

Medical Aspects of Estuarial Pollution. Angus McGregor. 3.1

Discussion 3.13

Industrialisation and the Ecology of Southampton Water. P.D.V.Savage. 4.1

Discussion 4.9

The Tidal Hydraulics of the Solent and its Estuaries. N.B.Webber. 5.1

Discussion 5.19

Sedimentary Processes within Estuaries and Tidal Inlets. E.J.Humby and J.N.Dunn. 6.1

Discussion 6.13

Planning and Pollution – A Case Study of some Economic and Social Pressures on Coastal Waters – South Hampshire. J.F.Barrow. 7.1

Discussion 7.18

The Pollution of Ports. D.R.Houghton 8.1

Discussion 8.6

Bacteriological, Biological and Chemical Parameters Employed in the Forth Estuary. R.W.Covill. 9.1

Discussion 9.57

Airborne Sensors for Monitoring Pollution. P.G.Mott. 10.1

Discussion 10.14

Techniques for Pollution Control in Estuarial Waters. D.W. Mackay. 11.1

Discussion 11.13

Factors Affecting Slick Formation at Marine Sewage Outfalls. J.R. Newton. 12.1

Discussion 12.13

Planning the Pollution Budget of an Estuary. R.E. Lewis and R.R. Stephenson. 13.1

Discussion 13.8

Experiences in Estuary Monitoring. T. L. Shaw. 14.1

Discussion 14.8

List of Participants. 15.1

Index 16.1

THE SOLENT AS A RECREATIONAL RESOURCE

Maldwin Drummond, J.P.

SUMMARY

1. The Solent is valued both as a unique stretch of sheltered recreational water with many natural assets and as the life line to the major ports of Southampton and Portsmouth. The rapidly improving lines of communication make the area within easy reach of the people and industries of London, the Southeast and the Midlands. At the same time, South Hampshire is one of the country's rapid growth regions, creating other demands such as increased sewage and waste disposal. Multiple use is essential and the conflicts between users must be resolved.

2. Recreation is viewed as a resource and the present situation is examined as a guide to future problems and opportunities.

GEOGRAPHICAL SETTING AND COMMUNICATIONS

3. The Solent is taken as that area of water sheltered by the Isle of Wight and enclosed by lines Foreland (Bembridge Point) to Selsey Bill, in the East and the Needles to Hurst Point, in the West. (Map 3)

4. Southampton is at the apex of the Solent triangle and the lines of communication radiate out from it, serving the port. Within twenty miles of the City, there are at present 1 500 000 people and the South Hampshire plan area alone (Map 1) is expected to expand to 1 000 000 by 1981 and to 1 400 000 by the year 2001 (H.C.C. and others, 1972).[(7)]

5. The completion of the M3 and A3 link, the M27 South Coast Motorway and the major works on the A34 to the Midlands, will improve the accessibility of the plan area Southampton/Portsmouth, from most parts of the country. Already within 100 miles, or just under two hours drive of Southampton, there are 18 000 000 people (Map 2).

6. The geographical location, rapidly improving lines of communication and population growth, associated with the South Hampshire Plan and the increasing container, ferry and tanker traffic to the Port of Southampton, puts a series of conflicting demands on the Solent. Portsmouth, still a very important naval base, contributes to this.

THE PRESENT SCENE—THE SOLENT AND ITS ESTUARIES

Planning

7. South Hampshire, which borders the Eastern Solent, contains the two large cities of Southampton and Portsmouth. It is an area of rapid expansion and population has been growing in the past few years at the rate of 12 000 a year. This growth puts pressure on land, roads, the

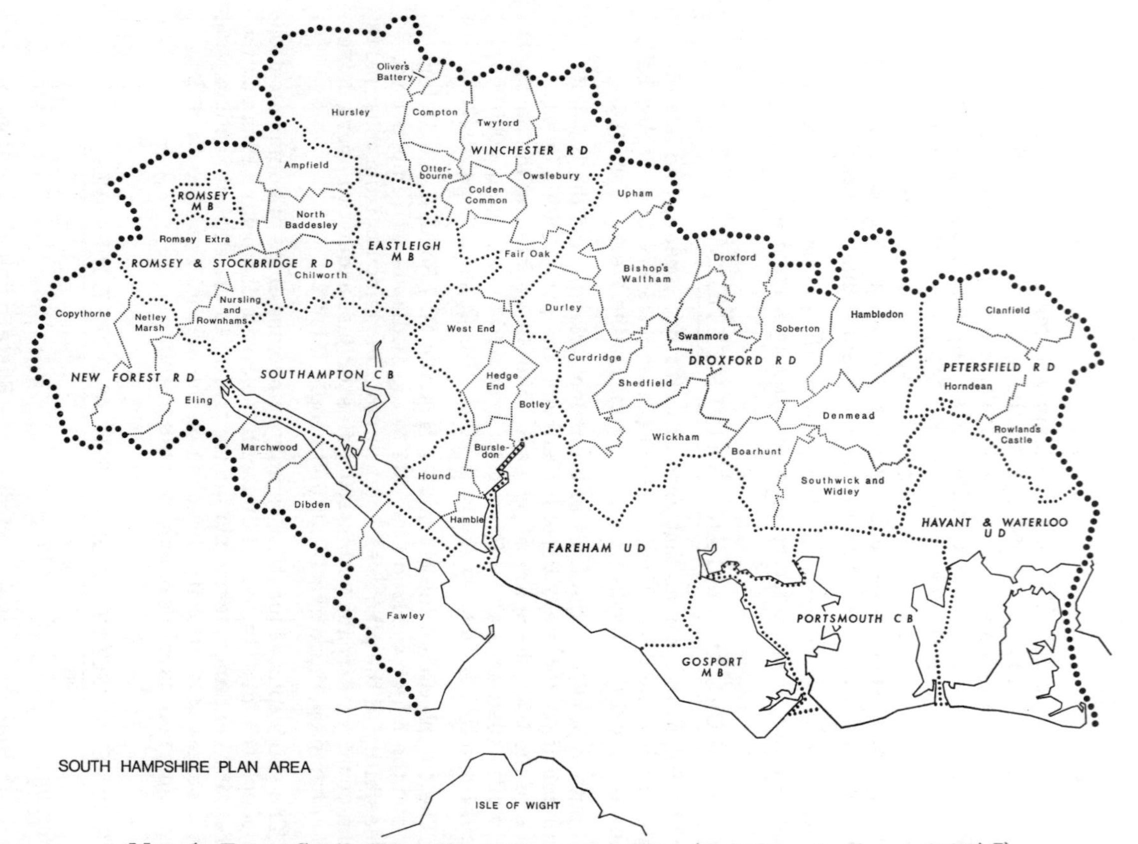

Map 1 From South Hampshire Structure Plan (H.C.C. and others, 1972)[(7)]

THE PORT OF SOUTHAMPTON

serves 25 million people within 125 miles radius

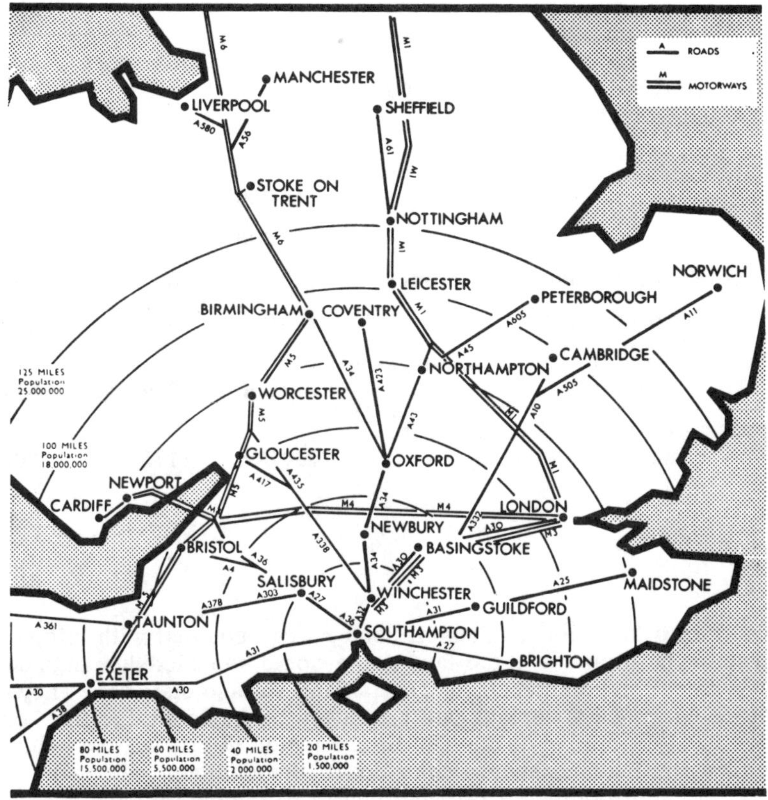

Southampton's central position makes it a first-class centre for the reception and distribution of merchandise of all kinds. Within a radius of 125 miles there is a population of 25 million, an area embracing London (under eighty miles away) and the industrial midlands, all linked to Southampton by regular rail and road services giving rapid and economic transport and distribution.

Map 2 Communications and Population (B.T.D.B., 1973)[(3)]

attractive countryside and coast, and so persuaded the three local government authorities, Hampshire, Southampton and Portsmouth, to come together in 1968 and work jointly toward the preparation of a preferred strategy that came out of selected studies. This was a considerable exercise in public participation and will, no doubt, be covered in more detail by J. F. Barrow, who headed the body responsible, the

South Hampshire Plan Technical Unit (Session 3). The plan area forecasts, including commitments, are shown below (Table 1).

SOUTH HAMPSHIRE PLAN FORECASTS
(H.C.C. and Others, 1972)(7)

	1966	1981	1991
Population	818 000	1 014 000	1 161 000
Employment	356 000	422 000	479 000
Housing (No. of permanent dwellings)	262 000	339 000	383 000
Transport (No. of cars)	145 000	370 000	473 000
Recreation—Visitors to the Countryside and coast	149 000	239 000	312 000
Education—No. of school children in maintained schools	130 000	175 000	209 000

Table 1.

8. The Hampshire County Council soon recognised that with this sort of pressure building up on the land there would be added demands on the Solent waters for commerce, recreation and sewage disposal. They therefore convened a working party, that led to the Solent Sailing Conference in the Spring of 1972. The final session of this will be held this year. The British Transport Docks Board, Southampton, had been worried by the problems of recreational craft before this and had assessed the problems of the port area from 1969 (B.T.D.B., 1969)(2) (Tables 2 and 3).

WORKING PARTY ON RECREATIONAL SAILING IN THE PORT OF SOUTHAMPTON (B.T.D.B., 1971)

		1969	1970	1971	1972
Craft Afloat		1 050	1 121	1 269	1 350
Percentage Increases:	1969-1970	6.8%			
	1970-1971	13.2%			
	1971-1972	6.4%			

Table 2.

RECREATIONAL SAILING IN THE PORT OF SOUTHAMPTON
(B.T.D.B., 1972)

Count of Recreational Craft underway in the Port of Southampton September 1972

Date	Time	Areas: Fawley/Hillhead/South Brambles/West Brambles/ Stone Point	Fawley/Dock Head	Rivers Test and Itchen	Total
Sept. 1972		71	9	4	84
Sat. 9th	1200	Wind N.N.W. Force 4	Sea Moderate,	Visibility 10 miles	
		84	49	5	138
	1400	Wind N.N.W. Force 4	Sea Moderate,	Visibility 13 miles	
		30	3	8	41
	1800	Wind N. Force 3	Sea Slight,	Visibility 11 miles	
Sun. 10th		152	54	12 and 21	239
	1200	Wind N.W. Force 4	Sea Moderate,	Visibility 15 miles	
		155	36	16 and 5	212
	1400	Wind N.W. Force 4	Sea Moderate,	Visibility 15 miles	
		49	5	2 and 2	58
	1800	Wind N.W. Force 4/5	Sea Moderate,	Visibility 15 miles	
Sat. 23rd	1200	200	50	30	280
	1400	120	20	20	160
	1800	180	26	13	219
Sun. 24th		170	130	9	309
	1200	Wind N.N.E. Force 3	Sea Slight,	Visibility 12 miles	
		90	145	13	248
	1400	Wind N.E. Force 3	Sea Slight,	Visibility 12 miles	
		45	3	4	52
	1800	Wind E.N.E. Force 2	Sea Rippled,	Visibility 9 miles	

Count of Commercial Vessels over 300 tons N.R.T. excluding Hovercraft recorded passing in and out at Calshot between 1200 and 1800 on the above mentioned dates

Saturday, 9th September 1972	9
Sunday, 10th September 1972	13
Saturday, 23rd September 1972	8
Sunday, 24th September 1972	8

Table 3.

9. The Solent Sailing Conference invited representatives from virtually all those who made demands on the Solent. Many of these uses were in conflict with one another. In addition, the commissioning authorities for the South Hampshire Structure Plan realised that it would economically and, indeed, environmentally be wrong to continue to discharge anything but fully treated effluent into the Eastern Solent by a multitude of outfalls (Map 3) and decided on a unified trunk sewage scheme for the plan area, with the outfall or outfalls in one place (Map 4). This decision followed investigations described in a report for the South Hampshire Advisory Committee (Watson, 1972).[19] The South Hampshire Main Drainage Board was set up to mastermind developments before their duties are taken over by the Regional Water Authorities (Water Bill, 1973).

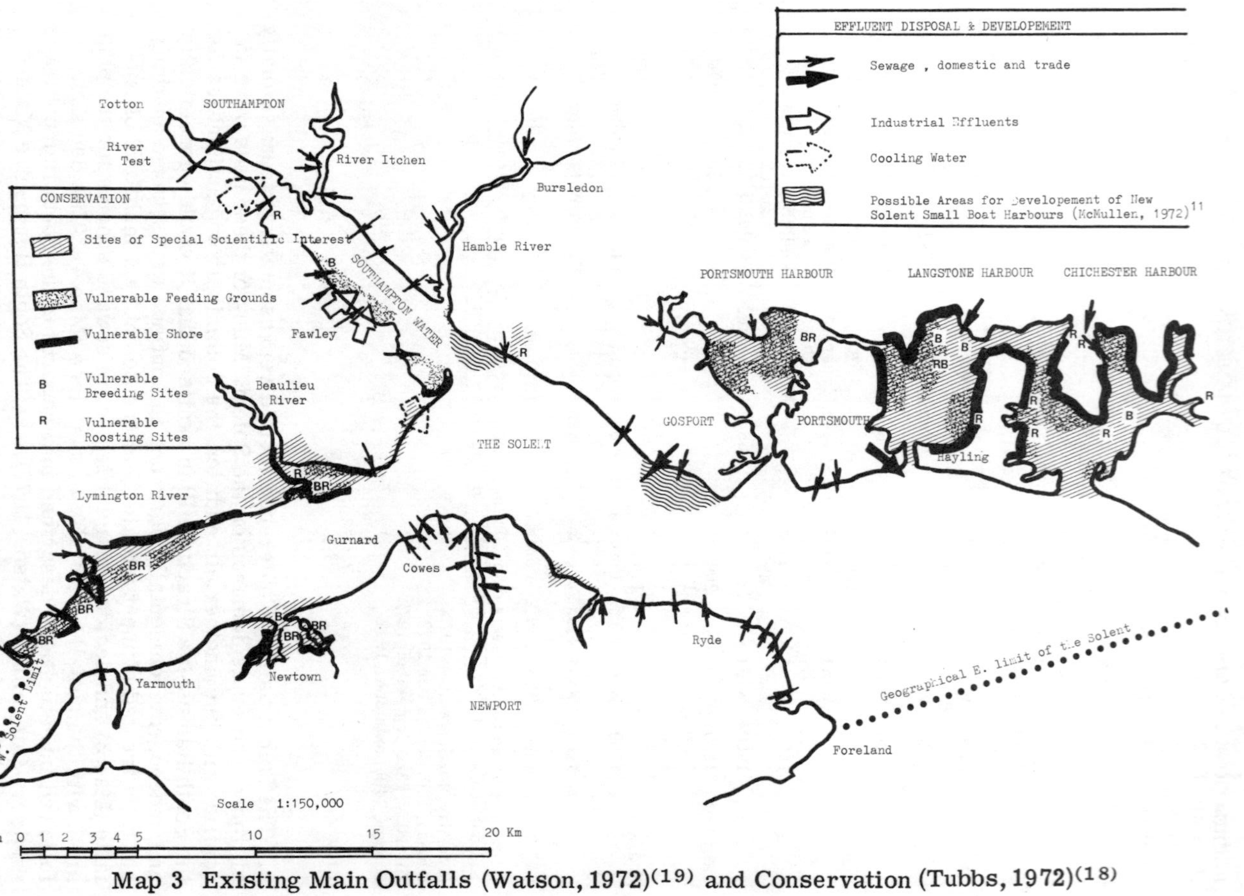

Map 3 Existing Main Outfalls (Watson, 1972)(19) and Conservation (Tubbs, 1972)(18)

	1a	1b	1ci	1cii	1ciii	1civ	1di	1dii	1diii	1ei	1eii	1eiii	1eiv	1ev	1evi	1evii	1eviii	1eix	1ex	1exi	2ai	2aii	2aiii	2aiv	2av	2b	2c	2d	2e	2f	2g	2h	2i	2j	2k	2l	2m	3
(iii) Power boat racing	X b	X c	X g							er, in conjunction with e(vii)	X	X	er, in conjunction with e(vii)	X	X	ould combine with (e) and cause conflict			X		X	X		X	X		X	X	X	X c		X	X					X 1
(iv) Cruising under sail	X b	X c	X h		X h						X	X		X	X		X	X	X		X c	X c	X		X c			X		X c								X 1
(v) Cruising under power	X b	X c	X h		X h						X	X		X	X		X	X	X		X c	X c	X	X c				X		X c								X 1
2b. Swimming				X e							X	X		X	X		X	X	X																			
2c. Sub-aqua activities	X b	X c	X k								X	X		X	X		X	X	X				X					X										
2d. Water skiing	X b	X c	X g								X	X		X	X				X			X	X	X	X		X		X	X		X	X			X	X	X
2e. Canoeing	X b	X c									X	X		X	X				X				X					X										
2f. Rowing	X b	X c									X	X		X	X				X		X c	X	X c	X c	X c			X										
2g. Wildfowling & gull's egging	X 1	X 1																X 1	X													X o						X o
2h. Bird watching	X 1	X 1																X 1	X				X					X			X o		X 1		X 1			X o
2i. Walking, rambling & quiet enjoyment		X c									X	X		X	X		X		X				X					X				X 1		X j				X 1
2j. Camping & caravaning											X	X		X	X				X														X i					X 1
2k. Picnicing & day visitor use by car to enjoy shore & sights at sea.				X k				X j			X	X		X	X		X		X													X 1						X 1
2l. Angling from the beach											X	X		X	X			X n	X									X										X 1
2m. Angling from a boat (other than charter angling)	X b	X c									X	X		X	X			X n	X									X										
3. SEMI-NATURAL ENVIRONMENT	X	X c		X 1		X 1					X m	X m		X m	X m			X ln	X				X 1	X 1	X 1			X			X o	X o	X 1	X 1	X 1	X 1		

TABLE 4 SOLENT CONFLICTS

NOTES

(a) Rogue dredgers have interfered with shipping.

(b) In the main dredged channel.

(c) In narrow estuaries and crowded harbours such as Lymington and Cowes. Disturbance to recreation by hovercraft noise (1965-1973).

(d) By prohibiting dredging in prohibited areas because of danger of fouling underwater pipes and cables.

(e) By digging pits below HW mark, causing danger and inconvenience to swimmers and beach walkers.

(f) Plastic cups have been found in the stomach of cod.

(g) By fouling pot markers and drift nets.

(h) By competing for quay and mooring space.

(i) By deformation of sea bottom and of fishing grounds by rogue dredgers.

(j) Visual damage if pylons near beach or camps or caravans too evident.

(k) Lobster and crab fishermen accuse skin divers of robbing pots.

(l) Can destroy habitat, physical damage, heat in the case of effluent and disturbance in case of shipping or recreation.

(m) Can cause eutrophication in harbours which can damage the recreational and natural resource.

(n) Cadland Creek outfall has caused considerable damage to bottom fauna and can produce oil taint. Marchwood on the other hand was responsible for spread of *Mercenaria mercenaria*.

(o) Both wildfowling and bird watching if not done with care can seriously conflict with semi-natural environment by damage and disturbance.

Multiple Use and Conflicts

10. The concern shown by the Planning Authorities over the future of the Solent in a time of fast growth and improving communications is well justified and matches public feeling. The Solent and its estuaries are supporting a high degree of multiple use. Some of these activities conflict with one another, others complement. It is essential to understand these conflicts (Table 4).

11. All these uses and demands impinge on one another, some doing damage, others enhancing, as is shown in outline.

Example 1: Effluent disposal v. Swimming.

12. Sewage disposal by the New Forest Rural District Council, under a scheme completed in the mid 1950's, discharging to a position near the head of Ashlett Creek, destroyed the desire of the people of Fawley to bathe in their nearest access point to the water, then a common use. In those days, there was little effective protest, as it was seen as community progress and the earth closet was the enemy. Today, with an extension to the load and services, the local authority have had to pipe the effluent to below high water mark in the Southampton Water itself. It is conceivable that the swimming use will return, though a heated, indoor pool is now a community desire.

Example 2: Commercial Shipping v. Recreational Sailing.

13. In the early 1950's a Dutch yacht, while racing, sailed dangerously close to the QUEEN ELIZABETH, giving cause for serious complaint by Trinity House. This changed the attitude of the then Harbour Authority, the Southampton Harbour Board, and they became labelled, unfairly, as anti-yachting. It could be said that this led to the close co-operation that was soon to be built up between the Harbour Master, Captain Andrew, and the Chairman of the Solent Clubs Yacht Racing Association, Air Commodore J. C. Quinnell. The result of this was a new look at the difficulties of yachtsmen and commercial shipping and the provision of a series of special yacht racing marks. This is a good example of a serious conflict being examined, understood and rationalized, to mutual benefit. Today, the fruits of that outcome are still evident, for Southampton Water and the Solent now harbour a record number of recreational craft, put at 21 700 in 1972 with a future already planned expansion to take this total to 30 900 (Smart, 1972).[16]

14. In the same period, there was a commercial pattern of movement that allowed a total of 76 029 ship movements, in and out, from the ports of Southampton (45 224) and Portsmouth (30 805) (Rogerson & Andrew, 1972).[15] (Maps 6 and 7)

15. It would be idle though to say that conflict does not occur. It does, and identification of the areas of conflict is apparent from the maps,

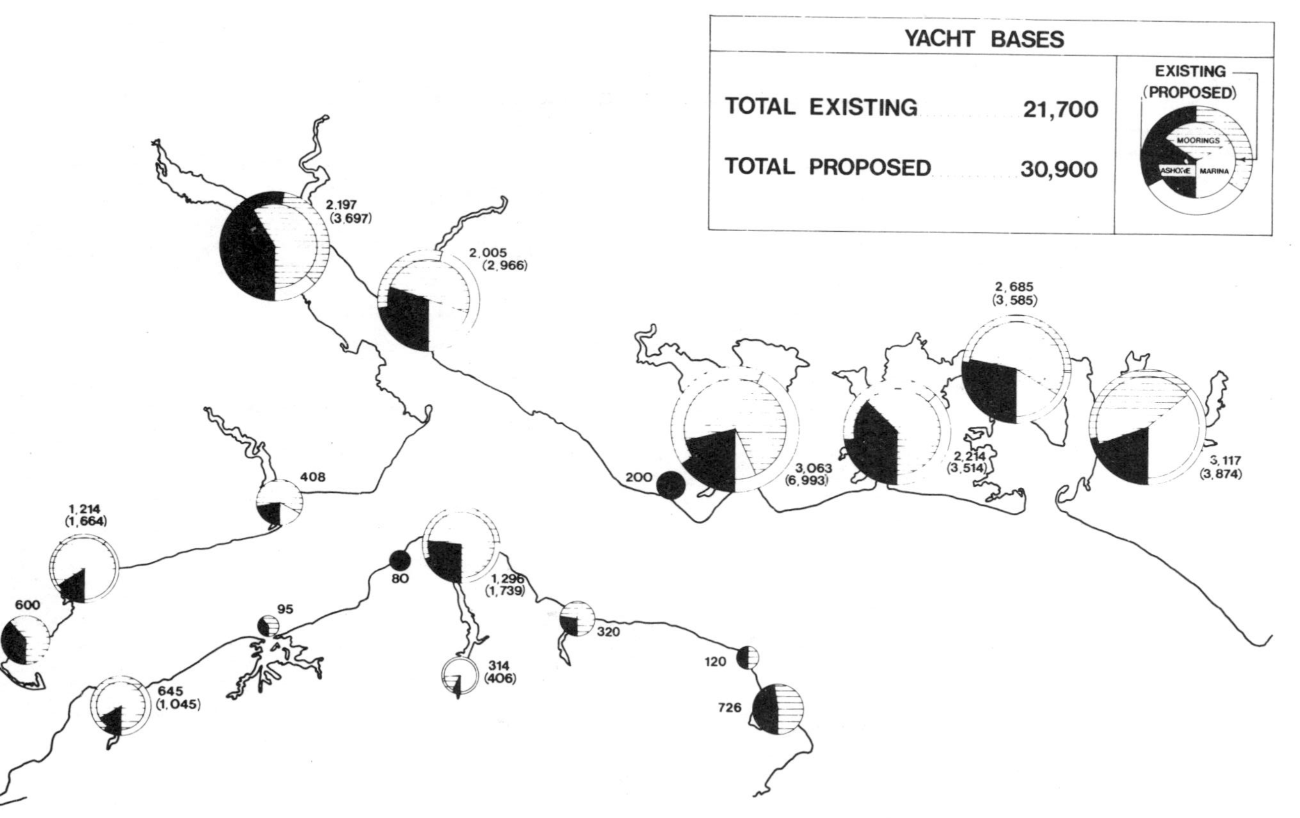

Map 5 Yacht Bases (Smart, 1972)(16)

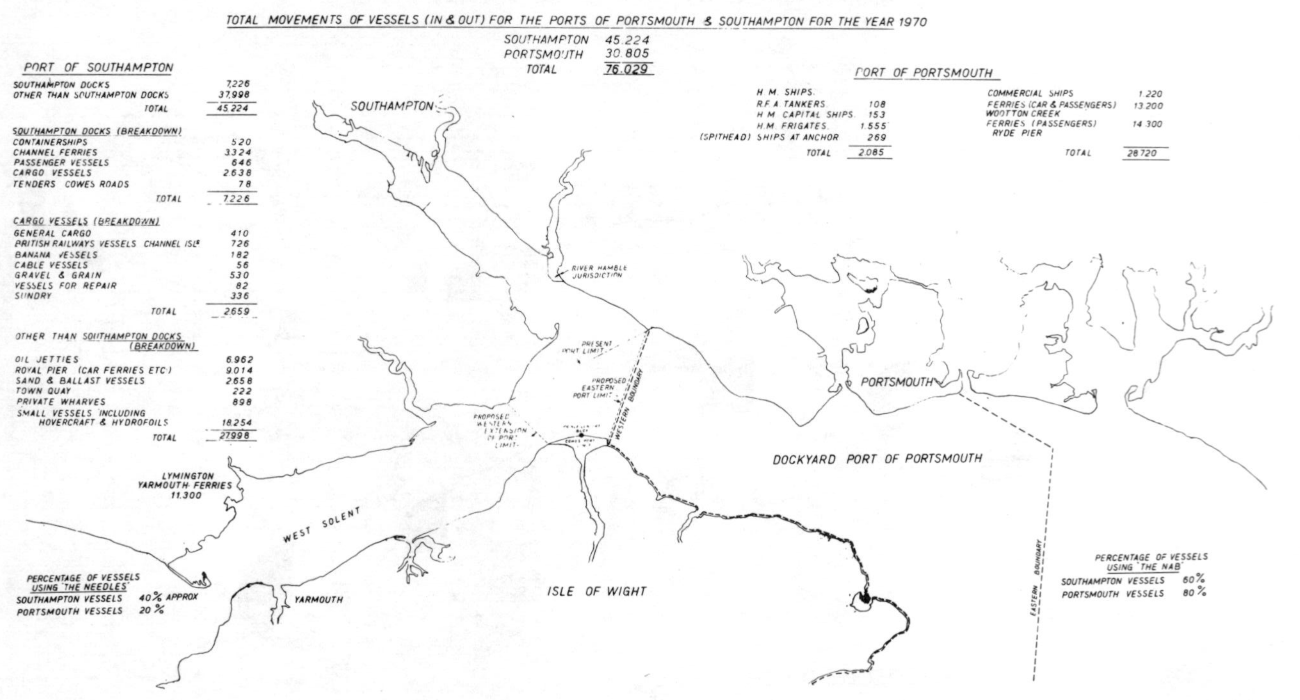

Map 6 Shipping Movements (Rogerson and Andrew, 1972)(15)

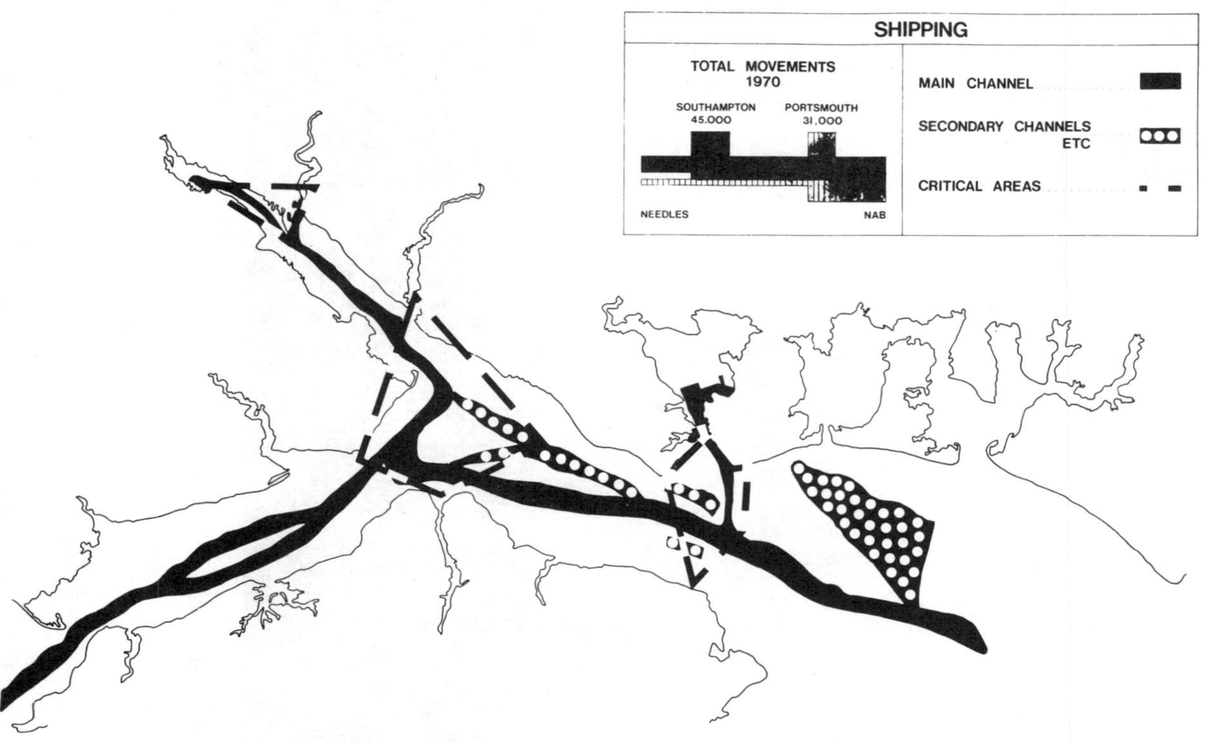

Map 7 Shipping Channels and Critical Areas (Rogerson and Andrew, 1972)[15]

for recreational craft do cross the shipping lane, for example (Map 7). It is clearly essential that management understands and recognises this pattern and that planning of new facilities ensures that the position is not aggravated. It is clear that narrow channels (approximately 300 meters at Calshot), accepting ships the size of the ESSO CAMBRIA, 250 000 d.w. tonnes, 374 m long and beam of 56 m, the FRANCE, 66 000 d.w. tonnes, 340 m long, with a beam of 37 m, and the QUEEN ELIZABETH II, 66 000 d.w. tonnes, length 315 m, beam 34 m (B.T.D.B., 1973), [(3)] must not become clogged by criss-crossing recreational craft (Fig. 1).

Photo by Beken of Cowes

Fig. 1 Solent Sailors

Commerce and Recreation

16. Multiple use of the Solent is essential though, and the physical characteristics of the area, in common with other estuaries, make it attractive both for recreation and commerce, and if the conflicts are understood, future planning and management can avoid serious clash.

The area is physically attractive to both these uses because:

(1) Sheltered water, for berthing, unloading and loading. The same characteristic is attractive for recreational sailing.
(2) Deep water for large ships in the main channels, without a large commitment to maintenance dredging. These comparatively narrow channels, which restrict commercial shipping, leave large areas free for recreational sailing.
(3) Good lines of land communication for both commerce and recreation (discussed in Geographical Setting and Communications).

Commerce, Recreation and the Semi-natural Environment

17. Both commercial and recreational use conflict though with the semi-natural environment by:

(1) The destruction of the habitat in order to provide facilities for commerce or recreation.
(2) By the disturbance of the habitat by, for example, landing on salt marsh shell spits and accidental damage to eggs during the nesting season. This is aggravated by numbers caused by improved facilities and better communication.
(3) By pollution caused by industry, shipping and recreation, for example, continuous low level pollution of Spartina grass can cause die back.

18. The Solent shores contain areas which are of considerable importance from the natural history point of view. The North shore, the Newtown River, Osborne Bay, the upper reaches of Portsmouth Harbour and the harbours of Langstone and Chichester, make particular claim.

19. The North Shore contains the Hampshire and Isle of Wight Naturalist Trust's reserves at Hurst/Pennington and Needs Ore. The whole Solent, particularly these two reserves, support 10 per cent of the total British population of the little tern, *Sterna albifrons*, and significant breeding colonies of the sandwich, *S. sandvicensis* and the common tern, *S. hirundo*. The North shore is an important over-wintering area for migrant and wintering populations of waders and wildfowl.

20. The Solent cord grass (*Spartina townsendii*) which colonized the area in the 19th century (started in 1870, near Hythe), is of great interest botanically, it being a hybrid between the indigenous *S. maritima* and the introduced species *S. alterniflora*. It is being physically eroded by both industrial and recreational development.

21. Langstone and Chichester harbours support significant populations of Brent geese, *Branta bernicla*, the shellduck, *Tadorna tadorna*, and the Godwit, *Limosa limosa* (Tubbs, 1972,[18] Pratt, 1973[13]).

22. But this natural richness adds to the value of the area from a recreational point of view, for the anchorage that is protected by featureless acres of anoxic mud and through whose doubtful atmosphere no

bird call is to be heard, is debased. In a survey carried out to examine the favourite harbours of those who keep their yachts at Lymington, it was found that Newtown River, the most undeveloped and natural harbour in the Solent, (in the Stage 1, Recreational River Development Phase, described later) was easily their favourite. Yarmouth, with full facilities, across wind and tide from the home port, would seem a more obvious favourite, but the attractions of the natural environment proved the more important (Drummond, 1972[(5)]).

23. Reduction and damage of that semi-natural environment affects the 'enjoyment' of the area and, therefore, the recreational resource, as is shown later.

THE RECREATIONAL RESOURCE

24. Olmstead, a far sighted American said in 1865...... 'It is a scientific fact that occasional contemplation of natural scenes of an impressive character, particularly if this contemplation occurs in connection with relief from ordinary cares, change of air and change of habits, is favourable to the health and vigour of man and especially to his intellect and understanding.'

25. One hundred and five years later the Jeger Committee in their report on sewage disposal, 'Taken for Granted', insisted,...... 'That everyone is entitled to find the relief from tension, which contributes positively to the good health of mind and body. Anything avoidable that detracts from personal relaxation and individual enjoyment of our incomparable coastline, should be eliminated' (Jeger, 1970[(9)]).

26. The Solent and its estuaries add to the pleasure and health of large numbers of people. The area has a lengthening season (April-December) and supports a very sizeable recreational community, as can be seen from the figures quoted above. The activities shown in the 'Summary of Conflicts', Table 4, and the yacht base map (Map 5) show this only too clearly, by giving the example of the present position for yacht accommodation in the Solent's fifteen recreational estuaries. Indeed, Bryer (Bryer, 1972[(4)]) has speculated privately that if all other constraints, such as conservation and shipping were disregarded, the Solent could support 44 000 yachts, excluding visitors. This is, of course, he says, inconceivable.

Recreational Development of Estuaries and Harbours

27. It is, however, worth looking at the way the 'deck acreage', as one water planner put it, grows. There are distinct steps from the natural, naked resource to the purpose built harbour, constructed on unprotected coastlines, as a result of pressure, profit, or community spirit.

STAGE 1—The raising of a river from one mooring to capacity for swinging moorings. with visitors accommodated against existing quays.

Dinghy space for tenders provided. The Newtown River is a rare Solent example.

STAGE 2—Conversion of swinging moorings in part or completely, to bow and stern moorings, either using buoys or piles. The provision of more dinghy space for tenders. Beaulieu was an example of this until the yacht marina was constructed in 1972.

STAGE 3—Marinas associated with bow and stern moorings on either side of the main channel. The Hamble River is at this stage now.

STAGE 4—The main channel in an existing or modified estuary used to capacity for servicing marinas along its banks, from navigable source to mouth. There is, as yet, no Solent example, as the Hamble is the most 'advanced' Solent River, with the exception of the commercial Itchen and Test.

STAGE 5—The provision of special man-made harbours on unsheltered coasts, as no estuary or sheltered water is available, or all possible harbours exploited. Brighton Marina is an example. There is no logical limit to the size of such facilities, for over 2 000 boats are accommodated in the U.S.A. at places such as Long Beach, California and Fort Lauderdale, Florida. A yacht basin can accommodate anything from 75-175 yachts per hectare (30-70 per acre) (Bertlin, 1972[(1)]). An investigation into the possibilities of Stage 5 in the Solent has been undertaken for the Solent Sailing Conference (McMullen, 1972[(11)]) and the report illustrates areas with potential from the hydrographic point of view (see Map 3).

Conflicts in the Use of Recreational Estuaries

28. As more facilities are provided, backed up by the planners requirements of adequate roads and car parking spaces (Hockley, 1972[(8)]), conflict arises with other users of the river. At Lymington, for example, the channel is shared with the British Rail car and passenger ferries (the new vessels will be nearly 65 m long, with a beam of 16 m and a draft of 2 m, and of necessity highly manoeuvrable). There are in prospect 1 500 yacht mooring spaces in the river (Drummond, 1972[(5)]). There is clearly a conflict here, though there is a plus, for if it were not for the ferry, the yachtsmen would have to pay for the dredging of the river. This is not the end of the story, however, for Lymington is now a thriving fishing port, with the Solent oyster 'explosion' in full swing (Key, 1972[(10)] and, therefore, rival demand for quay space and moorings at the head of the river (Fig. 2).

29. Conflict between the growing numbers of yachts themselves and the use of the river as an exit and entry channel, means that the recreational resource is being diminished by more of the same. The population, affluence and ease of travel is beginning to mean that leisure is being damaged by leisure, a problem foreseen by David Cobb, Chairman of the Solent Protection Society, and echoed by Hampshire's Planning

Photo by Maldwin Drummond

Fig. 2 Fishing Boats, Yachts and the Isle of Wight Ferry

Officer in his introduction to the Solent Sailing Conference (Smart, 1972[16]).

30. The water itself is the attraction, and is the vehicle that 'produces the relief from tension, which contributes positively to the good health of mind and body', to quote the Jeger Committee again. This is what people have to come to be by, on, or in, and for the purpose of well being. Stretches of recreational water are more and more important today, as tensions of modern society grow, aggravated really by over-population and, therefore, demand. It is, therefore, vital to conserve the recreational resource found in estuaries and sheltered waters. By so doing, we may well free a proportion of our hospital beds, increasingly reserved for mental illness.

31. The problem is that this same water has to bear the economic advantages of waste disposal, cooling and transport. The river authorities, and indeed their successors, the Regional Water Authorities, are freshwater orientated. Their view could be summed up that the brackish or salt water of estuaries is already contaminated as far as they are concerned and there are vast quantities of it to exploit. Their remit, the reason for their everyday existence inland, is not relevant here, yet. This is, of course, the need to provide drinking water and they will,

therefore, clean rivers and neglect estuaries and coasts. They will be tempted to relieve pressure on inland waters by using the coastal belt and, therefore, damage the recreational resource. The work, therefore, in the Solent of the South Hampshire Drainage Board is vital, and far-sighted. It is a change from such actions as that of Bolton, piping its sewage, for economic reasons, to the 'sink' of Liverpool Bay. It will not relieve Portsmouth's own load on its harbour or stop the Isle of Wight, yet, from thinking of discharging four and one-half million gallons per day (18 000 m^3/d) from their proposed new Gurnard outfall, situated between the world-famous sailing port of Cowes and their nearby bathing resort of Gurnard. It will not stop Lymington Borough, with the approval of the Ministry of the Environment, only screening their effluent from the Pennington outfall, which gives onto part of Britain's most prolific oyster bed.

32. But to return to the recreational resource. The Watson Report (Watson, 1972[(19)]) said that the oil-contaminated refinery cooling water outfall, discharging into Cadland Creek, Southampton Water, was equivalent to the demands on the receiving water of a city of 1.2 to 1.6 million. The effect on the recreational resource is shown by quoting from a letter to the Solent Protection Society in 1971 (S.P.S., 1971[(17)]).

33. 'On Saturday, 30th October, I arranged with a party of friends from Hythe, to fish in the Solent. Unfortunately it was very foggy and when we got down the river, as far as Cadland Creek, the fog became so dense that we ran onto the mud. It was a falling tide, so we settled down for several hours, to wait.

34. It was then that the horror dawned on us, as the water drained away with the tide. We were not on a mud bank but high and dry in the middle of acres of solid oil sludge. The fog started clearing from the river, but we then realised that we were in the sulphur steam from the refinery cooling plant.

35. The position was extremely hazardous. It was difficult to breathe and the fumes were choking. Fortunately the wind got up and cleared the fumes away, but it was very nasty at the time.' The writer goes on to say.... 'The reason I have mentioned the incident is that passing up and down the Southampton Water on the tide, one does not see the terrible state of the bottom. I for one never dreamed that it was so bad. Not a crab, not a piece of seaweed, not a living creature as far as one could see, for acres and acres'.

36. He did not mention that if he had succeeded in reaching his fishing spot and managed to catch a fine fish, he may have found on cooking it, that it was inedible through oil taint. Esso are now full out, trying to correct a system which through superceded design and old age, was letting 1 250 gallons per day of oil in suspension into the Southampton Water. The problems of rectifying this are difficult and Esso are well aware of their responsibilities. They have shown themselves consistently over the years not only as good employers but people who are

trying in every way to fit their plant, a necessary cuckoo, into rather a challenging New Forest/Solent nest.

37. Another example of damage to the resource is that of the Gosport boil. This is the town's discharge of virtually untreated sewage, via a long pipeline into the recreational waters of the Solent. This leaves a scar of dirty water for yachts to sail through, of nearly 1 000 meters at times.

Photo by David Mills

Fig. 3 The Gosport Boil

38. Contraceptives are notoriously difficult to disguise by treatment, and are so common in the Eastern Solent and off Cowes, that they are known as 'Spithead Salmon'. The naive blame the Navy for their presence afloat, and many an anxious parent has led her child from the water earlier than the water temperature demanded. Cowes' domestic sewage is disposed of into the harbour not 100 meters from the town quay. This was one of the arguments put out in favour of the proposed Gurnard outfall.

39. The Solent receiving water may be able to purify these effluents chemically, bacteriologically and biologically, but it stains the scene and reduces the ability of the waters of the Solent and its associated estuaries to produce 'relief from ordinary cares'.

40. The anxious public, looking for something to persuade pollution authorities, picked rightly on health, but wrongly emphasized the chemical, bacteriological and biological worries. The Public Health Laboratory Service Committee (P.H.L.S.C., 1959[14]) and the Medical Research Council (M.R.C., 1959[12]) were happy that the risk to health was minute. The plea really, however, was should we have to bathe in, sail on or look at, waters that appear or smell unhealthy. Little attention was paid to this because this was an aesthetic measurement and, therefore, subjective. It is easier to condemn the Winchester Rural District Council's Botley and Bursledon discharge, which gives into the Hamble, the Solent's most valuable recreational sailing port, as the River Pollution Survey (Welsh Office, 1970[20]) classed it as Grade 3 in their ten year river quality report. This denotes 'rivers of poor quality, requiring improvement as a matter of some urgency' and refers specifically to Hamble's Bursledon Bridge area, which is a reach crowded with marina berths.

41. A person out for the day for enjoyment and recreation has only his eyes and nose to go by. Aesthetic water quality measurement must use these for gauging the harm to the resource. They can add to this those needed for the invisible and demanded for public health reasons. Aesthetic criteria would include whether there are any visible solids or identifiable secondhand and unpleasant objects floating about, whether the water is unnaturally discoloured and if there is any foreign smell.

42. The new Regional Water Authorities should designate and publish their assessment of the aesthetic quality of their rivers, estuaries and enclosed waters used for public recreation. This should be done every year. The quality designation adopted by the River Pollution Survey would form the model and could be extended to cover estuaries and enclosed waters. Taking the Welsh Office's model, their subsection (d) in Classes 2, 3 and 4 form a useful base.

Class 2 (d) says rivers which have been the subject of complaints which are not regarded as frivolous but which have not been substantiated.

Class 3 (d) Rivers which have been the subject of serious complaint, accepted as well founded.

Class 4 (d) All rivers which have an offensive appearance, neglecting for these purposes rivers which could be included in this class solely because of the presence of detergent foam. (This last rider would have to be removed, as sailing on, swimming in, or sitting by foam, is hardly recreation.)

43. Regional Water Authorities would have to publish yearly their evaluation of the recreational waters status and stand by this judgment.

If challenged, they would have to ensure that the river for that year remained at the published rating. They would have to take every possible step to see that those offending were brought to book. The ratings would be checked nationally by the Department of the Environment, every ten years, as part of their river and, it is hoped, coastal water and estuarial studies. Such procedures would enable local people to judge whether upgrading and improvement was worth the financial burden or whether it was more expedient to use the receiving water to its purifying capacity— blow aesthetics. The Authority may decide to go beyond this, perhaps after public enquiry, and confirm its use as a waste carrier, or sink in the case of an estuary, taking steps to deny access for recreational purposes and, therefore, provide areas for this demand elsewhere. It would, however, be a public choice and there is little doubt the problem of river and estuarial pollution, as we know it today, would soon decrease. The cost, of course, would be great.

42. It would mean, however, that recreational needs would be given almost the same priority as drinking water. If local authorities looked at it in this way and stopped fouling these waters themselves, it would be easier to prevent recreational sailors from doing the same. Yacht disposal systems using the estuary or sea, do add to the damage, but not on a significant scale in comparison, at the moment. In due course,

Photo by Maldwin Drummond

Fig. 4 An Oiled Solent Castaway

Photos by Maldwin Drummond.

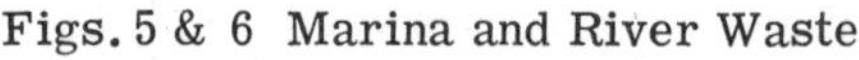
Figs. 5 & 6 Marina and River Waste

the 'pump out' station will be a feature of yacht harbours in estuaries, but again not before the major problem is tackled by the government and local authorities (Drummond, 1972[6]).

43. Commercial shipping is, at the moment, under no constraint to limit where they may pump out their sanitary tanks. They could do it in Southampton Docks, for the British Transport Docks Board have only powers over oil discharges.

44. Oil, as mentioned above, damages the recreational resource. It taints fish, mars the waterlines of yachts and ruins clothes of those who use the beach. Southampton has few enough bathing beaches, yet Weston Shore and Calshot, for example, are occasionally almost unuseable because of oil. The recreational sailor is not altogether blameless, to a minor degree, in the discharge of oily waste. No yacht with an engine should pump its bilge without ensuring that a non-toxic emulsifier is present. (Fig. 4)

45. Litter, too, is an increasing problem. The Solent shores are dotted with plastic disposable rubbish, thrown from ferries, ships and disposed of by yachtsmen. This can cause danger, as the rope round the propeller of the outboard motor demonstrates in Fig. 6.

46. Regional Water Authorities should take this problem on themselves. It is not unsolvable, as the Solent Protection Society showed, when after a certain amount of pressure, one of the major Solent ferry operators went back to china cups, as his identifiable plastic ones had become too much of an uncomfortable shore trademark. Harbour authorities can make life easier, by making more disposal points available and by a publicity campaign. An overloaded marina container is shown in Fig. 5. The Clyde River Purification Board, for example, advertises on the Clyde ferries that it is an offence punishable by a £100 fine to throw anything overboard. This may be effective.

47. The Solent and its estuaries are, perhaps, a long way from the quoted state of the Potomac which was said, at one time, to be 'too thick to navigate and too thin to cultivate'. They are, though, far from being free of those problems that do much to reduce their recreational resource and so damage their capability of contributing fully to the health and welfare of the nation.

RECOMMENDATIONS

1. That the effects of pollution on recreational water be fully recognized and standards approved to meet this problem in estuaries and enclosed waters. These standards should be strong enough to ensure that the quality of recreation, the recreational resource, of the Solent and other areas of enclosed water and estuaries of similar importance, is not marred. This could be drawn up on the lines of the River Pollution

Survey with a strengthened aesthetic requirement. In this effort, standards for each stretch of recreational water should be assessed and published by the R.W.A. It should act as a statement of standard, which would be initially open to public comment or objection and then, perhaps after enquiry, adopted as the required quality. Effluents should not be allowed to lower this stated standard either chemically, bacteriologically, biologically or *aesthetically*.

2. That the conflicts occurring within the Solent and its estuaries should be studied and conclusions drawn. This will enable multiple use of the waters to be achieved without undue damage to the resource. The study should form the basis for a water use plan with guide lines for management. (NB: This may well result from the Hampshire County Council's Solent Sailing Conference, 1973.)

3. That the Regional Water Authorities look into the problem of waterborne litter in estuaries (flotsom and jetsom) and take steps to effect an improvement by encouraging more convenient disposal points and by publicity aimed at both commercial and recreational users.

REFERENCES

1. Bertlin, D. P., Marina Design and Construction, Paper presented to the symposium on marinas and small craft harbours, Dept. Civil Engineering, Southampton University, 1972.
2. British Transport Docks Board, Report of the Working Party on Recreational Sailing in the Port of Southampton, 1969. Survey figures are updated yearly—Table 3.
3. British Transport Docks Board, Southampton Shipping Guide for February, 1973.
4. Bryer, Robin, Solent Capacities, A Personal Estimate, Solent Sailing Conference, Working Party discussion paper, 5-12-72.
5. Drummond, Maldwin, Conflicts in an Estuary—A Study of Conflicts in the Lymington Estuary, Dissertation, Department Extra Mural Studies, Southampton University, 1972.
6. Drummond, Maldwin, Pollution and the Yachtsman, *Yachting World Annual*, Edward Stanford, London, 1972.
7. Hampshire County, Portsmouth and Southampton City Councils, South Hampshire Structure Plan, Document for participation and consultation, South Hampshire Plan Advisory Committee, 1972.
8. Hockley, D. H. E., The Marina—In the Sailing Scene and the Environment, paper presented to the Symposium on Marinas and Small Craft Harbours, Department Civil Engineering, Southampton University, 1972.
9. Ministry of Housing and Local Government. Jeger Committee, 'Taken for Granted', Report by the Working Party on Sewage Disposal, H.M.S.O., London, 1970.
10. Key, D., The Stanswood Bay Oyster Fishery, Shellfish Information

Leaflet No. 25, Fisheries Laboratory, Burnham-on-Crouch, M.A.F.F., 1972.

11. McMullen, Capt. Colin, Notes on Provision of Sea Recreational Facilities in the Hampshire Area, Memorandum for the Solent Sailing Conference, 1972.
12. Medical Research Council, Sewage Contamination of Bathing Beaches, England and Wales, M.R.C. Memorandum No. 37, H.M.S.O., London, 1959.
13. Pratt, Norman, Solent Ornithology—Towards A Management Policy for the Hurst-Calshot Areas. Private discussion notes, first draft, January, 1973.
14. Public Health Laboratory Service Committee, Sewage Contamination of Coastal Bathing Waters in England and Wales, *J. Hyg.*, Camb. 1959, **57**, 435.
15. Rogerson, Cdr. K. F. and Andrew, Capt. J., Commercial and Defence Uses of the Solent and Southampton Water and Sailing in the Area. Paper presented to the Solent Sailing Conference, organised by the Hampshire County Council, Spring, 1972.
16. Smart, A. D. G., Introduction to the work of the Conference, Paper presented to the Solent Sailing Conference organised by Hampshire County Council, Spring, 1972.
17. Solent Protection Society (1971), Chairman's records of that date (Maldwin Drummond).
18. Tubbs, Colin, Biological Appraisal of the Coastline, Paper presented to the Solent Sailing Conference, organised by the Hampshire County Council, Spring, 1972.
19. Watson, J. D. and D. M., Marine Disposal of Sewage and Sewage Sludge, Summary report for the South Hampshire Plan Advisory Committee, 1972.
20. Welsh Office, Dept. of the Environment, Report on a River Pollution Survey of England and Wales, Vol. 1, H.M.S.O., London, 1970.

DISCUSSION

RODDA (Department of the Environment). Mr. Woodnutt, in the Conference Opening, referred to proposed changes in the administration of water 'affairs'. These should encourage more intensive studies and action to foster the development of estuaries for commerce, recreation and liquid wastes in such a way that health risks, pollution and nuisances are either avoided or kept to a minimum. Fish and wildlife generally should be protected. To do this large capital sums will be required.

The Author has set out the conflicts arising from the estuary's multiple uses and these show how the Solent is a good case for study and theme for this conference. Several of the major conflicts are a result of the demands of the 18 million people who live within 2 hours travelling time of the Solent, a fact which the Author emphasises.

Demands made upon the Solent should be met if possible, but they should be controlled so that Nature's balance is not upset. Ways and means should be developed to mitigate any undesirable changes which might occur. Restraints will have to be introduced to protect the environment, such as the control of waste discharges to preserve the wholesomeness of the Solent water, including some controls over marinas and any pollution they might cause.

Towards the end of his paper the Author has referred to some parameters for classifying the quality of water space according to nuisance values. These classifications should be extended to include some of the chemical and biological quality parameters which are included in the Department's River Pollution Survey. However, the setting of these classifications pin-points a difficulty as not enough is known about the chemical and biological nature of estuarine waters.

The Author comments on the possibility that estuaries might be given over to the large scale disposal of wastes. He refers to them as 'sinks' and comments that, as a consequence, recreational pursuits may be denied. I doubt if such a situation could be tolerated and suggest that it would be preferable to plan a strategy of balanced multiple use as a means of providing facilities for all who want to use estuaries.

DRUMMOND The remark about use of estuaries as sinks was meant to imply that the public should be invited to decide whether an estuary could be used for multiple use and then criteria produced for control of standards, or whether the authorities should impose the decision. When presented with the options and shown the costs, the public will opt to pay the extra needed to secure good environmental conditions. The public actually want to pay for it.

We ought to go full out for the establishment of aesthetic criteria. This will be difficult, and perhaps that is why people have shied off it. The general public have only got their eyes and their nose to go by. If they are going to recreate on these waters this is how they will monitor quality, and we must recognise this.

OFFORD (Rofe Kennard & Lapworth). Will not the proposed South Hampshire sewer outfall mentioned in the paper produce a bigger and better version of the Gosport Boil? Will it not be necessary in the long run to provide treatment at the landward end?

DRUMMOND It was hoped that the combination of treatment before discharge, and proper outfall design would get over the problem. The amount of thought and study given would hopefully give a good solution.

LITTLE (Brighton Marina Company, representing also the National Yacht Harbour Association and the ad hoc Committee on pollution control of the Ship and Boatbuilders National Federation). There is a need to maximise the aesthetic value of the waters under discussion. This must be stressed forcibly.

The Author implied that marina owners and operators were unlikely really to get down to the control of sewage discharged from boats unless forced to by legislation. Although this may be true in certain cases, members of the National Yacht Harbour Association would see to it that proper water pollution control facilities were provided, and that yachtsmen would use them, if only in their own self-interest.

DRUMMOND Local authorities must give a lead and clean up harbours, estuaries and coastal waters. Once progress has been made in this direction marina operators would be expected to respond.

JURY (Newport I. o. W. Borough Council, Chairman of the working party on water reorganisation for the Isle of Wight). It is a tremendous step forward to see all the constituent local authorities around the Solent beginning to act in unison in the field of sewage disposal. Unfortunately, a number of decisions on schemes must be taken immediately, before the new Regional Water Authority has time to get a grip on the problem.

There is a danger that large sums will be spent in the near future on expensive long sewage outfalls, only to have to add full treatment at a later stage. Someone in the Treasury or the Department of the Environment must decide whether full treatment before discharge is to be regarded as the only acceptable solution for the Solent.

DRUMMOND Full treatment will probably be demanded in the future, and anybody who thinks long sea outfalls without land-based treatment will be acceptable is deluding themselves. The coastal shelf will become our most valuable asset, for such activities as fish farming as well as for recreation. There will be tremendous effort in the next 25 years on improving our coastal waters. So we should ask Government to give a lead and come out in favour of full treatment on land before discharge.

SUTTON (Esso Petroleum Co. Ltd.) I speak as a citizen rather than as a representative of a company. The Author's paper illustrates very clearly that conflicts are a consequence primarily of increasing population and that this conference should touch on the question of population. The South Hampshire plan is based on the South East strategy plan which is not so much a plan as a prediction of what will happen if successive Governments continue to make no effort to influence the growth or movement of population. We should, as intelligent and enlightened people, include in any conclusions we reach some guidance to Government on this question.

CHARLESWORTH (Rural District Councils Association). Land-based treatment is to be encouraged. The effluent is then available for aquifer re-charge or other uses, and also standards of effluent are likely to be carfully watched.

WAKEFIELD (Coastal Anti-Pollution League Ltd.) asked for clarification of Table 4, Solent Conflicts.

DRUMMOND It is very important to try and analyse the various activities that are going on in an enclosed or semi-enclosed area of water, and to try and see where and if they conflict, because it is then possible to identify the problems which need to be resolved. An X in Table 4 means a conflict between two things. For example, there is a conflict between commercial use of the water and the extraction of gravel. This may not seem immediately obvious but the notes explain that rogue dredgers are interfering with shipping. So there is a problem straight away. What is to be done about people who have been given areas of the Solent in which to dredge gravel but stray into the main shipping lanes? This rather complex table illustrates the sort of steps that will have to be taken in order to look at the problem in the first place.

OAKLEY (J. D. & D. M. Watson). Treatment is a matter of degree. Seventy or eighty years ago a Royal Commission determined that for the circumstances at that time for discharge to inland waters a standard of 30 parts per million suspended solids and 20 parts per million B.O.D. was an adequate norm. This has come to be regarded as 'full treatment', but is not; it is not always adequate in inland waters, and higher standards are often requested. On the other hand, it is more than adequate in other circumstances and it is not always necessary for discharge to large bodies of open water.

We must realise what treatment to the so called 'Royal Commission' Standard does not do. It does not, for example, produce a sterile effluent, and he who drinks it, drinks it at his peril. Nor does it remove much of the nutrients present in the sewage, so that if one is interested in biological effects, 'full treatment' is not going to be adequate. What it does do is to remove or oxidise most of the organic matter having a biochemical oxygen demand, but it is generally recognised that oxygen depletion in coastal waters and estuaries is not often a very significant problem.

We must make a balance between what treatment we do on land, and what we leave to be done in the sea. In some circumstances, treatment on land to a 30 : 20 standard is proper; in other circumstances it is wildly extravagent.

DRUMMOND The table on conflicts will show the acceptability of any effluent, i.e. if it does not materially effect any one of the legitimate uses then the level of treatment is satisfactory. If however it does put somebody off bathing it does lessen the recreational resources. There is not sufficient recreational resource in this crowded island to permit this reduction.

CHARLESWORTH (Rural District Councils Association). Surely we want to find that degree and that level of treatment that suits the particular environment into which we are placing these wastes, and if we can find that for estuaries, the problem is solved.

LEGAL CONTROLS AND LEGISLATION

A. S. Wisdom, Solicitor

Clerk and Chief Officer, Avon and Dorset River Authority

INTRODUCTION

1. Amid the papers before this Conference which is considering in its various aspects the conditions, influences, problems and pressures of pollution in estuaries with particular reference to the Solent, it is appropriate that mention should be included about responsibilities for both causing and remedying such pollution within a legal framework. Which is what this paper is all about.

2. Before embarking on the categories of the legislation and enforcement agencies, it would be useful to know how pollution in estuaries arises and the various forms it takes in order to ensure that the legal remedies and legislative controls are aimed at the root causes and consequences.

3. Despite the vast degree of dilution afforded by salt water, the consequences of a pulluted tideway are apparent by the detraction occasioned to amenities and recreation, fouled beaches, risks to public health and bathing, and sustained injury to navigation, migratory fish, shell fish, plant life and so forth.

4. Causes of pollution in estuaries (and I suppose this covers most other tidal waters) include:—

(1) Polluted river waters entering the estuary from above the tidal limit. Since comprehensive powers already exist which enable River Authorities to control effluents discharged to inland waters and to prevent other types of pollution (such as oil) in non-tidal waters, this heading is outside the purview of the paper.
(2) Sewage and industrial effluents discharged directly to an estuary or coastal waters.
(3) Escapes of crude sewage, bilge contents, etc from vessels afloat.
(4) The dumping of waste materials off-shore.
(5) Discharge of sludge or industrial matter through long pipe lines.
(5) Escapes of oil from tankers and vessels at sea.
(7) Escapes of matter from off-shore installations.
(8) Toxic substances drifting ashore from wrecks.

5. At present we are living in and going through an era of continuous change and progress where existing institutions are being made 'more viable' in terms of area, functions and finance, and where current techniques and processes revolve in a continued cycle of improvement and innovation. Against this background and the modern mode of public opinion which has been educated through the various media of information

to visualise 'pollution' as a prohibited expression, it is timely and indeed inevitable that the legislation and administrative measures relating to pollution should be tightened up, amended and improved.

6. Over the last few years there has evolved a spate of reports and official documents recommending and urging reform in the pollution world. These include:—

The Report of the Technical Committee on the Disposal of Toxic Solid Waste (1970).
The Report of the Working Party on Sewage Disposal, Taken for Granted (1970).
Protection of the Environment, The Fight Against Pollution (1970)
Report of the Central Advisory Water Committee, Future Management of Water in England and Wales (1971).
Water and Sewage Services: England, Explanatory Memorandum on Proposals for Reorganisation (1971).
Circulars No. 10/72 and 18/72 giving the Ministers' Conclusions on the Report of the Working Party on Sewage Disposal.
Report by the Working Party on the Control of Pollution, Nuisance or Nemesis (1972).
Third Report of the Royal Commission on Environmental Pollution, Pollution in Estuaries and Coastal Waters (1972).
Consultation Papers on various aspects of Pollution issued by the Secretary of State.

HISTORICAL BACKGROUND

7. Prohibitions against maritime pollution have been on the statute book since early times. An ordinance of Henry VIII in 1542 imposed a penalty of £5 upon the casting or unloading of miscellaneous matter like filth, rubbish and ballast from vessels within a haven or port. Tin ore extraction in the West Country and its effects on ports was the subject of statutes in 1531 and 1535. Under Acts of 1745, 1814 and 1847 the ejectment of rubbish, earth, filth and solid matters from vessels or from the shore into harbours and docks was made an offence under fine.

8. The first comprehensive general statute dealing with pollution, the Rivers Pollution Prevention Act 1876, could be applied to such parts and points of the sea and tidal waters as determined by order of the local Government Board after a local enquiry and on sanitary grounds, but it was not an offence to carry sewage to tidal waters under statutory powers. Only six orders were made under that Act.

9. Under the Public Health Acts the discharge of crude sewage into tidal water has not been prevented, although at common law there is no right to discharge sewage to sea so as to cause a nuisance and a right to do so cannot be acquired by prescription.

10. The existing partway controls exercised by river authorities over effluent discharges made into estuaries and coastal waters stem from

the Rivers (Prevention of Pollution) Acts 1951-1961. Which are referred to later on.

11. Various Oil in Navigable Waters Acts have sought successively to restrict the effects of oil discharged from tankers, other vessels and shore-based installations into navigable waters, territorial waters and designated areas of the high seas.

PROPOSALS FOR FURTHER POLLUTION LEGISLATION AFFECTING TIDAL WATERS

12. The Government's intentions for re-organising the water services as envisaged in the Department of the Environment circulars 70/71 and 92/71 (and presumably, enshrined permanently in the Water Act 1973 by the time this Conference takes place) gave rise to a series of consultation papers issued by the Secretary of State which included several relating to forthcoming legislation on pollution. The first pollution paper was entitled 'Control of all discharges to tidal waters, estuaries and the sea' and pages 5-8 outlined in some detail the Secretary of State's proposals for revising the Rivers (Prevention of Pollution) Acts 1951 and 1961, particularly as relating to discharges into tidal waters. In a supplement to that paper it was affirmed that the Ministers would also take into account the recommendations set out in the third report of the Royal Commission on Environmental Pollution dealing with pollution in some British estuaries and coastal waters.

13. The second pollution consultation paper dealt with pollution from boats which has a somwhat confined recommendation affecting tidal waters. Another consultation paper has recently come out on oil pollution in estuaries.

14. Before the Water Bill (for the re-organisation of water and sewage) was first published there was a faint hope that it might incorporate clauses giving effect to all or some of the promised revisions to the Rivers (Prevention of Pollution) Acts 1951-1961, but the Bill appeared towards the end of January 1973 without reference to these revisions, although the Bill did provide for:—

(1) The existing pollution prevention functions of river authorities to be transferred to the new regional water authorities on 1st April 1974.

(2) The present responsibilities of local authorities for public sewers and sewage disposal works being taken over by the regional water authorities on 1st April 1974, subject to arrangements being concluded for the local authorities to continue discharging after that date sewerage functions apart from sewage disposal and certain sewers.

(3) Where in future new or altered outlets or discharges would be the responsibility of regional water authorities, they would become subject to regulations made by the Secretary of State providing that

any consents required for the outlets or discharges under the Rivers (Prevention of Pollution) Acts 1951-1961 must be granted (or deemed to be granted) by the Secretary of State.

(4) The Secretary of State being empowered to give directions for a joint committee to be set up by two or more regional water authorities for discharging their functions (relating to the restoration and maintenance of the wholesomeness of rivers and other waters) regarding an estuary situated within the areas of those authorities. There is, however, nothing to prevent the authorities from doing this on a voluntary basis.

(5) The net cost of water services (apart from land drainage) to be met by charges to be fixed by the regional water authorities by means of a charges scheme or by agreement. In fixing charges regard must be had to the cost of performing services, providing facilities or making available rights, by the authority.

(6) A new clause giving regional water authorities power to assist local authorities in taking appropriate action in the event of emergencies or disasters involving actual or prospective destruction of or damage to life or property. This has a possible context in combating a serious escape or spillage of oil affecting an estuary or area of coast.

(7) The requirement that each regional water authority should as soon as practicable after 1st. April 1974 carry out a survey of the water in their area, its existing management, the purposes for which it is used *and its quality* in relation to its existing and likely future uses, and prepare a report of the results of the survey. Then to prepare an estimate of future demand for the use of the water over a period of twenty years, following the completion of the report or such shorter time as the appropriate Ministers might direct. Finally, each authority must prepare a plan as to the action to be taken for securing more efficient management of water, including meeting future demands for water and its use and restoring and maintaining the wholesomeness of rivers and other inland and coastal waters. The reviews and plans have to be updated every seven years. In addition, each authority has to prepare programmes for discharging their functions over the next seven years subject to the approval of the appropriate Ministers.

15. This will enable a long-term policy to be established for estuaries by the regional water authorities.

LEGAL CONTROLS AND ENFORCEMENT AGENCIES

16. Next follows an explanation of the statutes concerned with pollution in estuaries (and this usually includes coastal or other tidal waters since the operation of the Acts are rarely confined to estuaries) and the public bodies responsible for enforcing the statutory provisions. Since

much of this law is under promise of revision, where possible a comparison is made between the present law and what the law may be after revision.

Effluent Discharges to Estuaries under the Rivers (Prevention of Pollution) Acts 1951-1961

Present Position

(1) Clean Rivers (Estuaries and Tidal Waters) Act 1960

17. New or altered outlets and new or substantially altered discharges of trade and sewage effluent made into the tidal waters within the seaward limits of most of the estuaries in England and Wales which are designated as 'controlled waters' and individually specified in the schedule of the 1960 Act are subject to control by the river authorities.

18. In each case the river authority may refuse their consent for a discharge or outlet, or grant it subject to conditions relating to the nature or composition, temperature, volume or rate of discharge in the case of an effluent, or relating to the point of discharge and construction of an outlet.

(2) Rivers (Prevention of Pollution) Acts 1951 and 1961

19. A river authority may apply to the Secretary of State for a 'tidal waters order' to enable them to exercise control over existing and new discharges of trade or sewage effluent (including outlets for such discharges) which are made into tidal waters. Only a river authority or other persons appearing to the Secretary of State to be 'interested' may apply. So far about fourteen orders have been made, whilst some applications for orders are under consideration. There is no provision for an order to be made to cover discharges to tidal waters existing before 1960, with the result that many discharges from public sewers and industrial effluents of long standing are outside the control procedure.

Future Position

20. The Government's intentions for introducing further legislation to revise the Rivers (Prevention of Pollution) Acts 1951-1961 are set out in the D.o.E. consultation papers (water pollution control papers Nos 1 and 5) and may be summarized briefly as follows:—

(1) S. 2 of the 1951 Act (prohibiting the discharge of poisonous, noxious or polluting matter to inland streams) will be applied to similar discharges to all tidal waters and the sea, except that matter dumped

in those waters will be exempted where the Minister of Agriculture, Fisheries and Food has granted consent.

(2) S. 3 of the 1951 Act (preventative pollution proceedings taken in the County Court) and S. 4 of that Act (cleansing bed of stream, cutting vegetation) may only be applied to particular parts of tidal waters, estuaries and the sea as are specified by the Secretary of the State by order after consultation with the Minister of Agriculture, Fisheries and Food and the regional water authorities. The order may be subject to negative resolution and might be amended from time to time. The argument for adopting this qualified procedure by order and bringing in only certain stretches of estuaries, etc. is that Ss. 3 and 4 are more applicable to inland rivers than to tidal waters, although (to quote the relevant paper) 'On the other hand the characteristics of a river in this context immediately above and below its tidal limit are often so similar that it would be difficult to justify confining the provision to freshwater...'

(3) S. 7 of the 1951 Act (consent required for post-1951 effluent discharges to rivers and for outlets for such discharges) will be applied to discharges to all tidal waters and the sea, again subject to matter dumped in such waters being exempt in cases where the Minister of Agriculture Fisheries and Food has granted consent.

(3a) There is a side effect from S. 7 (and this also arises under S. 1 of the 1961 Act—see the next para: (4)) in that it is an offence under S. 12 of the 1961 Act to disclose information supplied in connection with an application for consent. The Royal Commission on Environmental Pollution stated in their third report (p. 39, para: 108) that whilst S. 12 placed restrictions on the disclosure of information about effluents given to river authorities in administering control, or from samples taken for that purpose (as distinct from samples taken of river water where there is no restriction on the disclosure of information), the Commission did not consider that S. 12 should be repealed although wider disclosure should be made to responsible persons than there is at present of the nature and quantities of industrial effluents into tidal waters and estuaries.

(3b) The Secretary of State has also considered S. 12 in the somewhat wider context of whether the common law rights of riparian owners to abate pollution by action and injunction should or should not be replaced by a right to obtain compensation from the regional water authority where damage is suffered as a result of a discharge made in accordance with a consent. The Secretary of State set out his views in the consultation papers on water pollution control Nos. 1 and 5 which were issued before the publication of the Royal Commissions third report.

(3c) The Secretary of State's up to date thoughts on S. 12 have not apparently been clarified in subsequent writing, but it seems likely that S. 12 will not be altered and that a regional water authority will be required (in subsequent pollution legislation) to send copies of applications for consent to discharge to estuaries or the sea (includ-

ing their own proposals for new or altered outlets) to the Secretary of State and to defer a decision for a period in case the Department wish to indicate that they will call in the application for decision.
(4) S. 1 of the 1961 Act (consent necessary for pre-1951 effluent discharges to rivers) will be brought into effect so as to include similar discharges to tidal waters and the sea as regards discharges in existence prior to the revised legislation becoming law but the date of application of S. 1 to tidal waters will not commence until such date as the Secretary of State will appoint. The provisions referred to in para (3) above about applications and proposals being called in will similarly apply.
(5) The provisions of S. 5 of the 1961 Act giving river authorities the power to revoke or vary conditions of consents issued under S. 7 of the 1951 Act or S. 1 of the 1961 Act at periods of not less than two years will be incorporated with respect to consents relating to discharges to estuaries or the sea. Copies of the notices of variation or revocation will have to be sent to both the Secretary of State and the Minister, who will have power jointly to call in for determination a proposed variation or revocation.

21. The second pollution consultation paper, which referred to pollution from boats, proposes further legislation on that subject which affects both inland and tidal waters. As regards vessels fitted with sanitary appliances from which polluting matter does or may pass into the water, the amending legislation will:—

(1) prohibit the keeping or use in any freshwater stream of such vessels after 1st April 1979.
(2) until 1st April 1979 the regional water authorities would have power to continue working byelaws regarding boats fitted with sanitary appliances on freshwater under S. 5 of the 1951 Act as suitably amended.
(3) empower indefinitely the regional water authorities to make byelaws applying to boats or ships in tidal or coastal waters or, alternatively, giving them or local authorities power to apply to the Secretary of State for the application following 1st April 1979 of the general ban to specified parts of tidal or coastal waters.

22. In addition, the proposed legislation would enable regional water authorities:—

(1) to construct or contribute towards the cost of sanitary stations (either on-shore or on vessels afloat) for collecting polluting matter from boats or ships for disposal.
(2) to provide or contribute towards the cost of providing lavatories for use primarily by persons on boats not normally fitted with sanitary appliances.
(3) when imposing a charge for the registration of boats to include an element of cost for meeting points (1) and (2) above, with power to fix different charges for different classes of boats.

(4) to board and inspect any boat for the purpose of enforcing the above provisions.

Effluent Discharges to Estuaries under Public Health Legislation

Present Position

23. There is nothing in the public health legislation which prevents the discharge of sewage through pipeline or outfall into coastal waters, save the purely financial safeguard in that the sanction of the Secretary of State or Welsh Office is required for loans on capital expenditure for sewage disposal schemes. Large loans for new schemes involving coastal discharges will not be santioned if there is risk to public health or amenity.

24. The Public Health (Drainage of Trade Premises) Act 1937 as amended by Part V of the Public Health Act 1961 gives occupiers of trade premises a qualified right to discharge their effluent to a public sewer with the consent of the local authority. Certain substances which would adversely affect a sewer are prohibited and some industrial effluents, mostly of long standing, do not require consent. The result is that industrial discharges to estuaries and tidal waters through a public sewer, which escape the consent of the local authority under the above Acts, and do not require the consent of the river authority because the discharge originated before 1960, are not subjected to any controls.

Future Position

25. The Water Bill provides that as from 1st April 1974 the regional water authorities shall, in substitution for the local authorities, assume responsibilities for providing public sewers and sewage disposal works, subject to their making arrangements for the local authorities to discharge such responsibilities except for sewage disposal and also sewers receiving sewerage from two or more local government areas.

26. This means that the operation, maintenance and future provision of sewage treatment works and plant and the attendant responsibilities for sewage effluent discharges and outfalls will fall upon the new water authorities. When they propose to do anything regarding a discharge to an estuary or the sea, which requires consent under the Rivers (Prevention of Pollution) Acts 1951-1961, it is intended that they shall forward copies of their application jointly to the Secretary of State and the Minister of Agriculture, Fisheries and Food, who may call in the application within a stated period.

27. Under the Bill the water authorities will also on 1st April 1974 become responsible for enforcing the Rivers (Prevention of Pollution) Acts 1951-1961 the pertinent provisions of which Acts it is intended at some time thereafter (in accordance with the promised revision of these

statutes) shall be applied more strictly and generally to tidal waters and the sea. Accordingly, it will be increasingly incumbent upon the water authorities to keep their own house (more specifically—their tidal sewers) in order. This will take time and money to achieve, and might conceivably be carried out under a seven year functional programme of the type referred to in clause 22 (5) of the Water Bill.

Discharges Within Limits of Territorial Waters under the Sea Fisheries Regulation Act 1966

Present Position

28. Sea fisheries committees have a limited power of control over tidal pollution within their districts, in that they may make byelaws under the Sea Fisheries Regulation Act 1966, or previous repealed legislation for prohibiting or regulating the deposit or discharge of solid or liquid substances detrimental to sea fish or sea fishing within the three mile limit of territorial waters as long as the byelaws do not affect the powers of local authorities to discharge sewage.

Future Position

29. The Government propose (in their water pollution control consultation paper No. 1.) that, since the sea is believed to receive most of its pollution from rivers and estuaries, it would be irrational to control direct pollution of coastal waters independently of the control of estuaries and rivers. Local control by regional water authorities rather than by sea fisheries committees would be preferable because the water authorities will already have an organisation for the control of discharges to rivers and estuaries. It is therefore intended that control of discharges to the sea should be administered by water authorities and that the existing powers of sea fisheries committees to control tidal pollution would be ended.

Oil Pollution in Estuaries

30. The present system for dealing with oil pollution in estuaries has recently been reviewed in a consultation paper entitled 'The responsibility for dealing with oil pollution in estuaries' in the light that regional water authorities once set up will be given responsibility for control of pollution in estuaries.

Present Position

31. The Department of Trade and Industry under the Oil in Navigable Waters Acts (which will be consolidated under the Prevention of Oil Pollution Act 1971 sometime in 1973) have overall responsibility for, and pay for the cost of, clearing oil from the sea and estuaries beyond a mile from shore. Outside that limit the responsibility falls on local

authorities who receive a 50% grant towards expenses; there are some local variations, such as the Port of London Authority who deal with oil within their jurisdiction. Oil in inland waters is the concern of the river authorities.

Future Position

32. If fresh pollution legislation will give regional water authorities a duty to apply the Rivers (Prevention of Pollution) Acts 1951 and 1961 to tidal parts of rivers, estuaries and coastal waters, then those authorities will become responsible for maintaining and improving the wholesomeness of estuaries inside the areas not covered by the Department of Trade and Industry. To what extent local authorities would be involved with the clearance of oil from land and beaches or in an emergency is something which remains to be arranged.

Note on the Prevention of Oil Pollution Act 1971

33. The discharge of oil or mixtures containing oil from vessels or from land or as the result of operations involving the exploration of the sea bed and subsoil or the exploitation into (a) the sea within the seaward limits of the territorial waters of the United Kingdom, or (b) any other waters (including inland waters) within such limits and navigable by sea-going ships is an offence under the Prevention of Oil Pollution Act 1971. There are various defences to a charge in respect of oil discharges from a vessel, such that it was necessary to secure the safety of any vessel, or saving life, or of preventing damage to any vessel or cargo, or that the oil escaped in consequence of damage to the vessel or by reason of leakage and that as soon as practicable thereafter all reasonable steps were taken to stop or reduce the escape of oil.

Dumping Sewage Sludge at Sea

34. A few large cities dispose of sewage sludge at sea from purpose-constructed ships and recently under private Act powers a number of municipalities or joint sewerage boards have in hand proposals for building pipe-lines for the same purpose. Whilst there is no evidence to suggest that these methods are harmful, the risk remains, and studies and monitoring are being carried out by the Government.

Dumping Waste at Sea

35. As one of the signatories of the Oslo Convention on the dumping of wastes in the North East Atlantic, which applies to both territorial waters and the high seas, the Government have promised to introduce legislation which will enable the Minister of Agriculture, Fisheries and Food to control dumping off the shores of England and Wales. This will

be done on advice given by the Ministry's fisheries laboratories and control will extend into estuaries and controlled waters but not up to the tidal limits of rivers. Arrangements will be made for regional water authorities to be informed by the Ministry about proposals for dumping in tidal waters within the jurisdiction of such authorities. The powers which sea fisheries committees now have to control dumping through their byelaws would not be required in future.

DISCUSSION

SPEIGHT (Hampshire River Authority). Over the years the advice given by the Government to local authorities and river authorities has changed with the economic climate. In the past the Government has in effect said that standards should not be put too high as the country cannot afford to pay. We are now in an era of 'environmental concern', but will not the next economic crisis bring the same sort of advice again?

Would the Author comment on experiences of having consent conditions relaxed or varied by the Minister's Inspectors?

WISDOM I have not encountered many appeals. The opinion held by a Government Department is apt to be reflected in the financial restrictions by which Government Departments are influenced. They are not influenced, at the departmental level, by political decisions.

Fines imposed by the magistrates were often derisory, but recently, possibly as a result of public awareness of pollution, more realistic penalties have been imposed.

WAKEFIELD (Coastal Anti-Pollution League Limited). Is it not true that the River Authorities have been much more concerned with cleaning up the rivers than with safeguarding the condition of estuaries? If so, the withdrawal of the Sea Fisheries Committees powers will be a bad thing, as the new Regional Water Authorities are likely to adopt the attitudes of the River Authorities.

WISDOM In the past River Authorities have been more concerned with the quality of inland waters. Concern for estuary water quality is a later development. River Authorities, and before them the Catchment Boards, have only been in existence for a short time and it takes a few years to build up an organisation to carry out surveys. Most of them have now come to the stage where they know the condition of their inland waters and are starting to pay more attention to estuaries, but this is quite a late development and has not really happened until the last three or four years. The prospects under the Water Act 1973 are better because more emphasis is going to be placed on amenity, there are built-in provisions about amenities and recreation in the Act itself, and the new Water Authorities are going to be encouraged and required to clean up their estuaries. The Water Authorities will be only too keen to get on with the job provided they have the money and the staff.

SMITH (University College, Galway). It appears that the intention is that information about trade discharges to water courses will still be confidential under the proposed legislation, thus continuing the restriction under Clause 12 of the Rivers (Prevention of Pollution) Act 1961. It is important that the information should be generally available in order that the public may join in informed discussion on safeguarding amenities.

If the intention is to safeguard trade secrets, then surely it is up to the manufacturer to make sure that he does not dispose of substances to the river which will give away secrets. He should treat them in his plant. If he wishes to use the river, which is a public amenity, he should be prepared to say publicly what he is discharging.

WISDOM The position is that the River Authority cannot disclose to a third party details of a trade effluent sample they take from premises without the consent of the discharger. This is quite sound in appropriate cases. Current thinking is that it has been interpreted too strictly, but the Minister does seem prepared to alter the situation.

It should be stressed that this information is not kept from the River Authority. This is the body responsible for determining whether in the light of that information they believe that pollution would or would not be caused or that the discharge should be made elsewhere.

RIORDAN (Pfizer Chemical Corporation, Ireland). The local authorities are the bodies entrusted with the responsibility for pollution control. They have the technical knowledge to treat probelms in a balanced and reasonable way, whereas the public have not. In this situation it is not always helpful if local authorities disclose information which is likely to be exposed to irresponsible comment by the press and other news media.

CHARLESWORTH (Rural District Councils Association). The new Regional Water Authorities, the Chairmen of which are Government nominees, will be responsible for the purity of sewage effluents and for water abstractions and the control of all rivers and estuaries. It is most unlikely that any river authority is going to prosecute itself if one of its sewage works is producing an unsatisfactory effluent. A further restriction on capital expenditure on improvement of estauries will result from costs of water development being charged and recovered from the domestic or industrial water rate. There will be very strong opposition to considerable increases in water rates, and estuarial improvement is going to be very well down the 'shopping list'.

LITTLE (National Yacht Harbour Association). The Author said that there were difficulties in the way of imposing a ban on discharges of untreated sewage from boats by 1979. What are these difficulties?

WISDOM There are at present only two areas where there are restrictions on vessels discharging sink waste or toilet contents into river or sea waters. One case is the non-tidal Thames. That is because the Authority controlling the Thames had had its own private act pollution powers for the last 100 years. The second place is part of the Norfolk Broads where the River Authority has recently obtained powers to make byelaws. It took the River Authority nearly 5 years to get the powers through the Government Department concerned, and a lot of restrictions are placed upon them which make them less than completely effective. This rather leads to the supposition that this particular issue is being evaded. Wider pollution powers will be given generally to the new Water Authorities as soon as legislation is available for byelaws covering boats provided with sanitary appliances and kept or used on streams or territorial waters.

JURY (Newport I. of W. Council). To what extent are efforts being made to ensure a uniform level of environmental pollution control throughout the E.E.C.? Unless something is done, our home manufacturers will be adversely affected compared with their European neighbours.

WISDOM There are organisations compiling data on pollution and its control in various countries. No doubt codes and procedures will be suggested in the future.

GRIFFIN (Wallace, Evans & Partners). Will a joint committee of Water Authorities be an effective way of looking after the large estuaries?

WISDOM There is provision in the Water Act 1973 for joint committees for estuaries. They are being encouraged and there is no reason to suppose that they will not be successful. Co-operation can be achieved very conveniently and well. It has been happening for at least 20 years on various aspects of water affairs. The new Authorities will have far more money and resources than have the present River Authorities.

MEDICAL ASPECTS OF ESTUARIAL POLLUTION

Angus McGregor, M.A., M.D., F.F.C.M., D.P.H.

Port Medical Officer, Southampton Port Health Authority.

INTRODUCTION

1. As Port Medical Officer to the Southampton Port Health Authority I have responsibility for advising the Authority on all matters likely to affect the public health within the Authority's district or area of jurisdiction and in addition carry out certain other duties. The Authority's boundaries are wider than those of the Docks Authority and actually include all Southampton Water and its beaches and inlets and the middle part of the Solent between lines drawn from Stokes Bay to Ryde Pier on the east and from Stone Point to Gurnard Ledges on the west, but excluding the immediate coastal waters off Cowes to the shoreside of Prince Consort Buoy (Fig 1 page 3.8).

2. Some 25 000 ships entered the district in 1972 carrying over 600 000 persons and a similar number of ships and passengers left. In addition there are several thousand small craft permanently based within the district and still more use it. The duties of a Port Medical Officer include boarding incoming ships to prevent the importation of infectious diseases and other matters but in the context of the subject of this conference attention will be concentrated on the public health aspects of estuarial pollution. The next section of this paper describes the medical hazards that may arise from estuarial pollution and this is followed by a description of some work undertaken by the staff of the Port Health Authority into the state of the water and of its shellfish.

MEDICALLY SIGNIFICANT ESTUARIAL POLLUTANTS

3. The medically significant types of pollution which will now be considered in turn in relation to their source, medical effects and amenity effects are;

Bacterial and viral pollution
Chemical pollution
Waste
Heat
Noise

Bacterial and Viral Pollution

4. The major source of bacterial pollution of the estuary is untreated sewage, which comes not only from the major outfalls but also from direct human and animal defaecation on to beaches and tributary streams and from ships of all sizes. Not only does the Solent area generally receive a substantial volume of sewage—220 $10^3 m^3/d$ (48 million

gallons per day), of which 50 per cent is untreated[1]—but Southampton Water in particular is heavily polluted by both the many large passenger ships that use the port and by the thousands of small craft that make use of the nation's largest and best known yachting facility.

5. Untreated sewage may contain any of the bacteria and viruses pathogenic to man that will survive in fresh and sea water. In practise although the possible number of different types of organism is very great, only a few have medical significance, notably those organisms that multipy in the human bowel, particularly the Salmonella group and polio virus. Further, the dilution effect of putting sewage into an estuary is sufficient to reduce the likely dose to be obtained by gulping and swallowing sea water while swimming below the dosage at which infection will occur (i.e. below the 'infective dose') with two exceptions. Two organisms, *Salmonella typhi* (the cause of typhoid fever) and *Vibrio cholera* (the cause of cholera) both have very low infective doses. This may also be true of more common Salmonella varieties, e.g. *S. typhimurium*, particularly if faecal matter is swallowed in substantial particles, and of the polio virus, though here the evidence is particularly uncertain. The Medical Research Council 1959 report on sewage contamination of bathing beaches in England and Wales[2] explored this point extensively and concluded that there was no evidence of real risk of human infection. Work since then has suggested that the risk may be greater than the M.R.C. found but the point seems unlikely to be ever settled decisively, due to the relatively infrequent appearance of the organisms in this country.

6. However, while the hazard to swimmers or those suffering accidental immersion while boating is slight and the hazard from consuming fresh fish is negligible as cooking kills all the organisms concerned, there remains a serious potential hazard from the consumption of shellfish. Shellfish filter out and therefore concentrate organisms in sea water and are commonly consumed by man without cooking. Very small levels of Salmonella infection in the estuary may accordingly be associated with a high risk of human disease and evidence relating to this risk in Southampton Water is discussed in the next main section of this paper.

7. To put these diseases into perspective it is here necessary to describe them briefly. Typhoid fever is a serious infection carrying a fatality rate of up to 10% without treatment and as high as 2-3% with treatment. As it commonly runs an obscure course diagnosis is often difficult to make and delayed.

8. *S. typhimurium* on the other hand only produces a relatively mild gastroenteritis which may nevertheless be fatal to infants and the elderly.

9. Cholera is a serious disease with a fatality rate of 50% or more without treatment. This may be reduced to very low levels by early diagnosis and full treatment.

10. Poliomyelitis has almost disappeared from this country following the widespread use of vaccination. About 10% of unvaccinated persons contracting the disease become paralysed and some 2-10% of paralysed victims die.

11. Finally one other risk of infection from bathing in contaminated water must be mentioned. This is Leptospirosis or 'Rat Bite Fever.' This disease caused by the *spirochaetal leptospira* is endemic in rats and other wild animals. In man the disease presents as a severe fever with a fatality rate of up to 20% in the elderly. Due to the diluting effect the risk of contracting the infection is probably confined to inlets and rivers. However recreational activity in water seems now to be displacing occupation as the main source of infection (3).

Chemical Pollution

12. By and large chemicals toxic to marine life are also toxic to man so that the main consideration is the dose required to cause human disease and the likelihood of human exposure in practice. Consideration will be given in turn to the heavy metals, organochlorines and other insecticides, nitrates, radioactive chemicals and oil.

13. Of all the heavy metals in practice it is lead and mercury which present the most direct danger to human health. Both are discharged in substantial quantities from industrial processes directly or via sewerage systems into the estuary. Additional contamination results from the fallout of aerial particles on to the water, the major source being leaded motor fuels.

14. Whilst the risk from bacterial contamination can be overcome by cooking food this does not, of course, apply to heavy metal contamination. On the other hand eating habits vary enormously round the world and it has been estimated that the average daily intake of mercury in this country is within safe limits (4). It is nevertheless about 8 microgrammes, about 25% from eating fish, and as mercury has no physiological function can only be regarded as gratuitously unnecessary poisoning. Further while distant water fish have body loads of only 0.00-0.02 p.p.m., fish caught nearer home have loads of up to 0.5 p.p.m. This is still below the generally suggested safe limit of 1 p.p.m., but uncomfortably close. Chronic mercury poisoning may cause disease of the gums and kidneys, tremor, and mental disturbances. In severe cases this may be followed by frank mental illness and kidney failure. This type of poisoning has so far only been seen in industry.

15. Human lead poisoning has also not yet been reported from estuarial pollution, probably because although lead is present, levels are below the toxic point.

16. Organochlorine compounds find their way into the sea from agricultural use. They are persistant and unaffected by cooking so that consumption of contaminated fish directly poisons man. In man these

chemicals are cumulative but no evidence of human poisoning has been found from the effects of the low doses absorbed from fish and other foods. On the other hand, heavy dosage is known to attack the central nervous system causing mental disturbance, tremor, convulsions, respiratory failure and death.

17. Other insecticides and agricultural poisons are all also toxic to man but there is again no clear evidence of poisoning having occurred from the levels found in estuarial waters. Nevertheless it is necessary to reaffirm that these poisons have no physiological function and can only be regarded as at the least an unnecessary load on the human metabolism.

18. Nitrates are used extensively as fertilisers and find their way into rivers and the sea. Where heavy contamination of water used for drinking occurs, then there is a severe risk to the life of infants due to interference by the chemical with blood oxygen transport. Unfortunately, boiling the water only serves to concentrate the chemical still further. However, as sea water extraction and processing for human consumption is still rare, estuarial pollution is of only academic interest at the moment. Moreover the heat distillation processes now in use removes nitrates.

19. Radioactive chemicals are now such a clearly understood risk and so carefully controlled that there appears to be no risk from the known contamination of estuarial waters by these poisons. The same would not, of course, apply if there was any breakdown either of the control of known sources such as laboratories using isotopes or in nuclear reactors whether in power stations or on ships or submarines.

20. Finally, oil pollution is an ever present threat in Southampton Water due to the presence of the refinery and the very large number of tankers using the jetties. In practice spillage and leakage is kept to a remarkably low level considering the enormous volume of oil dealt with—around 18 million tons imported in 1972 and around 8½ million tons exported. Accidents do, of course, occur even in the best regulated circles but danger to human life appears only to arise from accidental immersion in oil covered water. This can give rise to oil inhalation and pneumonia. On the other hand oil contamination of fish renders them inedible and the general loss to amenity from oil pollution accidents can be enormous.

Pollution by Waste

21. Waste may have its origin either in industrial processes, though fortunately there are no major industrial users discharging into Southampton Water, or on a relatively minor scale by personal dumping of refuse—papers, wood, tin, etc. Though not directly damaging to human health this refuse is, of course, both highly destructive of amenity and a regular source of accidents. It is to be regretted that legal powers to control it are almost non-existant and even docks bye-laws are systematically flouted.

Pollution by Heat

22. While power station emissions are detrimental to most marine life they do not appear to have any effect on human health. However, locally they appear to have facilitated the growth of clams and as these are harvested for human consumption, this has led directly to an increased risk to human health.

Pollution by Noise

23. Finally it may not be entirely out of place to record the deteriorating situation with regard to noise pollution. Traditionally shipping and sailing have been relatively quiet activities but of recent years the increasing appearance of power boats, often towing water skiers, has caused high noise levels and substantial risk of accidents to bathers and small boat users. The appearance of hovercraft has added an intolerable further burden as these craft emit noise at a level which would not be acceptable in a lorry that violated our extremely lax emission limits. There is no way of legally controlling these craft and no evidence of any improvement. Whilst no one will resent the use of foghorns in the interests of safety, hovercraft do nothing to benefit the amenities or health of the estuary.

STUDIES OF PORT HEALTH DISTRICT WATER AND SHELLFISH

24. It has been known for many years that Southampton Water, the rivers entering it, and the Solent, though to a lesser extent, are all contaminated. Considerable investigations locally were made in 1953-1958 in association with the Medical Research Council investigation of sewage contamination of bathing beaches.[2] At that time it was found possible to recover *E. Coli (Type 1)* from the whole of Southampton Water and this can be accepted as clear evidence of contamination.

25. Knowledge of major sources of pollution such as Southampton's Main Sewage outfalls led my predecessor many years ago to advise against swimming in the River Itchen within the City or off the Weston Shore. Public notices have been displayed for many years and systematically disregarded. No hard evidence linking disease in the City with bathing in this contaminated water has ever been found, with the possible exception of two cases of poliomyelitis many years ago, in which the evidence was highly suggestive.

26. The increasing commercial interest in recent years in exploitation of the shellfish beds led to a re-appraisal of the health risk. It was first necessary to up-date the information on water contamination and to this end a series of samples was taken between July and December 1971 from 15 different points along the boundary of the Port Health District—9 from the eastern side and 6 from the western. The sample points are indicated in Fig 1, page 3.8 and described briefly.

Table I Results of analyses of Port Health Authority water samples (1971) Least and most contamined sample from each site

Site	Date	State of Tide	Water temp. °C	Free & Saline Nitrogen mg/l	Alb. Nitrogen mg/l	Nitrate Nitrogen mg/l	4 h Permang. mg/l	Coliforms per 100 ml	*E. Coli (Type 1)* per 100 ml
A	29. 7. 71	Young flood	23	0. 1	0. 08	0. 08	1. 4	50	8
	13. 12. 71	Ebbing	9	0. 18	0. 20	0. 91	0. 65	2 250	1 700
B	29. 7. 71	Young flood	24	0. 08	0. 13	0. 25	1. 1	170	17
	13. 12. 71	Ebbing	8	0. 30	0. 25	0. 82	0. 8	9 000	5 500
C	9. 9. 71	High tide	19	0. 10	0. 10	0. 37	1. 0	900	40
	10. 11. 71	Coming in	8	0. 13	0. 03	0. 49	0. 4	3 500	3 500
D	5. 8. 71	Ebbing	20	0. 03	0. 20	1. 32	1. 1	550	4
	21. 12. 71	Flooding	11	0. 26	0. 12	0. 82	0. 7	1 100	800
E	5. 8. 71	Ebbing	21	0. 26	0. 92	1. 40	13. 4	1 800	90
	16. 11. 71	High tide	Not taken	0. 18	0. 02	0. 58	0. 2	18 000+	16 000
F	5. 8. 71	Ebbing	19	0. 36	0. 35	1. 98	3. 3	18 000+	18 000+
	21. 12. 71	Ebbing	11	0. 26	0. 15	0. 66	0. 5	5 500	3 500

H	11.8.71	Young flood	20	0.02	0.17	1.24	0.6	16 000	5 500
	21.10.71	Flooding	15	0.17	0.08	0.62	0.2	1 100	40
J	11.8.71	Young flood	19	0.02	0.15	0.74	0.55	2 500	1 300
	29.12.71	H.W. + 5 h 40 mins.	8	0.15	0.05	0.41	0.40	1 300	350
K	11.8.71	Young flood	17	0.02	0.02	3.1	2.2	3 500	1 100
	21.10.71	Flooding	14	0.13	0.07	1.52	1.3	2 250	500
L	19.8.71	Ebbing	21	0.15	0.02	0.82	0.55	18 000+	18 000+
	1.12.71	H.W.+ 3h	10	0.99	0.12	0.1	0.55	3 500	2 500
M	19.8.71	Ebbing	21	0.18	0.08	0.91	1.05	16 000	3 500
	27.10.71	Flooding	13	0.30	0.46	0.91	12.6	9 000	1 700
N	27.10.71	Flooding	14	0.40	0.20	0.16	0.4	5 500	2 500
	1.12.71	H.W.+ $4\frac{1}{2}$h	10	0.48	0.03	0.1	0.65	18 000+	18 000+
P	2.11.71	Ebbing	15	0.12	0.12	0.74	0.4	0	0
	6.12.71	H.W.—2h	8	0.24	0.05	0.1	0.25	2 500	1 300
Q	25.8.71	Flood	19	0.02	0.13	0.25	0.8	250	20
	2.11.71	Ebbing	14	0.04	0.12	0.16	0.4	1 600	1 600
R	25.8.71	Rising	19	0.04	0.14	0.41	0.90	350	8
	6.12.71	H.W.—$\frac{1}{2}$h	18	0.13	0.02	0.1	0.30	2 500	250

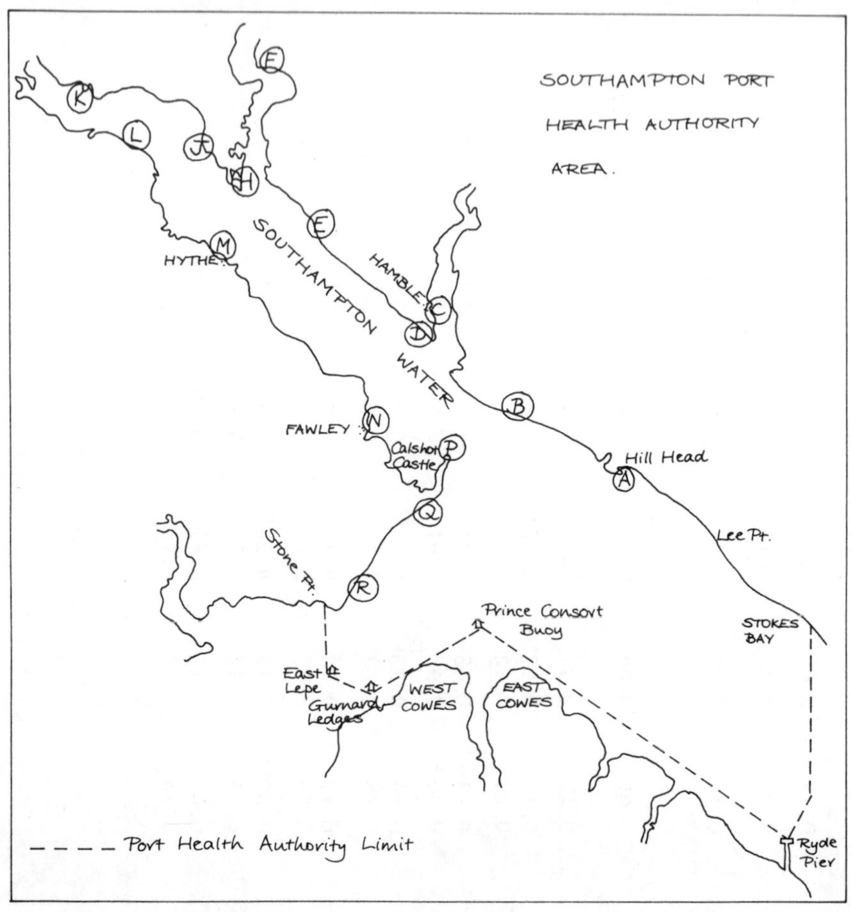

Points at which water samples taken (1971)

Description of Sampling Points

A	Shingle spit at Hill Head	J	Test River opposite Mayflower Park
B	Shingle spit at Solent Breezes	K	River Test at Redbridge
C	Pontoon in Hamble River	L	Test River at Cracknore Hard
D	Pontoon off Jetty	M	Hythe Beach
E	Weston Shore	N	Jetty at Fawley
F	Itchen River Nr. Northam Bridge	P	Jetty at Calshot
		Q	Beach at Stanswood Bay East
H	Itchen River at Dock Head	R	Beach at Stanswood Bay West

Fig 1

27. The results of analyses by Southampton City Analyst are set out in Table I page 3.6. They will be seen to include 2 sample results from each sampling point. These are the results of analysing the least and most heavily polluted of 4 samples taken.

Discussion of results of Port Health Water Sampling

28. There are very large variations in the results not only from site to site but from sample to sample at the same site. Further there appears to be no obvious correlation, contrary to expectations, between the number of Coliforms and the permanganate and nitrogen figures. Conceivably this lack of correlation may result from the majority of sewage pollution coming from treated effluent and *not* from crude sewage. No sample could be classed as grossly polluted but on the other hand all showed some evidence of pollution and departed markedly from natural sea water. Similar levels of pollution are of course seen in many of our rivers used for water supply extraction, as the results of sampling the Test and Itchen given in Table 2 below indicate.

Table 2 Results of sampling water in the Rivers Test and Itchen at the points of abstraction for public supply

1971 Bacteriological Analyses	River Test Coliforms per 100 ml.	*E. Coli (Type 1)* per 100 ml	River Itchen Coliforms per 100 ml.	*E. Coli (Type 1)* per 100 ml
Best samples	600	260	230	100
Worst samples	14 000	4 500	4 800	900
Chemical Analyses	Max. (mg/l)	Min. (mg/l)	Max. (mg/l)	Min. (mg/l)
4 h $KMnO_4$ value at 27°C	3.3	0.08	0.95	0.45
Free & Saline Ammonia	0.21	0.01	0.05	0.01
Albuminoid Ammonia	0.37	0.05	0.11	0.01

29. However, overall it was clear from the results of the water samples taken in the Port Health Authority area that there was always some degree of pollution of the water and that at some places it could rise to substantial levels. Accordingly it was considered necessary to test the shellfish.

3.10

PORT HEALTH AUTHORITY SHELLFISH SAMPLING

30. Shellfish sampling began in September 1971 as soon as the first water sample results began to come in and continued in 1972. Samples were taken mainly of clams but a few oysters were also tested. The results are given below in Table 3.

Table 3 Results of Shellfish Sampling

Date	Shellfish Type	Position taken	Result of Culture
16.9.71	Live Clams	Marchwood	No enteric or salmonella organisms. No *E. Coli (Type 1)* or *Clostridia* isolated.
25.10.71	Live Oysters	Stanswood Bay	A growth of *E. Coli (Type 1)*. No enteric or salmonella organisms isolated.
	Live Oysters	Stanswood Bay	
27.10.71	Clams	Marchwood	A growth of Coliforms. No salmonella isolated.
	Clams	Marchwood	
	Clams	Marchwood	
	Clams	Marchwood.	
26.11.71	Oysters	See Note 1.	A growth of Coliforms. No salmonella.
	Clams	See Note 1.	No *E. Coli (Type 1)*. A scanty growth of *Salmonella Amersfoort* was isolated.
16.12.71	Clams	See Note 2.	No growth of salmonella or *E. Coli (Type 1)*.
	Clams	See Note 2.	
10.1.72	Clams	Marchwood	No salmonella or *E. Coli (Type 1)* isolated.
	Clams	Marchwood	No salmonella isolated. No. of *E. Coli (Type 1)* 1/140g tissue
	Clams	Marchwood	
			No. of *E. Coli (Type 1)* per g body tissue
24.3.72	Live Clams	Newtown Relays	0
	Live Clams	Newtown Relays	1
	Live Clams	Newtown Lays	0
	Live Clams	Marchwood	5
			No. of *E. Coli (Type 1)* per g body
	Live Clams	Marchwood	5
	Live Clams	Marchwood	6
28.3.72	Live Oysters	Stanswood Bay	No. of *E. Coli (Type 1)* per g body tissue = 1
	Live Clams	Southampton Water —Town Quay	No. of *E. Coli (Type 1)* per g body tissue = 6
	Live Clams	Southampton Water —Town Quay	No. of *E. Coli (Type 1)* per g body tissue = 1
	Live Clams	Southampton Water —Town Quay	Growth of *Salmonella Senftenberg*. No. of *E. Coli (Type 1)* per g body tissue = 3
19.4.72	Live Clams	See Note 3.	No. of *E. Coli (Type 1)* per g of body tissue = 1. No Salmonella.

Table 3 Results of Shellfish Sampling (contd.)

Date	Shellfish Type	Position taken	Result of Culture
19.4.72	Live Clams	See Note 3.	No. of *E. Coli (Type 1)* per g of body tissue = 5. No Salmonella.
17.8.72	Live Clams	Marchwood	No. of Coliforms per 100 ml = 900 No. of *E. Coli (Type 1)* per 100 ml = 80
	Live Clams	Marchwood	No. of Coliforms per 100 ml = 25 No. of *E. Coli (Type 1)* per 100 ml = 13
	Live Clams	Gymp Foreshore	No. of Coliforms per 100 ml = 900 No. of *E. Coli (Type 1)* per 100 ml = 350
20.11.72	Fresh Mussels	Southampton Water	*E. Coli (Type 1)* = 1 per g of tissue No Salmonella organisms isolated
	Fresh Mussels	Southampton Water	*E. Coli (Type 1)* = 4 per g of tissue No Salmonella organisms isolated
	Fresh Mussels	Southampton Water	*E. Coli (Type 1)* = 3 per g of tissue Cultures yielded a growth of *Salmonella sp.*

Note 1 Oysters taken from commercial company lifting within district
Note 2 Clams taken from commercial company lifting within district
Note 3 Clams taken from commercial company lifting within district

Discussion of results of shellfish sampling

31. Shellfish growing in clean sea water would not contain any *E. Coli (Type 1)* at all. The finding of *E. Coli (Type 1)* in the majority of the samples confirms that the pollution of the water over the beds does in fact and not merely in theory lead to pollution of the shellfish themselves. Further the isolation of two different species of Salmonellae (*S. senftenberg* and *S. amersfoort*), both rather uncommon, demonstrates the efficiency with which shellfish can recover organisms from the sea even at very high dilutions.

32. The investigations therefore confirmed both that the water in the Port Health district is polluted and that shellfish are in turn infected to a degree which in some cases would be sufficient to cause human disease. The Authority's decision to take legal action to forbid the harvesting of shellfish for sale for human consumption anywhere within the Port Health district was fully confirmed. Finally it should be noted that as living shellfish clear themselves of infection if they are relaid

in clean water the Authority permits lifting subject to satisfactory evidence that the shellfish will be treated at an approved establishment.

CONCLUSIONS

33. The first section of this paper was a discussion of the medical hazards of estuarial pollution. In the second section it was shown that all these apparently highly theoretical considerations have real importance in the causation of human diseases. Not only did sampling confirm the presence of bacterial pollution throughout the port district but shellfish are in fact and not merely in theory picking up this infection. As shellfish are commonly eaten uncooked they therefore present a real hazard to human health.

34. The natural, if irrational, human proclivity to ignore dangers that can not be seen or do not have immediate consequences has resulted in practice in very large numbers of children and adults exposing themselves to infection by bathing in the polluted waters of Southampton Water. Further many adults are prepared to harvest shellfish in Southampton Water for sale to others and for their own use, despite both the legal prohibition and the reason for it—the serious risk of major disease.

35. Unfortunately, while public interest and concern in the problems of pollution of the environment is increasing, it has still not reached the point at which the government is compelled to interfere massively with the cheapest form of human sewage and industrial waste disposal—free discharge into the sea—by banning it. It therefore seems all too likely that the level of pollution will continue to rise and with it the direct danger to human health.

36. With the limited resources available to them the staff of Southampton Port Health Authority can do no more than attempt to protect the public's health from obvious dangers, working within the powers given by the existing law. There is neither the time nor the skill to investigate all possible dangers and it is vitally important that knowledge and information is accumulated systematically about the whole estuary, if all concerned, not least the public, are to be in a position to press for remedial action at national as well as local level.

ACKNOWLEDGEMENTS

I wish to asknowledge with thanks the devoted work of the Port Health Inspectors and the assistance received from the Director of the Public Health Laboratory, Dr. J. Graham, and the City Analyst, Mr. H. Dedicoat.

REFERENCES

1. Department of the Environment, Third Report of Royal Commission on Environmental Pollution. Command 5054. H.M.S.O., London, 1972.
2. Medical Research Council, Sewage contamination of bathing beaches in England and Wales. Memorandum No. 37. H.M.S.O., London, 1959.
3. Weekly Epidemiological Record, 26th January, 1973. No. 4. p. 59. W.H.O., Geneva.
4. Ministry of Agriculture Fisheries and Food, Survey of Metals in Food. H.M.S.O., London, 1971.

DISCUSSION

McGREGOR Since the paper was written, more evidence has come in that leptospirosis, which was thought to be a disease only contracted from stagnant waters, is probably becoming more dangerous in open water. (See para 11) Recent American work[3] showed that two thirds of cases in 1971 occurred in casual bathers. In the past this would have been considered extraordinary—it was a disease of those who worked in sewers or bathed in stagnant pools.

Turning to chemical pollution, locally there is no trouble at present. There are no significant releases of lead or cadmium or mercury, which are the three major poisons. Around the shores of Britain mercury contamination of fish is now clearly present, [4] but at not more than half the estimated safe limit. This is thoroughly unsatisfactory. The limit itself is based on a theoretical concept that the human body can tolerate certain levels of metallic poisoning for an indefinite period, when in fact there are no long term results to draw upon at the present time. Further, it does not seem likely that levels of contamination will drop. Experience suggests the opposite. Lead has now been recovered from shellfish from all European coastal areas.

Samples of clams collected recently were found to contain *Vibrio parahaemolyticus* which is also the cause of food poisoning in man. Mussels sampled in the River Hamble were not infected, but had such an oily taste that they were not edible.

Notices are displayed banning the lifting of shellfish. This should be enough to protect people from infection. But in practice illegal lifting of shellfish for direct sale does go on. It is almost impossible to control this activity. The trawlers come in from the open sea, at irregular times, take the shellfish they need and depart again. The trawler has only got to get near to a jetty for a quarter of an hour, get rid of its oysters and clams on to a lorry, and then disappear at high speed.

In the past it has not been possible to show medically a danger to health because one has concentrated on the sort of danger swimmers are exposed to in taking a casual dip. This has always produced a most difficult problem of proof. Thousands of people visit the coast. Some of them take ill, some of them do not. They return home. All have bathed at some stage.

With the increasing pollution of Southampton Water the point is being reached at which we can expect to see genuine seaborne outbreaks of disease. From this point of view then, the only satisfactory solution is land-based treatment of all sewage. Even though it does not remove infection completely it does at least reduce it.

The present situation is not satisfactory. There is now clear evidence that shellfish are concentrating dangerous human disease organisms from the water and there is considerable danger that deterioration in conditions is occurring.

HAMILTON (Institute for Marine Environmental Research). Is sufficient attention paid to the possible long-term effects of low levels of pollution, for instance by mercury and lead? Processes of adaption and compensation are possibly taking place within the ecological system of which man is a part. Medical aspects of this do not appear to be receiving attention.

On another topic, how significant are counts of *E. Coli* and other bateriological organisms? Very wide ranges of values are quoted. Have they any practical significance?

McGREGOR It is correct that there is some adaptation to heavy metals. The point is that they nevertheless have no physiological function. What is the point in learning to adapt to a poison? Medically speaking interest in the past has been in the acute overdose situation. Lead is an example. This is because it has been a problem in the industrial field where people working in factories handling lead have been subjected to massive overdoses and massive lead poisoning has resulted. In industry low levels of lead poisoning have also been detected, and the effects observed. Until very recently there has been no attempt to evaluate the effects of microscopic doses.

It is only in very recent years, as life expectancy has increased, that we have begun to tackle the diseases of old age. With the eradication of the old killers, we are looking more and more at the insidious poisons that produce disease in the sense of less than full health rather than those that shorten life. All the evidence is that cumulatively over a long period these poisons are adverse. This is a very general statement which cannot be made much more specific at the moment.

E. Coli is found in human faeces in enormous quantities at all times. It is not an organism that normally causes disease in man. It is used as an indicator because presence of *E. Coli* type 1 for practical purposes is evidence of human faecal contamination of the water. The significance of individual figures is therefore very doubtful. Tide, wind and weather all affect the actual levels at any one point in a water system as big as Southampton Water on a single sampling day. Even a multiple sample does no more than give a very rough idea. The sample results shown in the paper show only that infection is present, and no more than that.

SMITH (University College, Galway). From personal experience of work in this field, a major difficulty is to obtain reliable information on the history of the very large number of minor infections, such as those of the ears, nose, and throat, or infection of cuts. To do any work on the etiology of these diseases is extremely difficult. In highly faecally contaminated sea water there is surely a high incidence of what we might call minor infections. This would manifest itself most frequently at the level of cuts that take just a bit longer to get better. This is a real medical problem in our waters. It is extremely difficult to prove. There is a subjective experience that there is a risk, but it is not possible to produce any factual evidence.

To what extent do the beaches represent a reservoir of contamination? Does sewage contaminated sea water deposit organisms on the beach, and the beach then act as an infective source? These kind of problems have not been investigated. What is known about the distribution of multiple antibiotic resistance factors which have been reported from studies of beaches? [6,7] Does sea water represent a reservoir of antibiotic resistance against enteric bacteria? The organism *Vibrio parahaemolyticus* has only recently been looked for. Therefore, none of our present thinking uses experience of sampling this organism. As well as being a cause of food poisoning in human beings, it is an organism which kills shellfish, lobsters and other fish.

There is also the problem of synergistic reactions—the differences in effect of substances found alone or in admixture. We do not know enough about the medical bacteriology of sewage contaminated sea waters.

Is there any inter-relationship between oil pollution and microbial pollution? One of the effects of oil pollution is probably to decrease dissolved oxygen levels in the water underneath the oil, thereby producing anaerobic conditions. In these circumstances sewage borne pathogens like *Clostridium perfringens* may be significant. These conditions will be most common in the head of inlets and mud flats, etc. Money for research into these problems is badly needed.

McGREGOR There is no doubt that a fall into grossly polluted water would turn a cut septic, or at least impair the healing of it. With the open sea, the problem is much less clear cut, and there is a lack of medical knowledge. The classical work in public health and environmental control was done before the discovery of bacteriology. The fact that it cannot be proved now that some of these infections in water are causing human disease in some terms is not an obstacle to taking action. There is a clearly established risk to human health through shellfish.

This is now shown beyond any doubt. Other risks to health are definite possibilities and in other countries with other weather conditions have been proved beyond any doubt. From the medical point of view there is no excuse for waiting. Action must be taken now to remove the contamination entirely, or at very least to contain it at present levels.

BAKER (Field Studies Council Oil Pollution Research Unit). Oil degradation by bacteria is almost entirely an oxygen-requiring process. Experiments to measure oxygen diffusion through oil films in places like rock pools show that there is an oxygen depletion under the oil film, but in the open water with a thin oil film this is unlikely to be significant. It is only important where oil is stranded in sheltered pools and on beaches. When oil gets into muddy sediments, which is often the case in estuaries, it remains there for many years, because the sediments are anaerobic naturally to quite a large extent. The oil may be degraded to a slight extent to use up remaining oxygen. In these anaerobic sediments therefore the oil remains for the longest period of time.

The rate of oil degradation is so variable that generalisation cannot be made. Thin oil films or small droplets of dispersed oil in the water usually degrade rapidly but oil stranded on muddy shores and salt marshes may take a very large number of years to degrade if it sinks into the sediments. Also the bacteria that degrade oil require nutrients such as nitrates and phosphates, so that the degradation rate may sometimes be controlled by the level of these nutrients. As sewage contains these nutrients, a bit of sewage pollution may actually aid the oil degradation process.

McGREGOR The area under discussion is Southampton Water and the Solent where mud abounds and oil spillage occurs. These are the cirsumstances under which oil degradation takes the longest possible time. Recreation areas are largely bounded by these flats. Hence the worry.

WAKEFIELD (Coastal Anti-Pollution League Ltd.). Human life could probably be maintained until everyone lived to 100 or even more, if enough energy and resources were diverted to that one end, but there are many other methods of getting killed or dying. The motor car is probably the biggest killer of all. We have got to preserve amenity values. Is it not better to live a short, but interesting and full life with beauty around us, than to live in an environment which is completely protected and maintains our life for a very great length of time?

McGREGOR There is no basic conflict. It is not suggested that the infections that are obtainable from our waters are by and large lethal. They are not. Typhoid is one which carries a certain mortality, nowadays low because of successful antibiotic treatment. The others by and large are lethal only to infants and old people. Elimination of these infections is needed not because of their effect on the length of life but because of their effect on its quality. This is agreed. Clearing the Solent/Southampton Water complex as a source of danger and disease to human beings is important as a factor influencing quality of life. Most of the infections contracted from eating infected oysters, for example, will cause illnesses lasting two to three weeks only. Mortality will be rare. However, it is not an improvement in the quality of life to get food poisoning when this can be avoided.

The quality of life is important and therefore to remove the oil nuisance is just as important as to remove infection. At the last resort these environmental factors affect mental health. Mr. Drummond used the phrase 'recreate' for recreation. This is medically speaking an impeccable objective—to restore mental balance by recreation.

ARNOLD (Welsh Office). Why has no mention been made so far of die-off of bacteria? Is it not true that disease organisms die rapidly in the alien environment of sea or air, away from the human body? The references by Mr. Smith to a 'reservoir' of pollution in the beach seems contrary to this.

Has enough thought been given to estimating actual *E. Coli* levels at the end of long outfalls where detention times in the outfall may be very long particularly when the pipe is running below design discharges?

McGREGOR The die-off concept is a useful simplification. It is quite clear that generally speaking the organisms that infect man are peculiar to man and cannot exist with any comfort outside his human environment, but it is such a sweeping generalisation that it has little practical value when applied to the sort of conditions under discussion. Viruses survive in water. Poliomyolitis is essentially a bowel infection. Almost all organisms that cause gastroenteritis, diarrhoea and vomitting illnesses will survive in water. Sea water is much more close to the human internal environment than plain water, and so the organisms fare better. Most of the disease organisms, it now seems, will survive quite comfortably in sea water. In Southampton and on the South coast in general the sea water is relatively warm so survival may be longer here.

With reference to beaches, we assume too readily that disease organisms die away. It seems much more likely that many of them can survive for quite extended periods given a protective environment. For example, solid human faeces protects the organisms inside for a long period. The die-off theory does not have the applicability that it was once thought to have.

JURY (Newport I. of W. Borough Council). None of our present sewage treatment works aim specifically to remove pathogens. Must attention now switch to this problem? If so, this represents a major change in approach.

McGREGOR There is some reduction in microbial population in a sewage treatment works. Full treatment is the first step. When this is universally practised, perhaps pressure for sterilisation of effluents will be appropriate. At present it is a side issue.

TUCK and HAWES (C.E.G.B.). The author states in para 22 of his Paper 'While power station emissions are detrimental to most marine life they do not appear to have any effect on human health'. Could he please explain what these detrimental effects are?

McGREGOR The water temperature at the point of emission is high enough to discourage many forms of marine life while encouraging others.

6 Smith, H. W., Incidence of R* *Escherichia coli* in coastal bathing waters of Britain, *Nature,* **234,** 155-156, 1971.

7 Feary, T. M. et al, Antibiotic resistant coliforms in fresh and salt water, *Archives of Environmental Health,* **25,** 215-220, 1972.

INDUSTRIALISATION AND THE BIOLOGY OF SOUTHAMPTON WATER

P. D. V. Savage, B.Sc., M.I.Biol.

Central Electricity Generating Board, Marine Biological Laboratory, Fawley, Southampton.

INTRODUCTION

1. This paper is concerned with the descriptive biology of Southampton Water. A detailed knowledge of the changes that are occurring in the fauna and flora will allow a realistic assessment to be made of the effects of industrialisation. The work described is being carried on mainly by the C.E.G.B. Marine Biology Laboratory at Fawley, Southampton University, and a number of local laboratories.

2. In looking for the effects of industrialisation in an estuary, the complexity of the natural system presents immediate problems. Often there are very wild and large changes in the whole range of physical and chemical parameters. Some of these are effects of human activity, many originate naturally. In comparison, the magnitude of permanent trends is small.

3. In considering the effects of industrialisation on an area such as Southampton Water one is faced with a number of difficulties. In general there is a lack of 'baseline data', that is, information collected over the years before industrialisation had much impact. There have been two surveys carried out in connection with the development of the Power Stations at Marchwood and Fawley.[1,2] These have provided information on the ecology of Southampton Water back to about 1955 and have permitted some useful comparisons to be made. The C.E.G.B. Marine Biological Laboratory was opened in 1969 and has concentrated since its inception on obtaining a detailed picture of the ecology of the area, considering not only biological but chemical and physical factors as well.

4. Since 1950, the main industrial developments along the western shore of Southampton Water have included the oil refinery at Fawley, a variety of chemical plants, Marchwood Power Station which commenced operation in the mid-1950s, and a 2000MW Power Station at Fawley of which the first generating set was commissioned in 1969.

5. In addition to the potential environmental problems which may be posed by the use of seawater for industrial purposes, such as for discharge of wastes and toxic materials and for cooling purposes, the associated major civil engineering preparations and work required for developments on such a scale are themselves likely to influence the local fauna and flora to varying degrees.

6. Reclamation, of course, will reduce the area available to bottom-living fauna and flora. Land has been reclaimed on an extensive scale along the shores of Southampton Water and into the Test Estuary. Recent reclamations at Western Docks have involved about 40 ha, whilst at Dibden about 2500 ha are currently being consolidated.

7. Dredging can produce effects in a number of ways. Fauna and flora may be affected by changes in the nature of the substrate following dredging activities and any resulting change in shore level may then be critical. Shorter term effects include smothering of organisms by settling suspended matter and possible reduction in light penetration into the sea, although these are unlikely to be permanent unless dredging is protracted in a particular area. Ecological changes might well follow dredging and reclamation works should they be on such a scale that general water circulation patterns were affected.

8. Industrialisation has led to an increase in the number of jetties, sea-walls, piles and other structures in Southampton Water, and these are all surfaces which may and indeed have been colonised by a variety of animals and plants. At the turn of the century, the area consisted mainly of a muddy estuary with comparatively few man-made structures. Since then there has been a progressive development of communities of fauna and flora living above the bottom and accounting for a significant part of the overall biomas. This in turn will have led to marked changes in the composition of plankton with a succession of species through the year.

9. The increase of shipping movements through the area, associated with the development of the Port of Southampton generally as well as industrialisation, has also provided the opportunity for the accidental introduction of animals and plants from other countries. Several of these have become well established over the years.

10. Concurrent with the effects of industrialisation are those associated with an increase in population.

11. Increase in the sewage inflow will affect the ecology of the area. It is difficult to separate the effects of these different influences. For instance, with power stations, one is not concerned solely with the effect of temperature as the water may have been polluted by a number of other sources beforehand. When that warm water leaves the power station the levels of other materials that might be present may have changed. This is a very complex situation which makes interpretation of the effects of individual industries rather difficult.

12. A recent paper by Professor Raymont[3] covered certain aspects of water quality, including organic matter, phosphates and nitrates, trace metals and also thermal effects. All of these are possibly affected by a number of the different industries that are found on the waterside. The conclusion that Professor Raymont reached is one which is shared by many workers in the area, that although at the present time Southampton

Water perhaps cannot be regarded as a grossly polluted area it is certainly one where continuous monitoring of the situation is necessary, particularly of toxic metals and other substances which may be concentrated in the tissues of marine organisms.

FLORA AND FAUNA OF SOUTHAMPTON WATER

Plankton

13. The real test is what actually lives in the area.

14. Data for plankton exists back to the 1950s with Raymont and Carrie[2] conducting a detailed survey of zooplankton over a period of 6 years to assess the effects of Marchwood Power Station on local fauna. The marked seasonal changes in zooplankton density which were found in Southampton Water and the differences in zooplankton production from one year to another emphasise the need for continuing observations over a number of years. In the initial period following the commissioning of Marchwood Power Station there was an increase in the zooplankton population in the area. Certain changes appeared in the composition of the plankton which, it was felt, may have been assisted by increased temperature. Most obvious was a marked increase in late spring/summer/autumn populations of barnacle Nauplii terminated by *Elminius modestus*. Acartia tonsa and the planktonic harpacticoid, *Euterpina acutifrons* were species of copepod also possibly favoured by increased temperatures.

Clams

16. There is evidence that the American hard-shell clam, *Mercenaria mercenaria*, has colonised the area over the period of increasing industrialisation in Southampton Water. The species first appeared in the region in 1956 at Lee-on-Solent, and in Southampton Water itself the first specimens were found around 1957. Since then it has colonised wide areas. Between Southampton and Marchwood in the Test Estuary, it has become the predominant bivalve. However, workers have looked at the age of specimens that were found when these first specimens were reported and it is possible that the first introductions, whether they were accidental or otherwise, were made prior to about 1935/1936. It is an immigrant species introduced from North America where temperature conditions are warmer than those that would have been experienced in Southampton Water at that time. In the years following its initial introduction, the animal would not breed every year. It would breed only when the natural summer temperatures approached those normally experienced in the United States. The opening of the Marchwood Power Station in 1955 created in a local area around the power station outfall conditions of warming which obviously favoured this particular species. It is from this period that the rapid rate of colonisation and density increase can be traced. Professor Raymont[3] quotes

figures of densities of clams based on Mitchell's work of the order of $100/m^2$ in the Marchwood area. In the localised area of warming near the power station, the number is in the order of 160 specimens of commercial size per m^2. Densities of young clams of a centimetre or so in size have been quoted up to $4750/m^2$. These densities are many times higher than those one would expect in other areas that are commercially rich, so from the point of view of this clam the Marchwood area of Southampton Water is certainly a conducive environment.

17. It is interesting to look at the temperatures at which this animal breeds because this will have an important influence on the rates at which it will extend its range. Literature from the U.S.A. [(4,5)] shows that there this animal breeds at temperatures between 23°and 30°C. Ansell, [(6)] who worked on the clam in the early 1960s, found that clams taken from Marchwood would breed at around 18°C. Mitchell's work, reported in Raymont's paper, [(3)] has shown that the maximum densities of clam larvae in the Marchwood and Town Quay area of Southampton Water were found when the water temperatures were about 18 or 19°C. This work is being continued by Coughlan at the Fawley Marine Biological Laboratory.

18. It appears then that an introduced species has, over the course of a number of years, lowered its breeding temperature by 3 or 4°C.

19. Turning to benthos in general, in 1971 an opportunity arose of working with the Consulting Engineers to the South Hampshire Plan, Messrs. J. D. and D. M. Watson, to carry out the first major survey of the benthos of the lower reaches of Southampton Water and the Solent. [(7)] It was possible for the first time to look at the structure of the communities at 116 different stations. This is a good example of getting information in advance of a major industrial development—in this case the siting of a new large sewage outfall. Over 100 different invertebrate species were found.

20. The slipper limpet, *Crepidula fornicata,* another immigrant species, was found over 90% of the total area of the survey.

Oysters

21. This survey in 1971 also enabled an assessment to be made of the abundance of native oysters, *Ostrea edulis,* in the Solent, particularly in Stanswood Bay where the outfall of Fawley Power Station is situated. Although a few live specimens were found a year or two before, the first detailed surveys were carried out during 1971, by C.E.G.B.'s Fawley Marine Biological Laboratory and the Ministry of Agriculture, Fisheries and Food [(13)]

22. Grab and dredge samples were taken in 1971 and 1972. It was estimated that between 8 and 9 million oysters were present in this area. Heavy commercial fishing which was carried out during 1972 removed, it is estimated, 100 or 110 tonnes of oysters. This represents

about 2 million individuals and means that about one quarter of the population was removed in three or four months. There was obviously a case of gross over-fishing. For this reason a close season was imposed and a public enquiry was held into the writing of a Several Order for this fishery. This has now been granted, which means that there will be considerably more control over what is a very valuable natural resource. A cooperative of local fishermen now have the rights to fish for oysters in this area. They also have the obligation to maintain the fishery in a satisfactory state, and in that connection it is interesting to note that recently several tonnes of cultch (cockleshell) were scattered over the bed in this area to encourage the resettlement of more larvae to improve the fishery. The fishing cooperative is hoping in the long run to remove 400 tonnes of oysters in a year.

23. The reappearance of oysters in Stanswood Bay is interesting from the point of view of any assessment of the effects of industry on the area because, although Stanswood Bay is just outside Southampton Water as such, any serious effects of pollution from Southampton Water would have an effect on an area adjacent to the mouth of this waterway. Yet from about 1968 or 1969 when it is thought oysters first settled in Stanswood Bay to give the stocks that were assessed in 1971, conditions there must have been reasonable for settlement.

24. The growth of the oysters in the area, however, is not particularly good for producing fattened oysters. It would appear that the area of the Stanswood Bay fisheries will be a good nursery for oysters from which stock will be removed to other estuaries to fatten up. This is a normal procedure, as often an area which is good for settlement of larvae is not a good one for subsequent growth and fattening.

25. It is thought that the larvae which have produced most of the Stanswood oysters have originated in the Beaulieu River and that the restocking and success of oysters in the Beaulieu River in the late 1960's was one of the factors which brought about the revival of the oyster fishery in Stanswood Bay. Hydrographic conditions are such that currents between the Beaulieu River and Stanswood Bay produce a movement of water from the Beaulieu River which will carry larvae into this area.

26. *Crepidula* are abundant in the area, and their shells are one of the surfaces preferred by oyster larvae for settlement. Man too can increase the surface area by a supply of cultch. Another factor which is important is that the number of predators, such as crabs, that normally prey on young oysters is at a comparatively low level in Stanswood Bay. Water quality conditions in the area support the settlement and growth of oysters and this provides a good index of the general state of water quality. The colonisation commenced at about the time that the first of the 4 generating sets of the power station was commissioned at Fawley in 1969. The situation seems to be that a major industrial concern injecting large volumes of warm water into the area has had no

apparent ill effect at all on the development of what has turned out to be one of the largest oyster fisheries in the country.

Fouling organisms

27. The Fawley Laboratory has been engaged on studies into fouling animals and plants. These are important from the C.E.G.B.'s point of view because fouling animals settle in culverts and the water systems of power stations creating a number of problems. Coughlan[8] has looked at the abundance and diversity of fouling communities in Southampton Water, in the Solent, and as far as Langstone Harbour. He has found that the greatest diversity and biomass of fouling occurs at the head of Southampton Water rather than the Fawley end. One would expect the pollution loads to be higher at the Southampton end of the Water around the Royal Pier, but the fact that such diverse fouling communities can exist there is some indication of the general water quality.

Introduced organisms

28. Introductions to the area include the American clam, already discussed, the Japanese Sea Squirt, *Styela clava,* several zooplankton species and the Slipper Limpet, *Crepidula fornicata.*

29. Introductions are another index that can be used to some extent to gauge the ability of an area to support marine growth. If pollution has reached a certain point then it is unlikely that new species will become established in the area. The most recent species to appear is the Japanese seaweed, *Sargassum muticum,* which was first recorded off Bembridge, Isle of Wight, early this year. Already the weed has spread to the mainland and specimens have been found in Portsmouth Harbour. It is thought that the weed may have been introduced accidentally with imported oysters or may have entered on shipping. Experience from the north-west coastline of the USA and Canada has shown that the weed is able to colonise new areas rapidly and create a number of problems. Biologists from Portsmouth Polytechnic are monitoring its distribution locally and a programme of eradication by means of manual removal has been started.

Wood borers

30. Work has been done on the distribution of wood borers[12], which are often mentioned in connection with warming effects. In 1955 the C.E.G.B. started a comprehensive monitoring programme using wooden battens that were exposed at different sites in Southampton Water. Borers of the genus *Limnoria* predominate. It was found that for about 3 years after the commissioning of Marchwood Power Station there appeared to be an upward trend in the total number of borers that were found but that there was no further increase in activity after this initial period.

31. A great variation is found from year to year when one examines the total number of borers in the battens. It appears that the activity of the wood borer may be influenced not so much by local warming as by the presence of old pieces of timber which are themselves infested with *Limnoria* and which will infest any new structure in put the area.

32. The wood borer monitoring programme was extended in 1968 to examine the possible effects of Fawley Power Station. Although the numbers vary from year to year, there appears to be no upward or downward trend. One may conclude that the warm water is having no effect on the total number of wood borers or wood borer activity.

Temperature conditions

33. Marchwood Power Station started in 1955 and Fawley was opened in 1969. Fawley Power Station which produces 2000 MW requires around 70 m^3/s (50 million gallons per hour) of seawater for cooling purposes. This represents only about 4% of the water available to the station. The seawater becomes warmed by about 8°C during passage through the condensers. Temperature measurements have confirmed that the warming that occurs is localised and stabilised. It is restricted to a comparatively small area around the power station outfall. A particularly detailed study was carried out in connection with the development of Marchwood Power Station in the middle 1950s. During zooplankton surveys it was found that at this location, where the temperature rise was about 6 to 7°C, the difference declined rapidly so that 300 metres away from the outfall the temperature had already dropped to within 3°C of the ambient temperature (Raymont and Carrie).[2] Jarman and de Turville[9] looked in detail at thermograph records from 15 stations in Southampton Water and the Solent for the three years from 1965-1967. They examined the effects on the general water temperature in Southampton Water of Marchwood Power Station and Fawley Refinery. (It is important to mention the refinery because most people consider that the only warm water outfalls in this area are from the Power Station, but a considerable quantity of warm water can be discharged by other concerns using water cooling.) He found that the general warming that occurred in the area over these three years was of the order of 0.6°C. When these figures were compared with mathematical model predictions using a mixed segment model, described by Downing,[10] it was found that there was very close agreement.

34. It was obviously important to record temperature changes found close to outfalls, especially when looking at effects on marine life. The Fawley Laboratory has been looking in detail at the temperature situation existing close to power station outfalls using infra-red techniques, sensitive thermistor probes and also ordinary temperature/salinity bridges. Again it is confirmed that the warming that occurs is localised.

35. It is important to compare these temperature changes with the range of temperatures occurring naturally in Southampton Water. The temperature range at the estuary head is from about 0°C in the winter up to 22°C in the summer. At Calshot the minimum temperatures are slightly higher, ranging from about 4°C to 20°C. Animals living in Southampton Water have to deal with a very wide range of temperatures under natural conditions. On mudflats, where the shellfish are living, temperature range over the course of 24 hours can be as much as 10°C from a frosty night to a warm afternoon.[11] These are changes which the animals have to deal with under natural conditions. This point is important because the variations in environmental temperatures that occur in British waters through a year are much greater than will be found in many tropical and subtropical areas and some parts of the U.S.A. In a tropical situation maximum water temperatures may be around 27 to 28°C, changing at most by only a few degrees in a year. As far as inherent ability of those animals to adapt to any change in temperature is concerned, the increases in temperature found near particular power stations in this country are likely to be so small that there will be little or no change in the general ecology.

36. There are therefore geographical and biological reasons why the observations that one might make on the effects of power stations in one country may not apply to the situation in another. Also, in Britain the maximum demand for power is in winter, hence cooling water discharges are highest in winter. In many areas, particularly in the U.S.A. the reverse is true because air conditioning requirements produce the peak demand. When the amount of water available for cooling in such large areas is at its lowest the power stations require the maximum supply of water and may raise temperatures in the receiving waters much closer to the upper temperature tolerances of particular species concerned.

CONCLUSION

37. It is long term data which will be of prime value in the interpretation of the changes brought about by industrial activities. Obviously any investigation must be broadly based. The work of the Fawley Laboratory includes, as well as the biological survey work, studies on chemistry and water quality, seawater and substrate temperature measurements and biochemical investigations on trace metal uptake by organisms. The concentration of metals in a number of local shellfish is being monitored.

38. Studies of this type carried out over long periods will eventually show whether industrialisation is affecting the ecology of local waters, and will also allow some separation to be made between these effects and those resulting from natural changes.

REFERENCES

1. Pannell J. P. M., Johnson A. E. and Raymont J. E. G., An Investigation into the Effects of Warmed Water from Marchwood Power Station into Southampton Water, *Proc. Instn. Civ. Engrs.*, 1962, **23,** 35-62.
2. Raymont J. E. G. and Carrie B. G. A., The Productivity of Zooplankton in Southampton Water. *Int. Revue ges Hydrobiol.*, 1964, **49**(2), 185-232.
3. Raymont J. E. G., Some Aspects of Pollution in Southampton Water, *Proc. Roy. Soc. (B)*, 1972, **180**, 451-468.
4. Nelson T. C., On the Distribution of Critical Temperature for Spawning and for Ciliary Activity in Molluscs, *Science,* N.Y., 1928, **67,** 220-221.
5. Loosanoff V. L., The Spawning of *Venus mercenaria,* 1937, *Ecology,* **18,** 506-515.
6. Ansell A. D., *Venus mercenaria* (L.) in Southampton Water, *Ecology,* 1963, **44,** 396-7.
7. Barnes R. S. K., Coughlan J., Holmes N. J., A Preliminary Survey of the Macroscopic Bottom Fauna of the Solent with Particular Reference to *Crepidula Fornicata* and *Ostram Edulis, Proc. malac. Soc. Lond.*, 1973, **40,** 253-275.
8. Coughlan J., Unpublished work at the Fawley Laboratory.
9. Jarman R. T., De Turville C. M., Industrial Heating of Southampton Water, *The Dock and Harbour Authority,* 1971, **51,** (606).
10. Downing A. L., Association of River Authorities Yearbook, 1966.
11. Spencer J. F., Temperature Studies on the Intertidal Zone of an Estuary. Symposium on Marine Biology, Central Electricity Research Laboratories, Leatherhead, 1969 (Ref. RD/L/M 269).
12. Hockley A. R., Holz und Organismen, pp. 457-464, Duncker and Humblot, Berlin, 1965.
13. Key, D., The Stanswood Bay Oyster Fishery. Shell Fish Information Leaflet No. 25. Ministry of Agriculture, Fisheries and Food, 1972.

DISCUSSION

HAMILTON (Institute for Marine Environmental Research). Are there any differences in the physiology or metabolism of *Mercenaria* which have been observed between the present population of Southampton Water and those in U.S.A. waters? Such differences could be due to adaptation to the environment.

COUGHLAN The *Mercenaria* found locally are certainly able to breed at lower temperatures. Also, whereas in waters of the U.S.A., areas of lower salinity appear to favour growth, the opposite is found here. *Mercenaria* at Lee-on-Solent grow faster than those at Netley, which in turn grow faster than those in the Test Estuary at Hythe/Marchwood. It is possible that growth at the head of Southampton Water and in the Test Estuary is inhibited by unspecified 'pollution'.

WAKEFIELD (Coastal Anti-Pollution League Ltd.). The Author discussed new fauna which had arrived since the introduction of power stations, but did not say whether any had disappeared. Have any species disappeared?

SAVAGE The mollusc *Mya* has been largely replaced by *Mercenaria,* particularly in the Marchwood area, apparently because *Mercenaria* is more tolerant to pollution, although other factors may have been operating. This is a good example of replacement of a species.

SOULSBY (Hampshire River Authority). Would the Author, as a professional marine biologist, comment on the possible effects of discharging untreated sewage through properly sited outfalls?

SAVAGE This problem should be approached by considering changes likely to occur in nutrient levels. Measurements of the basic nutrients, phosphates, and nitrates, have shown no increase during the last three years, although sewage discharges have increased. The evidence suggests that Southampton Water has some further capacity to absorb more nutrient from increased sewage effluent quantities.

ALLEN (The Ecologist). Is there any evidence that ouftlows from power stations have imposed an unusual thermal stability on the estuary, or that species tolerant of marked influctuation are intolerant of this stability? In other words, power station outflows might smooth natural temperature fluctuations, which hitherto were sufficiently severe to make the environment inhospitable to all but a few pioneer species, so that these would then be replaced by other species.

The Author remarked that the instance of invasion by exotic species is an indication of water quality. Obviously quality must be higher than if there were no species there at all, but it might be too low for indigenous species to survive. To what extent have invading species replaced the indigenous ones? Are data adequate, before the period of industrialisation, for us to know this?

SAVAGE In the area immediately round the outfalls, where temperatures are raised 3 to 4°C, there have been changes in densities of species, particularly of *Mercenaria*. Changes in habits would also be expected. Elsewhere, differences are too small to produce detectable changes.

Almost certainly there is not enough information from the period before industrialisation. It is agreed that the appearance of new species does not necessarily indicate that water quality is good. Obviously species can be introduced that are more tolerant of pollution and therefore they will colonise simply because they are more tolerant, such as *Mercenaria*.

Another example is *Sargassum muticum* which was first found off Bembridge, Isle of Wight, at the beginning of this year. The fear is that it will grow on such a scale in certain areas that it will take up space which would be occupied by bottom animals. The work carried out by

Coughlan[7] and his colleagues for the South Hampshire Plan in 1970/71 showed how the distribution of large fauna was determined to some extent by what grows on the bottom. The same applies to *Crepidula*. If this gets a hold, the main problem to the oyster fisherman is that it is occupying space that would be better occupied by oysters.

These are the sort of changes which can occur. Time will tell whether this might happen with *Sargassum*.

OAKLEY (J. D. and D. M. Watson). How does pollution show itself? Is it a luxurious growth of one or two dominant species? Possibly the very complexity of estuaries gives some protection against the effects of pollution. This question is central to the theme of the conference.

SAVAGE One characteristic of pollution in a system is that there is a decrease in the diversity of species and an increase in some species which will take over the space previously occupied by other species. This is an indication of a deterioration of the situation.

READ (Napier College of Science and Technology, Edinburgh). A biological study has recently been started in the Firth of Forth. In the Forth certain areas of the intertidal zone and inshore waters are grossly polluted with untreated mixed domestic and industrial sewage. This condition has given rise to plant and animal communities of the intertidal zone that are characterised by a low species diversity and, in the case of rocky shores, an increase in the number of individuals of certain 'tolerant' species and species which are not normally members of these communities, such as certain spionid and capitellid worms.

On the question of nutrients, conventional sewage treatment does not reduce by very much the discharge of nutrients in sewage effluents. Such treatment is aimed only at a reduction in organic load and suspended solids, unless a tertiary process of nutrient stripping is installed. The introduction of such conventional sewage treatment will therefore, assuming no change in the manner of discharge, affect the biota through changes in suspended solids and organic load.

Does the Author consider that the introduction of full treatment of sewage discharged into the Solent had has any measurable influence on the biota?

SAVAGE I believe that insufficient data exists at this stage to make such an evaluation.

SMITH (University College, Galway). Returning to the question of effects of temperature, it is found in work with bacteria, and possibly also with animals, that it is not change of temperature that matters, but rate of change of temperature. This should be taken into account in any studies.

SAVAGE I agree with you that the rate of change of temperature is extremely important. We are particularly interested in the situations where sea water is heated rapidly on passage through condensers and where warm discharges flow across the shore or along the sea bottom.

HOWELLS (Natural Environment Research Council). On the question of biological diversity of estuaries, it has been generally accepted that the number of species within the euryhaline, that is the changing stability part of an estuary, is less than the number found in the limnetic, freshwater, part of the river, and also less than that in the sea. The two graphs of number of freshwater species and of marine species cross in the estuarine zone, and this is where the total number of species is at a minimum. Because there is ecological 'space' in that part of the estuary, the abundance of individuals of a reduced variety of species will be very great: there is reduced variety, but a very large biomass.

These estuarine organisms are particularly adapted to withstand wide changes in environmental factors: they accept or adapt to large changes in salinity, changes of water level, and large natural ranges of temperature, for example, mudflats act as very efficient heat storage banks during the middle of the day if exposed, and then act as heat exchangers at night.

Furthermore, estuaries are areas where vast quantities of nutrients go into the aquatic system, whether or not there are people there. There is also a tremendous sediment load.

All of these things vary, by day, by season, and year by year and the organisms that survive and succeed in those situations have a built-in adaptability achieved by a high degree of tolerance, by physiological regulation, or by genetic selection so that there may be groups or varieties which are hidden away in odd corners and which can recolonise places that may have been depopulated by a pollution or other catastrophic incident.

All these kinds of variation produce the same sorts of stresses that arise when an estuary is polluted. The development of an estuary in terms of human population produces, not new nutrients, but an added load to the fluctuating levels of nutrients that are there normally. The same situation applies to heat, and many trace metals. For example, the estuaries in Devon and Cornwall have much higher levels of heavy 'toxic' metals than will be found in places which are industrialised and polluted. The local species in these estuaries can tolerate these high levels; they survive and they even prosper. The difficulty then is to distinguish the effects that have been introduced and the effects which are natural. The major problem is that we do not have any good or simple measure of the biological changes that go on in estuaries. We do not have the right sort of background knowledge so that we can distinguish the signal due to a polluting influence from the general noise situation generated by natural events.

Another of the complicating factors is that most estuaries are used for a very wide variety of purposes, and all sorts of things are going on at the same time. In the laboratory it is possible to test what are the effects of heat or of depleted oxygen or other single factors, but it is not at present practicable to assess the effects of a whole mixture of natural and introduced agents fluctuating during the day or a 24 hour period. A further problem is how to distinguish these effects, not only on an individual or on a species, but on a population, and even going further, on the whole community of inter-acting populations living in an estuary. It is for this reason that we are faced with a very real problem of finding some practical way of assessing the effect of pollution on an estuarine area.

MITCHELL (Water Resources Board) (written contribution). The author suggested that the establishment of the American hard-shell clam, *Mercenaria-mercenaria,* in Southampton Water was probably due to elevation of temperatures by Marchwood Power Station outfall promoting ideal conditions for breeding. However, recent evidence suggests that *Mercenaria* was firmly established in Southampton Water before 1956, when the first generating set of Marchwood Power Station was commissioned.

Heppell(14) suggested that the population of *Mercenaria* in Southampton Water and the Solent may have been derived from larvae carried across the channel from small populations in Brittany, or that they were introduced with oysters imported to the Langstone Harbour oyster beds. Ansell(6) suggests that the population became established from accidental introductions via the galleys of transatlantic vessels using the Port of Southampton. From observations on the annual rings of individuals, Ansell *et al* (15) suggest that *Mercenaria* was introduced into the upper reaches of Southampton Water some time prior to 1936. However, evidence now indicates that *Mercenaria* was in fact deliberately introduced into Southampton Water in 1925 (Smith, personal communication). A number of live clams were sent over from America to be tried as eel bait, but as they did not prove to be any better than the traditional bait the unused portion of the shipment, 'a few dozen:, were laid in the River Test at a point which was opposite to the present site of Marchwood Power Station— the actual location would be beneath the docks built since that time. In support of this evidence Cove (personal communication) states that *Mercenaria* was quite common in the River Itchen at Woolston from as early as 1935. It is locally reported that in the second war this species had become sufficiently well established to support a small fishery—the clams being sold to American servicemen passing through the port.

It was also suggested by the author that *Mercenaria* had competed out the native clam *Mya arenaria.* While it is true that *Mya* is now almost completely absent from Southampton Water, it is more likely that the cold winters of 1947 and 1962/63 devastated the population— large beds of dead *Mya* (i.e. empty shells) occur in several areas of Southampton Water, some of which are only on the fringe of the present extent of the *Mercenaria* population. *Mercenaria* would probably be far less affected by these cold winters as this species has a good tolerance to lowered temperatures and hence the population would have expanded to fill the niche left by Mya.

Waugh[16] quoting Ansell shows that the mortality of *Mercenaria* in Southampton Water during the severe winter of 1962/63 was 4.6% which was considerably lower than the native species of *Cardium edule* and *Venerupis decussata* with mortalities of 62.5% and 55.9% respectively. *Mya* is not mentioned here, so presumably the numbers of this species had already been much reduced by the severe 1947 winter. Support for the fact that extreme winter temperatures do affect this species is recorded by Newell[17] who states that in the Whitstable area *Mya* is now rare where it was formerly common before the cold winter of 1962/63.

The maintenance of species and habitat diversity were suggested as suitable criteria for a clean industry. However, this could be misleading, for certain forms of 'pollution' may actually lead to increase of species diversity by creating new habitats. One example of this was given by Mackay (Paper 11) in his paper where he gave an example of a shore in the Clyde which would normally be gravel with a restricted fauna which had become overlain with organic mud from sewage and which now supported a rich invertebrate fauna with associated feeding waders. Another example is found in Southampton Water where the fauna around Marchwood Power Station outfall is more diverse than other areas in the estuary. This may in part be due to the effects of the warm water, but is more likely to be caused by the physical effect of high flow preventing siltation and thereby increasing the area of clean surfaces available for the establishment of encrusting organisms.

TUCK and HAWES (Central Electricity Generating Board) (Written contribution). No mention was made in the paper nor in the subsequent discussion of salmon and migratory trout. However, Southampton Water continues to be a migratory route for these fish, to and from the River Test. This in itself is a good indication of the quality of the estuary, as well as a tribute to its managers—the Hampshire River Authority—who have maintained the high standard of water quality for these fish to be able to continue their migrations despite the extensive industrialisation in the recent decades.

REFERENCES

14. Heppell D., The Naturalization in Europe of the quahog, *Mercenaria mercenaria* (L), *J. Conch,* 1961, **25**, 21-34.
15. Ansell A. D., Lander K. F., Coughlan J. and Loosmore F. A., Studies on the hard-shell clam, *Venus mercenaria,* in British Waters. I. Growth and reproduction in natural and experimental colonies. *J. Appl. Ecol.*, 1964, **1**, 63-82.
16. Waugh, G. D., In the effects of the severe winter of 1962/63 on marine life in Britain. Crisp D. J. (ed.) *J. Anim. Ecol.*, 1964, **33**, 173-175.
17. Newell G. E., In, The effects of the cold winter of 1962/63 on marine life in Britain. Crisp D. J. (ed.) *J. Anim. Ecol.*, 1964, **33**, 178-179.

THE TIDAL HYDRAULICS OF THE SOLENT AND ITS ESTUARIES

N. B. Webber, B.Sc. (Eng), C. Eng., M.I.C.E., M.A.S.C.E., M.I.W.E., M.I.Mun.E.

Department of Civil Engineering, University of Southampton

INTRODUCTION

1. Few maritime areas of the world are cherished so much by so many different types of people as the Solent and its adjacent coastal waters (Fig. 1), and deservedly so, for as the authors of a recent book [(1)] have observed: "It is this feeling of inexhaustible variety—the kaleidoscope of the changing scene—that is the attraction of the Solent. Even the fact that it is approaching saturation point is in its favour, because if it has got thus far and yet retained its magic and gentle charm then it will endure for as long as there are yachtsmen to sail and navigators to make landfall from distant parts."

2. The tidal features are by no means the least of its fascinations and, indeed, the complexities are such that the Admiralty Tide Tables [(2)] include a special table entitled 'Swanage to the Nab Tower' which sets out the hourly variations at 18 places in the area, so great is the deviation from the normal shape of tidal curve and the behaviour at the standard port (Portsmouth).

3. Tidal hydrodynamicists [(3)] and hydrographers [(4)] have studied its characteristics, but it is certain that they would be the first to agree that it still retains many of its mysteries. And the present author must confess, at the outset, that there is very little that he can add by way of new material or ideas. This paper, therefore, takes the form primarily of a brief outline of present knowledge concerning the tidal hydraulics with particular emphasis on considerations of pollution.

GEOMORPHOLOGY

4. During the Pleistocene period, which embraces the last two million years, world sea level is known to have risen above and fallen below the present level, and in the last recession, about 17 000 years ago, it was about 100 m lower. The greatest and fastest rise in the geological record reached its peak about 6 000 years ago. The cumulative incursion of the sea flooded low-lying coastal lands in every part of the world. At about this time, or a little earlier, the Straits of Dover were formed.

5. The coastline was many miles to the south of the present Isle of Wight and a great river, which Clement Reid [(5)] has called the Solent River, ran parallel to the coastline and discharged to the sea, somewhere

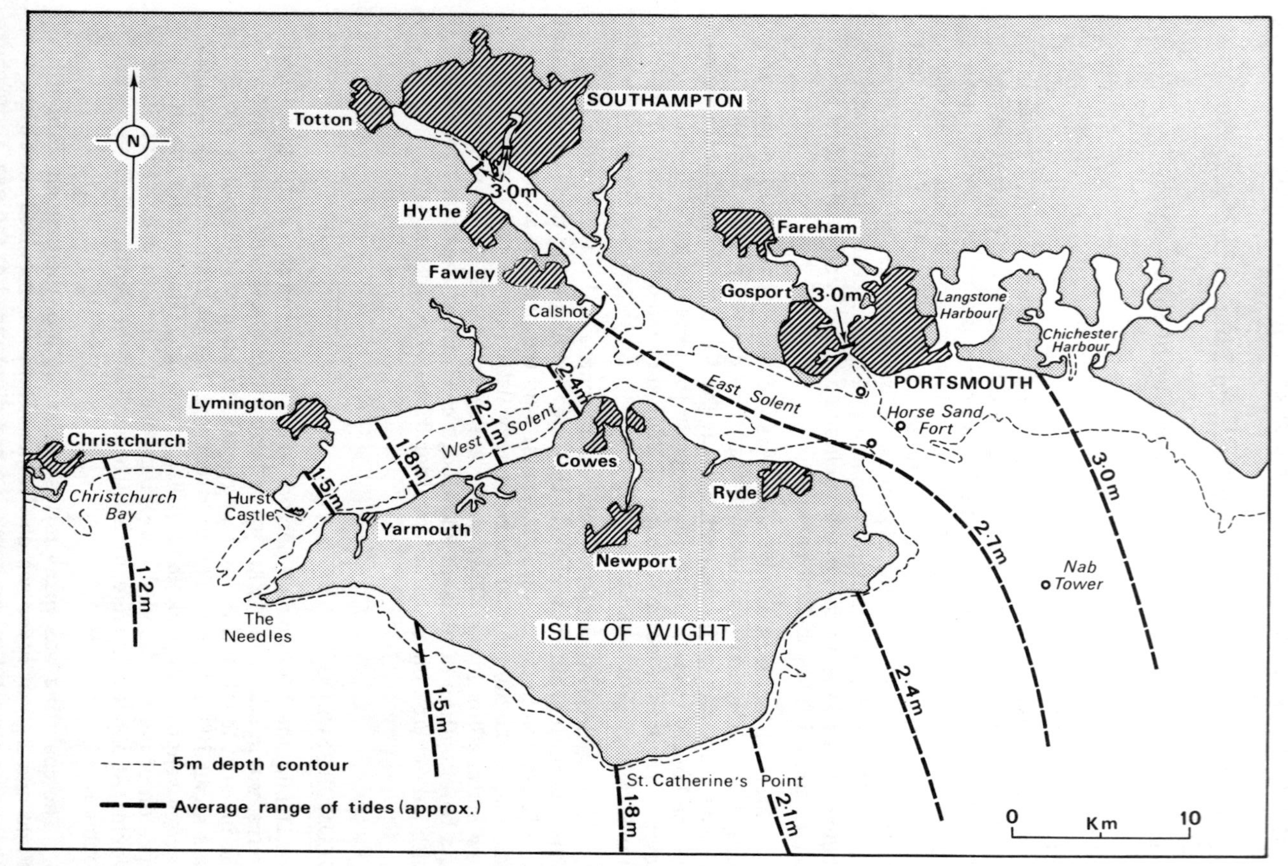

Fig. 1. The Solent system

south of the present Littlehampton (Fig. 2). The tributaries, Test, Itchen, Avon, and the Dorset rivers drained much larger areas than is the case now. There is ample evidence for all this from boreholes in Southampton Water and elsewhere.

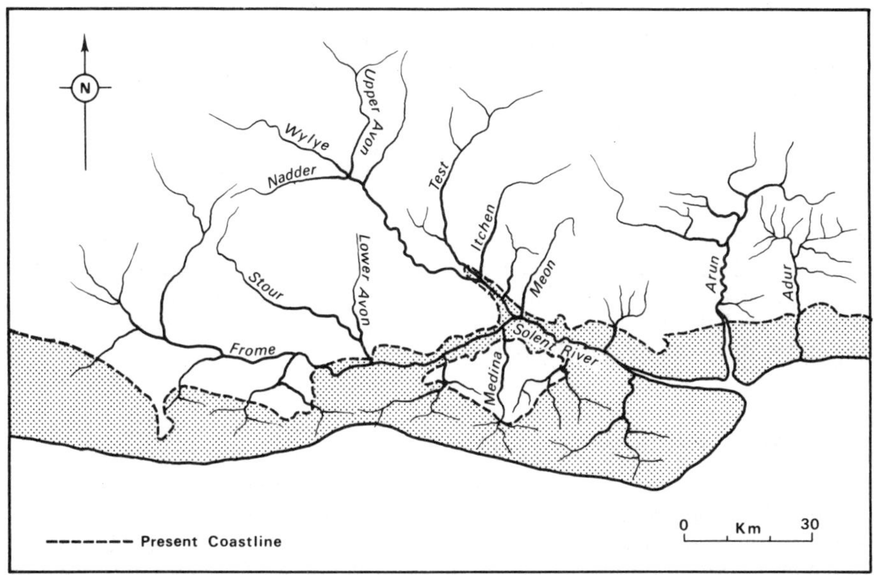

Fig. 2. The 'Solent River'. (Clement Reid, Mem. Geol. Surv.)

6. With the continuing rise in sea level, the soft chalk ridge on a line between the Needles and Dorset was eroded and the sea broke into the valley of the Solent River forming a semblance to the present Christchurch Bay. Though at this early stage the Isle of Wight was severed from the mainland, it was probably at first only cut off by small channels and marshes and was sometimes an island, sometimes part of the mainland as sea level varied. The final isolation took place fairly recently, perhaps no more than 4000 years ago.

7. Clement Reid [(6)] summed up these last stages of geological development in an adress to the British Association at Portsmouth in which he said "Geology helps us to understand the origins of one of the most important harbours of the world. The Solent, Spithead and Southampton Water are parts of an ancient submerged valley system. The magnificent waterways thus formed are now slowly silting up, but that the process is not more rapid is due to the happy accident which diverted so much of the drainage of the ancient Solent to the open sea. Had it been otherwise, instead of the present fine waterways we should have had a series of alluvial flats and sand banks such as now block the lower reaches of the Thames".

8. A 'happy accident' is a very apt way to describe the evolution of the sheltered and attractive waterways, part coastal and part estuarine, that we now all enjoy. And we can speculate, from experience of 'developed' estuaries elsewhere, on the likely levels of pollution that would exist had not geological fate treated us so kindly.

9. During the past 2000 years or so sea level has remained fairly constant, but with small eustatic changes, which, in southern England, amount to a relative subsidence of the land of the order of 0.1m per century [(7)]. There do not appear to be any pronounced natural changes currently taking place in the Solent area, although, at the western end, the profiles of the Shingles Bank and Hurst Spit have exhibited some variation.

10. Of course, the configurations of the upper reaches of Southampton Water and Portsmouth Harbour have been altered significantly by man's activities in recent years, and these are discussed later, but it is most unlikely that there has been any noticeable effect on the tidal regime of the more open waters of the Solent.

TIDAL BEHAVIOUR IN THE ENGLISH CHANNEL

11. For the Solent the open sea is the English Channel, but for the English Channel the open sea is the Atlantic Ocean. It is therefore of some value to examine the tidal characteristics of the English Channel since these are the controlling factor with respect to the Solent.

12. The co-tidal lines for the English Channel are shown in Fig. 3. These are lines where it is high water at the same time—the pecked lines indicate the average tidal range. It will be noted that high water in the eastern portion occurs at about the same time as low water at Penzance, also that the range is a minimum half-way along the Channel in the vicinity of Poole Bay.

13. This pattern of behaviour is indicative of a degree of resonance with the semi-diurnal oceanic tide and, in fact, the natural period of oscillation of the English Channel is about 10 hours. Higher tidal ranges occur on the French side and, apart from the pronounced local effect of the Cherbourg peninsular, this is attributable to Coriolis force.*

14. A crude but instructive analogy can be drawn with the oscillation of water in a rectangular tank or bath (Fig. 4). At the transverse section there is maximum mass transport of water (maximum velocity) but no rise and fall of the surface. Thus, although the range of tide to the west of the Isle of Wight is relatively small there is considerable water movement (maximum currents of 1.2 m/s or more). It is a further feature of a standing wave oscillation that the amplitude increases

* The force induced by the rotation of the earth which has the effect of deflecting a tidal stream to the right in the northern hemisphere.

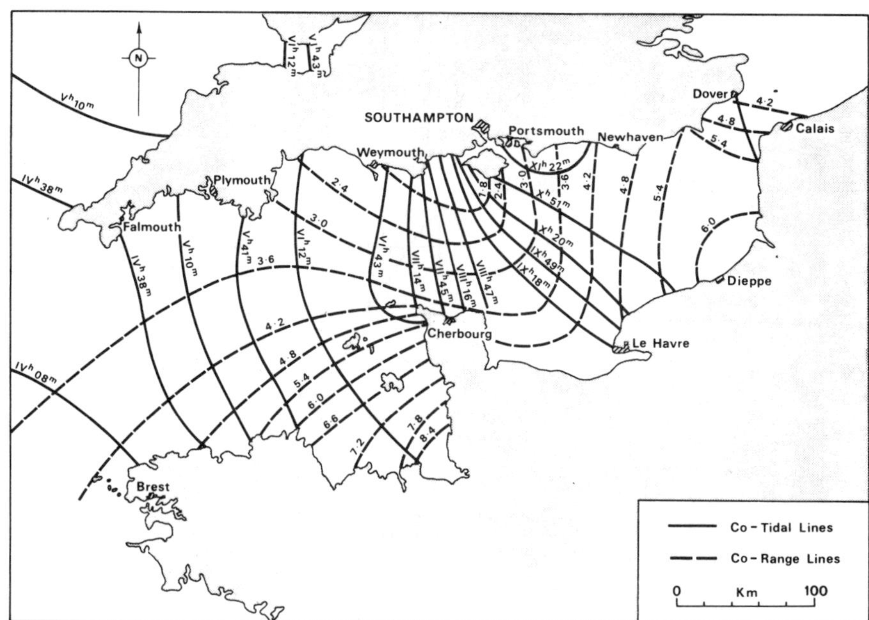

Fig. 3. Co-tidal lines for the English Channel

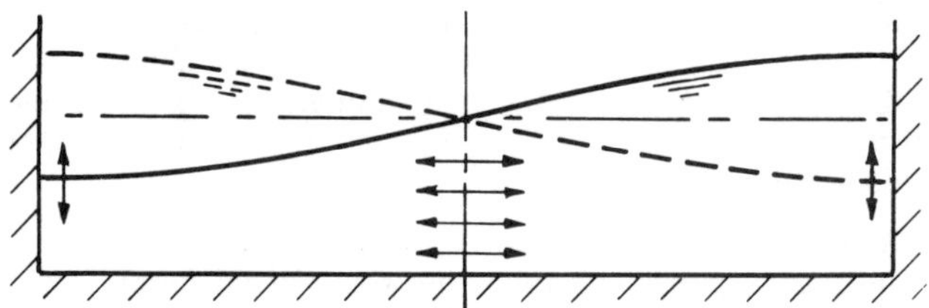

Fig. 4. Standing-wave oscillation

rapidly with distance from the nodal point or line. This is indeed so in the area with which we are concerned. In Fig. 5, the tidal curves for springs for the Christchurch Bay area and the Nab Tower (about 8km east of the Isle of Wight) have been superimposed and it will be noted that over a distance of only some 80km the range of tide is nearly doubled; furthermore, the co-range lines of Fig. 1 indicate that the major change occurs in the West Solent over a distance of only 16km. The same relationship would be found at neaps, and, in fact, at neaps the tidal range is typically one-half that at springs.

15. In regions, such as the vicinity of a node, where the amplitude of the semi-diurnal (M2) constituent is weak, combination with a powerful

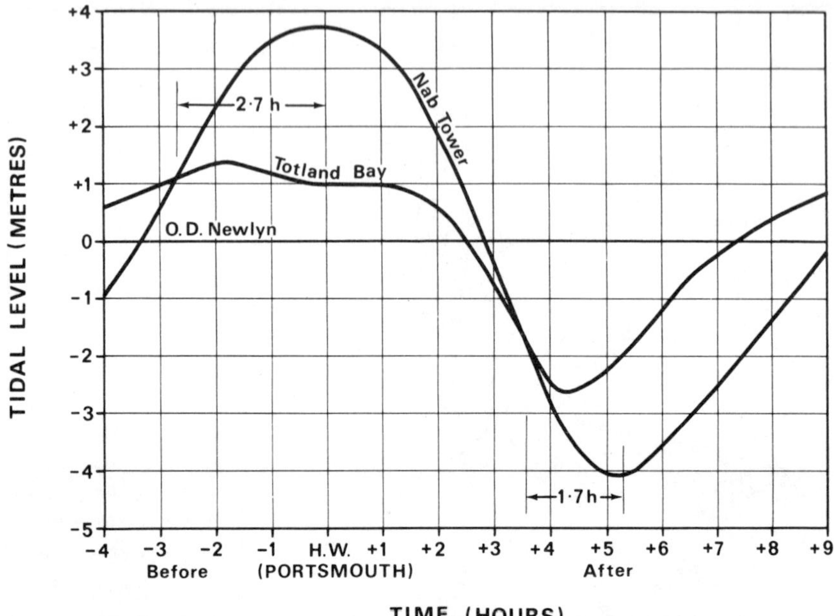

Fig. 5. Tidal curves (springs) for Totland Bay and the Nab Tower

shallow-water constituent (M4) is liable to result in double high or double low waters if the phase difference is favourable. Configurations and depths in the Christchurch Bay area are certainly conducive and we find that double high waters do indeed occur, whereas in Spithead and further east they do not, although the tidal curves exhibit a prolonged period of near high water.

16. The important end conditions for the Solent system are thus defined. These would exist, although perhaps in modified forms, were the Solent there or not.

TIDAL BEHAVIOUR IN THE SOLENT

17. What happens inside the Solent is dependent upon the conditions outlined in the previous Section and on the hydraulic characteristics of the waterway system. The latter include the configuration (length, width, depth, alignment) of the various estuarine components and the resistance to flow which they offer. Clearly, there is some inter-dependence since it is the ebb and flow of the tide which influences the form of channel.

18. It will be seen from the tidal curves (Fig. 5) referred to earlier that there is an east to west gradient of water commencing at about 2 hours before high water Portsmouth so that the ebb is running fast

whilst water levels within the system are still rising. This means that the last of the flood enters estuaries, such as the Medina, from the easterly direction rather than the western. The reversal to a west to east gradient occurs at a corresponding but lesser time before low water Portsmouth.

19. Characteristically, with the closer proximity to a tidal node, and because of the smaller sectional area of channel, tidal streams in the West Solent are much faster (up to 2 m/s) than in the East Solent (up to 1.2 m/s) and the open waters of Spithead and beyond. In the narrows at Hurst Castle, which have been maintained by scour to a depth approaching 60m, the maximum tidal stream at springs is recorded as 2.2 m/s. These values have been taken from Admiralty data [8] and refer to the average current over a depth of about 9m; surface currents would be higher.

20. The appreciable velocities and consequent high frictional loss in the western arm are evidenced by the considerable hydraulic gradient which on a spring tide is as much as 0.9m in 16km. In fact, the cross-sectional dimensions and the discharges during peak flows are approaching those of the largest river in the world, the Amazon, which is estimated to have a discharge of 100 000 m^3/s. Incidentally, 1 m^3/s is equal to 19×10^6 gal/day and it is perhaps a little unfortunate that it is the latter unit that engineers generally employ for defining sewage

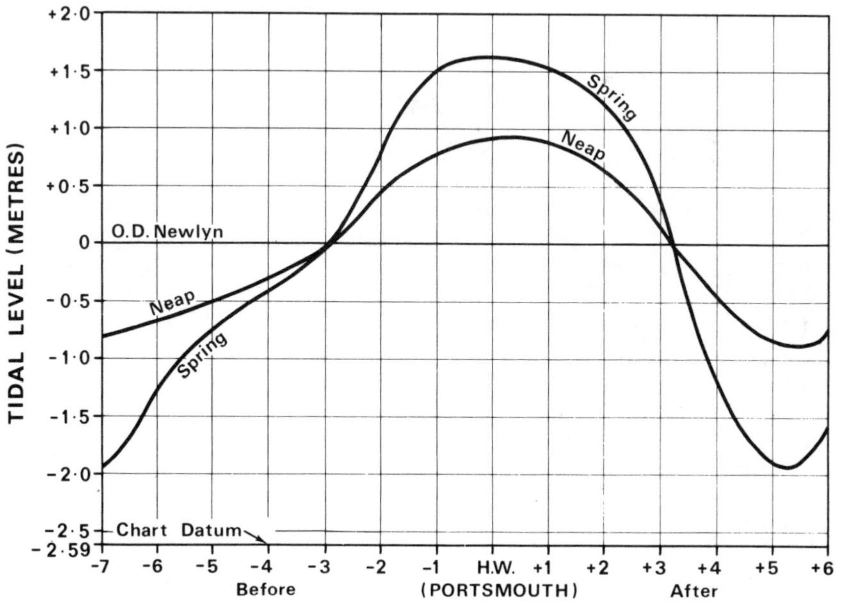

Fig. 6. Tidal curves for Cowes

discharges, tending to give the layman a false impression of their magnitude relative to tidal flow.

21. The general form of the tidal curves for the Solent region is of interest and the curves for Cowes (Fig. 6) are probably as representative as those for any other place.

22. There is a long period of 'stand' around high water, extending to about 3 hours. In the estuaries, this is of advantage to those that are dependent on adequate depth of water for the movement of vessels or the enjoyment of a sporting activity.

23. The effect of the high water stand is to shorten the period of ebb to around 4 hours instead of the customary 6 hours. The flood has a normal duration but there is an intermediate near stand of about 1 hour commencing at about 2 hours after low water. In consequence, contrary to normal estuarine behaviour, the ebb currents are faster than the corresponding flood and therefore provide a beneficial flushing of silt and contaminants in a seaward direction. However, as with tidal flow generally, the locations of the dominant ebb and flood currents may well differ, being influenced by configurations and particularly those in the approaching flow direction.

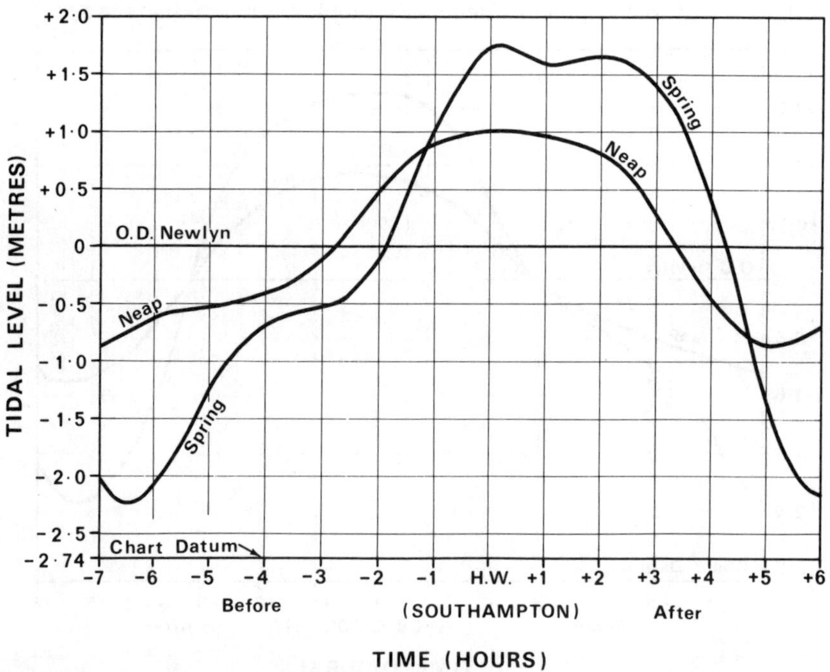

Fig. 7. Tidal curves for Southampton

24. In the tidal curves for Southampton (Fig. 7) these characteristics are seen to be accentuated and the double high water feature is pronounced. A harmonic analysis of this curve shows that it is composed of some 25 constituents and that the sixth diurnal (M6) rather than the quarter diurnal (M4) is the governing factor in the second high water, which is perhaps attributable in part to some resonance effect within Southampton Water itself.

25. Certainly the unique shape of the Southampton Water tidal curve demonstrates how the various influencing factors have combined to make the Solent tidal phenomena some of the most complex and difficult to explain in the world.

TIDAL VOLUMES

26. The tidal prism, that is the change in water volume between high and low water, for the Solent system is enormous. At neaps, the tidal prism is a minimum and for Hurst-Gilkicker it is of the order of $300 \times 10^6 m^3$, of which about $1/6$th is accounted for by Southampton Water. This represents a cubic block of water with sides nearly 0.7km flowing in or out every 6 hours or so. During spring tides the volume is about twice this amount.

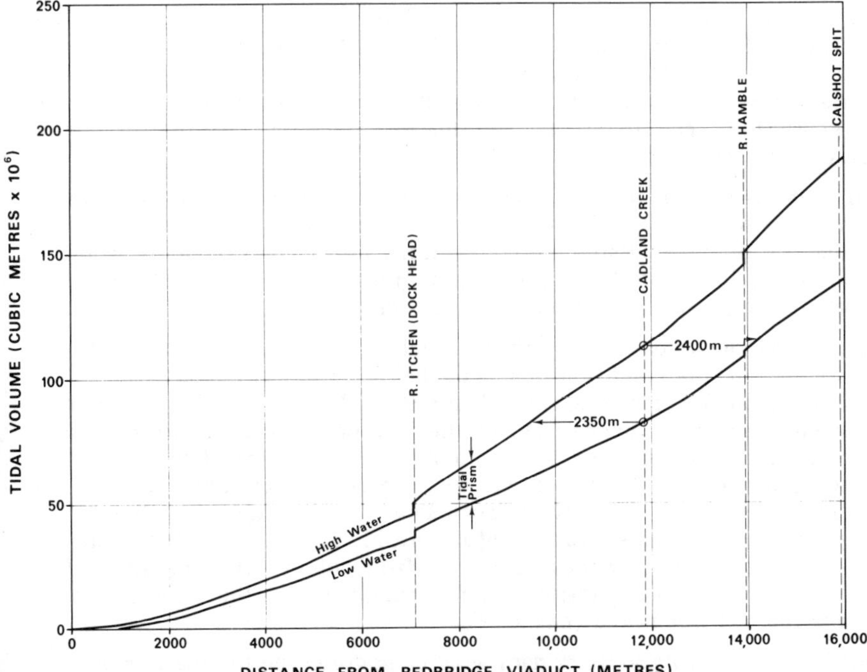

Fig. 8. Tidal volume curves (neaps) for Southampton Water

27. The tidal volume curves are of interest, because, besides indicating the volumetric flush, they give an indication of the average particle travel within estuaries. For example, Fig. 8 shows the tidal volume curve for Southampton Water for neaps. A particle released at Cadland Creek would travel 2.4km upstream on the flood of a neap tide and about the same distance seaward on the ebb. Of course, this presupposes that the velocity distribution is uniform, which it is not, but it does give a useful indication of potential effluent behaviour on the basis of simple desk calculations. It is no substitute, however, for well-directed field investigation.

FRESH WATER INFLOW AND SALINITY

28. By definition, an estuary system must have a source of fresh water and the greater proportion is usually introduced at the head of the system. This inflow of water is important from the point of view of the net flow seawards, the behaviour and distribution of currents, particularly in the upper reaches, and the sedimentary characteristics. Also, salinity is an especially valuable parameter from the pollution point of view because it serves as a conservative tracer indicating the degree of exchange with the open sea.

29. In the case of the Solent (Hurst to Langstone) the estimated annual inflow is of the order of $870 \times 10^6 m^3$ and this is equivalent to $1.2 \times 10^6 m^3$ per tidal cycle, with marked variations according to the time of year. Of this quantity, probably about 10 per cent is in the form of sewage effluent, discharged into the estuarine or coastal waters with or without treatment.

30. The largest rivers, Test and Itchen, enter at the head of Southampton Water and account for about 45 per cent of the total inflow. Even if one considers Southampton Water alone this is a very small proportion (about $1/75$th) of the neap tidal prism and, of course, for the Solent as a whole the proportions would be much less.

31. In the open sea the salinity is about 34 parts per thousand and in the Solent this condition is approached. The lower reaches of Southampton Water are fairly well mixed (salinity about 31 parts per thousand) but in the upper reaches there is pronounced stratification. For example, in the upper Test estuary, it has been found, in some places, that surface currents are always directed seaward. More interestingly, opposite to Marchwood Power Station there has been observed a three-layer stratification in which the fresh water from the river and sewage works outfall at Millbrook overlies the warmed water from the power station, which in turn overlies the colder saline water from the seaward reaches (9).

32. Some valuable salinity data have been collected for the Itchen Estuary by the Hampshire River Authority and it has been found that

the estuary seaward of Northam Bridge is partially mixed whereas up-river it is stratified. The typical salinity contours at high water for an average tide are shown in Fig. 9. From these data it is computed that the retention time for effluent discharged at the head of the estuary is about 4 days, depending on the fresh-water inflow. This pollutional aspect is further discussed in the next Section.

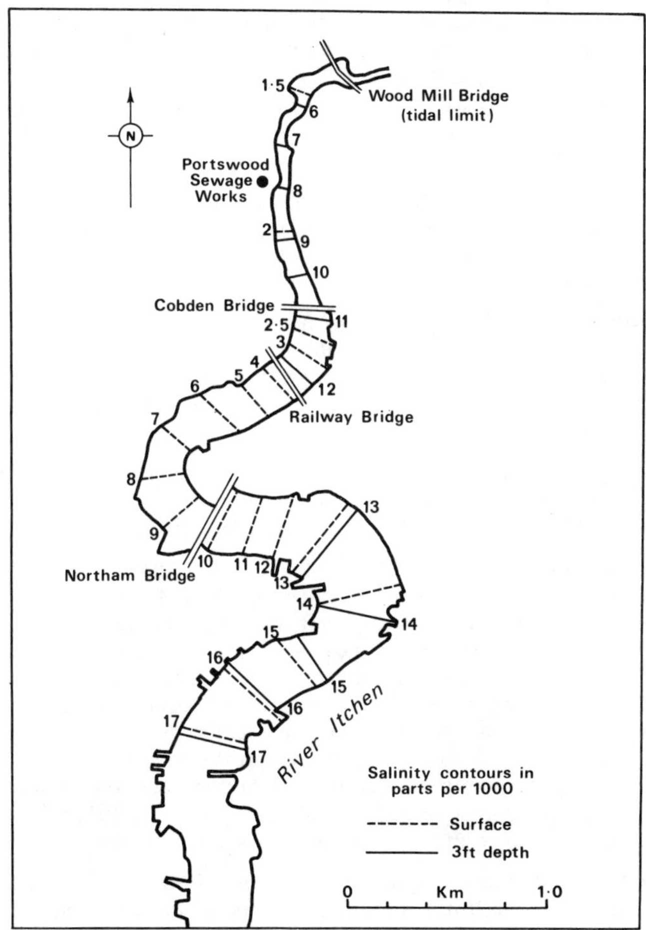

Fig. 9. Salinity contours for the Itchen Estuary at high water

CHANGES DUE TO MAN'S ACTIVITIES

33. Both Southampton and Portsmouth have been developed as ports because of the sheltered berthage afforded and the natural deep-water channels leading to them. Nevertheless, extensive channel deepening

had been undertaken over the past 100 years or so and large areas of intertidal mudflat have been reclaimed for port facilities and industrial usage.

34. A glance at a part of Murdock Mackenzie's chart dated 1783 (Fig. 10) of Southampton Water reveals how very rural the scene was then in comparison with the present day (Fig. 11). Nevertheless, because of the practice of 'cut and fill' the tidal prism has not been greatly altered; for example, the recent extension of the western docks has only reduced the spring tidal prism for Southampton Water by about one per cent. The estuary regime is probably only marginally changed although slightly higher water levels may now be experienced in the upper reaches of the Test.

35. In the case of Portsmouth Harbour, a quite large area (about 200 ha) of the upper estuary was reclaimed about three years ago. There was no compensating dredging and the tidal prism for the harbour has been reduced by about 6 per cent.

36. Recent projects in both Southampton Water [10,11,12] and Portsmouth Harbour [13] have been the subject of hydraulic model investigations and there is no doubt that the hydraulic model is a useful tool in predicting the effects of major civil engineering works such as those described.

37. But of greater significance in the present context is the introduction of municipal and industrial effluents on a large scale, not all of which have been treated to the standard that is desirable.

38. Recently, using basic data kindly provided by the River Authority, a study [14] was made of the pollution status of the Itchen Estuary and the likely effects of further abstraction just upstream of the tidal limit (Wood Mill) for water supply purposes. It was found that the two sewage works (Portswood and Woolston) gave rise to moderate pollution and that there was an 'oxygen sag' in the vicinity of Northam Bridge. There are, however, proposals to improve the standard of the effluent in both cases. It was further concluded—although the requisite data for this calculation were barely adequate—that further limited abstraction for water supply purposes would not prejudice the condition of the estuary provided that the effluent standards were improved.

39. In this necessarily over-simplified study, based on the method of Pearson and Pearson [15], salinity was, as referred to earlier, the tracer for mixing whilst ammoniacal nitrogen identified the sewage, the latter parameter being regarded as a more reliable indicator in the present case than B.O.D. which was extremely variable.

THE PROPOSED SOLENT OUTFALL

40. As is described in another paper, the South Hampshire Plan envisages the provision of a trunk-sewerage system which will collect

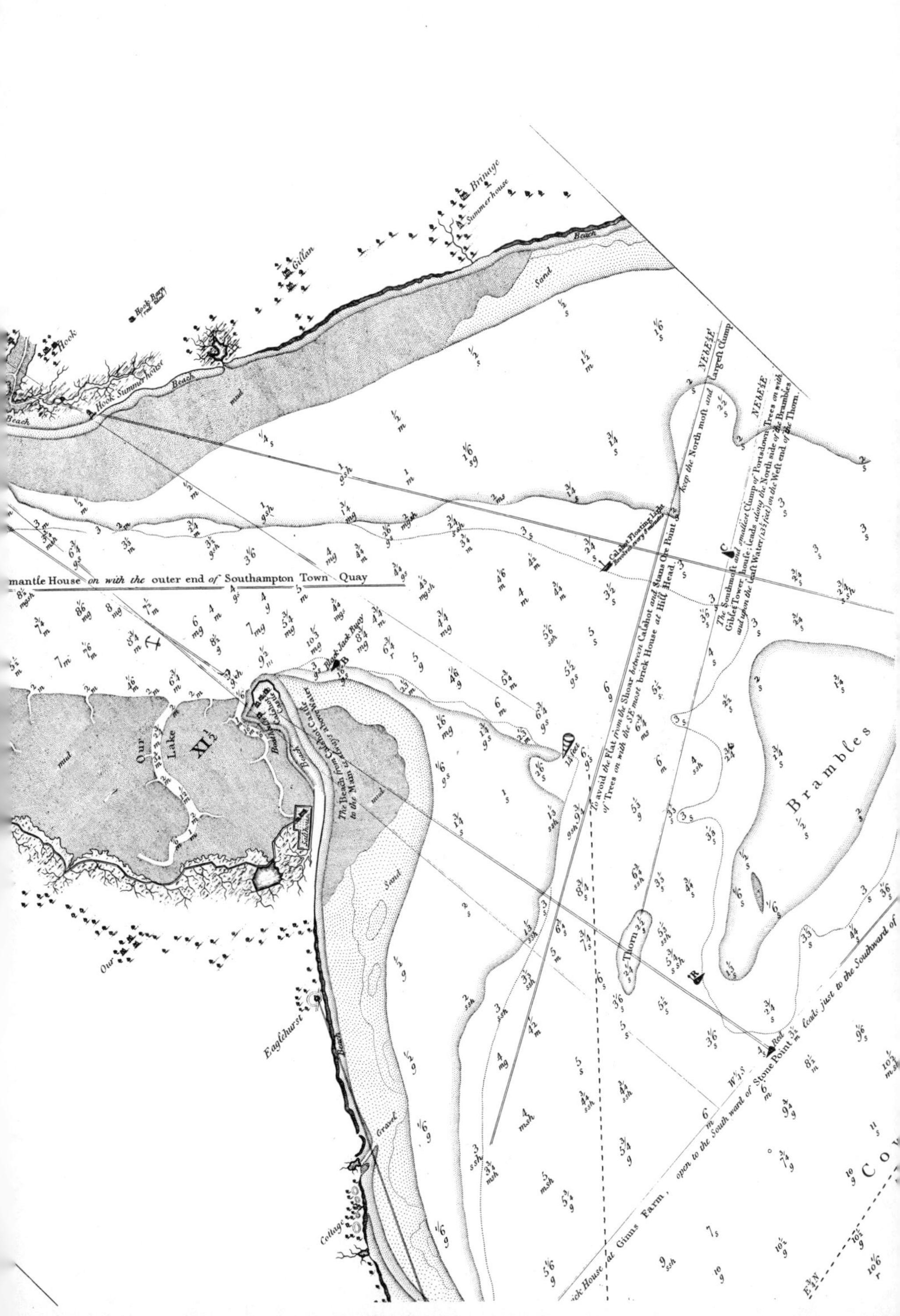

Brinage
Summerhouse
Gillan
Hook Buoy
Hook
Hook Summerhouse
Beach
mud
Sand
Calshot Floating Light
mantle House on with the outer end of Southampton Town Quay
NE b E ½ E
Keep the North most and largest Clump
The Southmost and smallest Clump of Portsdown Trees on with Gibbet Tower house; leads along the North side of the Brambles and upon the least Water (3½ feet) on the West end of the Thorn.
To avoid the Flat from the Shoar between Calshot and Stans Ore Point
of Trees on with the S.E. most brick House at Hill Head
Black Jack Buoy
Our Lake
Calshot Castle
The Beach from Calshot Castle to the Main is dry at Low Water
Brambles
Thorn
Red
Eaglehurst
Gravel
Cottage
ick House at Gims Farm, open to the South ward of Stone Point leads just to the Southward of

(Southern Newspapers Ltd.)

Fig. 11. Southampton Water, looking towards the Solent

the foul drainage from a large part of the South Hants region, principally between Eastleigh and Fareham—an area that is to be extensively developed and is in need of rationalisation with respect to the sewerage services.

41. In 1969, the Working Party on Main Drainage commissioned an investigation [16,17] involving the University of Southampton's Solent Model for the purpose of examining the behaviour of sewage effluent discharged from four possible outfall sites. These were located in minimum 9m depth of water at some distance from the shore off Hamble, Hook, Hill Head and Browndown, respectively. The discharge of effluent was simulated to scale by means of a dye, heated so as to provide the correct density effect, and its dispersion behaviour was observed. It was found that the Hamble and Hook outfalls would result in contamination of nearby shorelines whilst the Hill Head and Browndown outfalls gave rise to considerable dispersion, with the former having the better performance. For neap tides the sewage field was found to extend 13km in the longitudinal direction, and, of course, a very much greater distance at springs.

42. The above work, of an essentially very preliminary nature, was followed in 1970/71 by a far-ranging and most interesting investigation conducted by the consulting engineers, J.D. and D. M. Watson, concerning the marine disposal of partially treated sewage. Its objective was to ascertain the most suitable sites for outfalls and the appropriate methods of discharge. In their report [18], presented in 1972, it was concluded that an outfall at Horse Sand Fort (Fig. 1) was a feasible proposition.

43. During the course of this investigation a considerable amount of field work was undertaken, principally in the area East Solent to the Nab Tower, and the data collected constitute a valuable addition to our knowledge of the tidal hydraulics of the area. Some of the most interesting hydrographic features observed were a net north to south drift of surface current across the East Solent and a residual flow of water through the system from west to east, the latter being deduced by K. R. Dyer, now of the Unit of Coastal Sedimentation, Taunton, who directed research by A. R. de Mesquita [19].

44. In connection with this investigation a two-dimensional mathematical model, embracing the area East Solent to the Nab, was formulated by J. M. Williams of the Hydraulics Research Station, Wallingford. The model is based on the velocity and direction of water movement at points 1km apart for half-hourly intervals throughout the tidal cycle for any required wind speed or direction. The introduction of appropriate dispersion coefficients enable computations to be made as to the likely extent of the sewage field.

45. The South Hampshire Main Drainage Board, which is now the statutory authority for the area, has decided that a full-treatment works should be constructed and that the outfall should be in the East

Solent, discharging only purified effluent. The mathematical model referred to earlier has been applied to a study of the dispersion from this outfall. C. A. Brebbia and R. A. Adey [20] of the Department of Civil Engineering, University of Southampton, have also studied the same problem using a mathematical model based on the finite element technique.

CONCLUSIONS

46. The Solent, its tributary channels and adjacent waters, constitute an estuarial system of considerable complexity. It would require much more than a single paper to adequately discuss the tidal hydraulics and the author is more than conscious of the omissions and shortcomings of this present effort. For example, the network of waterways that constitute Chichester and Langstone Harbours have not even been mentioned.

47. From the scientific viewpoint, the Solent waters have received nothing like the attention that has been given to other important estuarial systems such as the Thames, Mersey, Tees, and so on. The reason is, of course, that these latter estuaries have given rise to considerable public concern as to their condition and there has been a consequent willingness to finance costly investigation. A useful start has, however, been made in respect to the Solent system by the formation, with the encouragement of the Natural Environment Research Council, of an inter-disciplinary technical panel. The task of this panel is to assess the present state of knowledge concerning the physical, chemical and biological aspects; their report should be of value to planners, engineers and others with immediate practical problems, and it should also serve to draw attention to the important gaps that need to be filled.

48. We all wish to see these delightful waters preserved free of pollution and there is a need for monitoring and research on all the various aspects, not least the hydraulic. Investigations for specific engineering projects have provided useful data from time to time but are really no substitute for long-term purposeful study. Scientists and engineers are eager to contribute but financial support is essential.

REFERENCES

1. Hay, D. and J., 'The Solent from the Sea', Stanford, London, 1972.
2. 'Admiralty Tide Tables—1973', **1**, Table III, pp. XX-XXI, Hydrographer of the Navy, 1972.
3. Doodson, A. T. and Warburg, H. D., 'Admiralty Manual of Tides', pp. 223-226, H.M.S.O., London, 1941.
4. Macmillan, D. H., 'The Hydrography of the Solent and Southampton Water', Ch. 4 of F. J. Monkhouse (Ed), 'A Survey of Southampton and

Its Region', Camelot Press, Southampton, 1964.
5. Reid, C., 'The Geology of the Country around Ringwood', Mem. Geol. Survey, H.M.S.O., London, 1902.
6. Reid, C., 'Discussion on the Former Connection of the Isle of Wight with the Mainland', Rpt. Brit. Assoc. Adv. Science, 1911, p. 384.
7. Rossiter, J. R., 'Sea-level observations and their secular variation', *Phil. Trans. Roy. Soc. Lond.*, 1972, A, **272**, 131-139.
8. 'The Solent and Adjacent Waters—Pocket Tidal Stream Atlas', Hydrographer of the Navy, 1962.
9. Pannell, J. P. M., Johnson, A. E. and Raymont, J. E. G., 'An Investigation into the Effects of Warmed Water from Marchwood Power Station into Southampton Water', *Proc. Inst. Civ. Eng.*, 1962, **23**, 35-62
10. Wright, W. and Leonard, R. D., 'An Investigation of the Effects of a Proposed Dredging Scheme in Southampton Water by means of a Hydraulic Model', *Proc. Inst. Civ. Eng.*, 1959, **14**, 1-18.
11. Webber, N. B., 'An Investigation into the Behaviour of the Estuarine Cooling Pond for the Proposed Power Station near Calshot', Report to C.E.G.B. (Unpublished), 1961.
12. Webber, N. B., 'An Investigation of the Hydraulic Effects of the Proposed Scheme for Extending Southampton Docks', Report to Brit. Trans. Docks Board (Unpublished), 1966.
13. 'A Study of the Effects of Reclamations on Portsmouth Harbour and Its Approaches', Report by Hydraulics Research Station, Wallingford (Unpublished), 1959.
14. Snell, M. R., 'An Investigation of the Itchen Estuary with particular reference to its Pollution Status', Report No. 11/72, Univ. of Southampton, Dept. of Civil Eng., 1972.
15. Pearson, C. R. and Pearson, J. R. A., 'A Simple Method for Predicting the Dispersion of Effluents in Estuaries', Proc. Symp. on New Chem. Eng. Problems in the Utilisation of Water, 1965.
16. Webber, N. B. and Helliwell, P. R., 'Hydraulic Model Investigation for Sewage Outfall Sites in the Solent Area', Report to S. Hampshire Plan Advisory Committee (Unpublished), 1969.
17. Helliwell, P. R. and Webber, N. B., 'The Use of the Solent Model for an Investigation into Sewer Outfall Location', Proc. Symp. on Math. and Hyd. Modelling of Estuarine Pollution, H.M.S.O., London, 1973.
18. 'Marine Disposal of Sewage and Sewage Sludge', Report by J. D. and D. M. Watson (Consulting Engineers) to S. Hampshire Plan Advisory Committee, 1972.
19. Mesquita, A. R. de, 'Studies on the Mean Flow of the Western Solent', Univ. of Southampton M. Phil. thesis, 1972.
20. Brebbia, C. A. and Adey, R. A., 'Mathematical Model for the Dispersion of Effluent in the Solent', Report No. 2/73, Univ. of Southampton, Dept. of Civil Eng., 1973.

DISCUSSION

WEBBER The paper is a very broad review of the tidal hydraulics of the Solent system and, by way of introduction, the opportunity is taken to examine, in a little more detail, two specific problems which are relevant to pollution—namely, the assessment of tidal flow in the Solent and the dispersion of contaminants within one of the estuaries.

Conditions in the western arm of the Solent are the most favourable for discharge assessment. The waterway here is about 16 km in length, reasonably straight, with an average width of about 4700m, and with a maximum depth, on average, of 16m. There is considerable difference in tidal range at the two ends (Fig. 1), which can be as much as 1.5m during springs, and the resulting difference in water level is conducive to fast tidal currents. More recently, tidal data have been obtained for Thorness Bay near the eastern end of the West Solent affording confirmation of the considerable hydraulic gradients which exist.

All this suggests that the West Solent can be treated as a hydraulic channel subjected to time-varying flow. A very preliminary and over-simplified start has been made on this problem. Using the Manning equation and three alternative values of the friction factor n (0.02, 0.25 and 0.03) the discharges have been computed for a spring tide.

For an n value of 0.025 the maximum discharge is about 90 000 m^3/s in the ebb direction and occurs around the time of H.W. Portsmouth. This is a very large discharge indeed and when one bears in mind that the mean discharge of the R. Test, the largest river which flows into the Solent, is only around 12 m^3/s one can gain some impression of the immense volume of water that is being transported.

Of course, during neap tides the flow is much less and, inevitably, with reversing flow there are periods when the discharge falls to zero, but the duration is relatively short. In any event, although there may be no net flow at a cross-section there will be circulatory motions in existence which will mean that the water is never completely still. In the central channel, the particle excursion per tide is of the order of 20 km at springs and about half this amount at neaps.

Integration of the discharge curves yields the total quantity flowing in each direction and hence the residual flow. But more basic data and a more rigorous approach, taking into account meteorological, accelerative and storage effects are needed before this computation can be made with any degree of confidence.

The other problem—the dispersion of contaminants—concerns Southampton Water and is more directly related to pollution.

Southampton Water and its tributary estuaries, Test, Itchen and Hamble, are shared between many interests—a large residential population, industry, commercial shipping and recreation—all of which must co-exist, paying due regard to each others' needs.

Sewage effluent is discharged to the estuary at a number of points (Map 3, p. 1.6), the most significant of which are Millbrook (treated) Slow Hill Copse (treated) and Ashlett Creek (untreated), together with the effluent entering via the Itchen Estuary. In addition, industrial effluent is discharged at Fawley and there is a cooling water circulation at Marchwood Power Station (about 32 m^3/s) and at Esso Refinery (4 m^3/s). The quality of the sewage effluent is known but that of the industrial effluent, though monitored, is not publicly available.

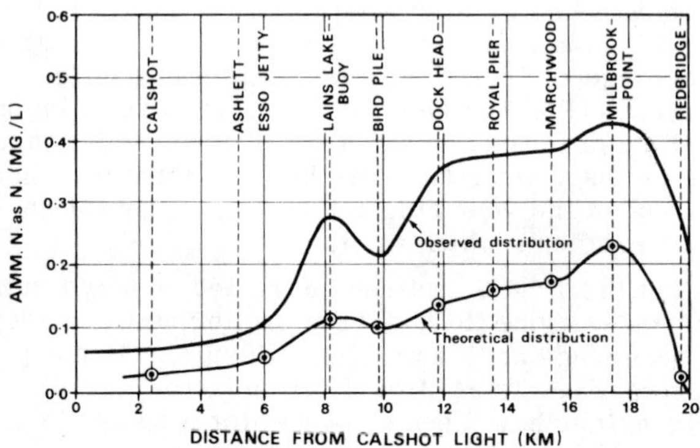

Fig. 12. Salinity and ammoniacal nitrogen profiles for Southampton Water.

Fig. 12 shows a salinity and an ammoniacal nitrogen profile for the navigation channel of Southampton Water obtained for the period of the high water stand.[21] Salinity is, of course, an excellent indicator of interchange with the open sea and it will be noted that the estuary is fairly well mixed in the lower reaches with some stratification in the upper. As referred to in the paper, ammoniacal nitrogen has been taken as the indicator for the degree of pollution in preference to B.O.D. which was found to be extremely variable. It will be observed that ammoniacal nitrogen does show a significant increase in the vicinity of the industrial complex; also that in this vicinity there is a slight reduction in salinity.

From the various relevant data—salinity, average fresh-water discharge, pollution load (excluding Fawley)—it is possible to compute,[22] using the simplified approach of Pearson and Pearson,[15] the theoretical ammoniacal nitrogen profile (Fig. 13). The observed and theoretically derived curves can be compared and it will be noted that there is discrepancy in magnitude but that the general shape is not dissimilar. It was thought worthwhile to introduce a correction coefficient to bring the theoretical curve broadly into line with the observed behaviour, although the over-simplifications (steady state conditions, vertical mixing, conservative tracer) and unknown factors (e.g. industrial effluent) suggested that this was a bold step to take.

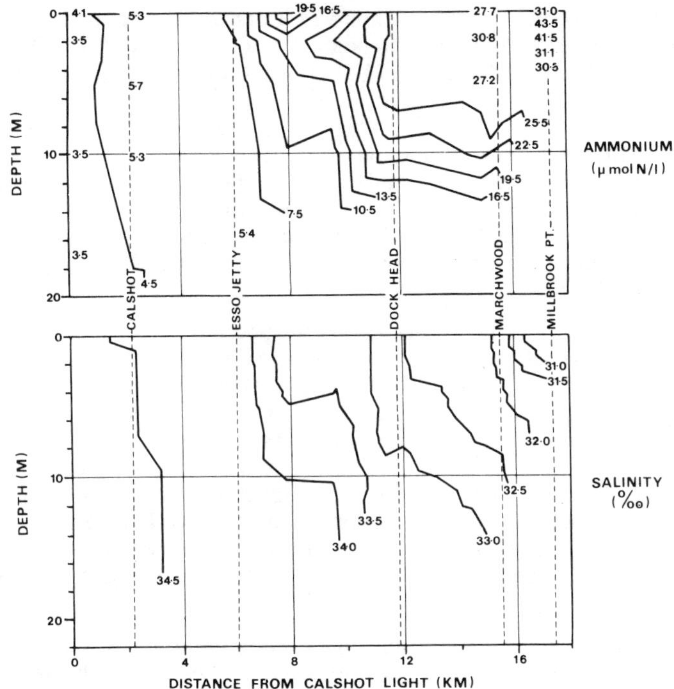

Fig. 13. Ammoniacal nitrogen profile for Southampton Water—observed and theoretical.

Having established this rather crude mathematical model it was now possible to vary the basic parameters of fresh water inflow, effluent discharge and standard of treatment in order to predict the pollution status of the estuary in the future, taking account of increased population and water abstraction. Inserting what seemed to be likely figures for these parameters it was found that there would be an increase in contaminant properties, but that the general levels would be within acceptable Royal Commission limits. For various reasons, such as higher standards of treatment than allowed for, the prediction might well err on the pessimitistic side.

Although one would readily agree that this procedure is open to serious criticism, it does incorporate most of the important parameters and has as its objective the prediction of the future state of health of the estuary, which must surely be the aim in any long-term environmental planning. As and when additional and more detailed data become available so will application of the more sophisticated mathematical techniques be justified.

OAKLEY (J.D. and D.M. Watson). There is a continuing controversy over the direction of residual flow through the Solent. Would Mr. Webber comment on the influence of wind and barometric pressure on the movement of water?

WEBBER Wind and barometric pressure do have a significant influence on the mass transport of water in shallow seas. Often the two are linked in that high winds are coincident with relatively low barometric pressure. Water will flow to a zone of low pressure and the effect of wind is to produce a mass transport of water on account of the tractive stress exerted on the surface.

The consequences for a restricted channel open to the ocean at both ends are perhaps best illustrated by reference to the Straits of Dover. Bowden [(23)] has shown that there is a residual flow of water in accordance with the relative barometric pressures and the direction of wind. Most of the time, therefore, the residual flow is from the English Channel to the North Sea. But, occasionally, there are spectacular reversals of residual flow, as for example during the disastrous North Sea tidal surge of 1953, when an enormous volume of water flowed from east to west, raising water level significantly as far west as Newhaven. In the Irish Sea a similarly directed (south to north) long-term residual flow has been identified.

With regard to the Solent the situation is more complex, since there are other influencing factors, of a hydraulic nature, which could mean a long-term residual flow from east to west contrary to the prevailing wind. Nevertheless, the evidence from tide gauges in the Solent area shows that strong winds can cause quite a significant set-up over distances of only a few kilometres. Further research on the residual flow and other aspects of the tidal hydraulics in the Solent area is currently in hand.

MITCHELL (Water Resources Board). Can a biologist offer a comment on the question of direction of residual flow in the Solent? There is some evidence that suggests that the massive oyster bed in Stanswood Bay is the result of a good stock of oysters breeding in the Beaulieu River [(24)] (see Savage's paper). The larvae were probably carried out of the Beaulieu River and transported with the residual current, from the west to the east. Other evidence for a residual current from the west to the east in parts of the Solent comes from distribution of the animal *Mercenaria mercenaria*, the hard shell clam. This animal first introduced into Southampton Water now has a distribution outside Southampton Water which is entirely on the mainland side of the East Solent. In other words the larvae are carried out of the Southampton Water into the East Solent. There is no evidence of any establishment of this animal in the West Solent, although some of the sediment appears to be suitable for this species. Both these examples suggest that there is a residual current from the west to the east.

WEBBER Dr. Mitchell has made some most interesting observations concerning the influence of residual flow on the spread of marine species.

There is, of course, a residual flow out of Southampton Water represented by the fresh-water inflow, which, as has been stated in the paper, is a very small proportion of the tidal prism. The distribution of velocity across the estuary mouth at Calshot is not known and there may well be an imbalance favouring ebb movement eastwards along the mainland shore. Certainly, this is the direction of littoral drift, such as there is.

With regard to the migration to Stanswood Bay, a velocity traverse undertaken some years ago in connection with a service crossing between Stansore Point and Egypt Point, a distance of 2.75 km, entailed detailed current meter observations at 10 representative stations in the cross-section. The results indicated a residual flow in a westerly direction in the northern portion and to the east in the southern, but the nearest station to the northern shoreline was about 400 m distant, so it is possible that further inshore different conditions could prevail. Again, the direction of littoral drift is easterly so that this may be the deciding factor.

Further information concerning the importance of current and other physical factors with regard to the migration of a particular species would be helpful in this context.

CAMP (J. D. and D. M. Watson). It might be helpful if some explanation were given regarding the desire to ascertain the direction of residual flow in the Solent. As investigations into the feasibility of disposing of a major new discharge of municipal sewage via a sea outfall to the Solent developed, it became apparent that some indication of the maximum residence time within the Solent was desirable so that a check on nutrient build-up could be made. Dyer's theories of a net movement east to west were worrying until it was realised that Dyer's data indicated changes in direction of residual movement possibly with a five-day period. Subsequently it was shown that, to satisfy standards for coliform concentration at the shore lines it was necessary to site the terminal point in the Horse Sand Fort area.

In one of his studies the Author showed predictions of ammoniacal nitrogen for 1970/80 and 1980/90, which suggested a doubling of the load in Southampton Water. How was this figure derived?

WEBBER If improved standard of treatment are adopted then the levels predicted would be unduly pessimistic, but, in any event, as stated earlier, they would be within Royal Commission limits.

The increased load predicted for Southampton Water is largely accounted for by greater effluent discharges reflecting forecast population increases.

SUTTON (Esso Petroleum Co. Ltd.). In paragraph 27, discussing movement of the tide in Southampton Water, starting from Cadland Creek presumably because it is the centre of the industrial discharges there, a particle released would travel 2.4 km upstream on the flood and the same distance seaward on the ebb. Why would not anything discharged from the industrial centre oscillate forever between the Hamble and Hythe? This does not happen. For instance, occasional oil spills eventually disappear.

WEBBER The method of predicting effluent travel from a consideration of the shape of the tidal volume curves does represent an appreciable over-simplification, since it assumes uniform velocity distribution in the transverse direction. Nevertheless, in the absence of velocity, float or tracer observations, it does give indicative information on a comparative basis.

Eddy current and turbulent mixing ensure that contaminants are flushed out of Southampton Water in the course of tidal cycles. A preliminary study is currently in hand directed at obtaining some estimate of the degree of exchange with the more open waters of the Solent. It requires data concering salinity variation throughout a tidal period and at the moment this information is fairly limited.

CHARLTON (University of Dundee, Tay Estuary Research Centre). Several speakers have raised the question of residual tidal currents and their bearing on the problem of outfall placing and design. We tend to imagine that the West Solent could be treated as a uniform channel, but when tidal flow is studied in such a channel, flood and ebb channels are frequently found, although bathymetric surveys may not indicate this. There may be a greater flow in the North in one direction than in the South and vice versa. This is very much so in the Tay Estuary which has some areas similar to the Solent. Moreover, at the turn of the tide, there are induced very considerable rotations in the bodies of water. For instance, just inside Hurst Castle a mass rotation would be expected on the high tide, so that material coming out of Yarmouth Harbour crosses the channel to the Lymington side. And likewise, there may be separate bodies of water along the whole channel which are rotating independently and perhaps in opposite directions to each other. These form very powerful mixing agents in such a body of water.

WEBBER Predominantly ebb and flood channels are certainly to be found in the Solent area, and, characteristically, at the turn of the tide, reversal of flow occurs at the margins before that in mid-channel, inducing the large scale eddies referred to. But there is no evidence that sediment discharging from Yarmouth Harbour crosses the Solent to the Lymington side and this would hardly be expected since the Harbour discharges when the current in the main channel is ebbing quite fast.

The valuable work of the Tay Estuary Research Centre is a good example of the merits of an inter-disciplinary approach to marine studies and it has been duly noted by those currently engaged in the preparation of 'A Scientific Survey of the Solent'.

WAKEFIELD (Coastal Anti-Pollution League Ltd.). In 1958/9 when float tests were carried out for the Gosport outfall, it was said that the residual flow was from west to east, and that sewage discharged at Gosport would pass quickly, in a zig-zag fashion, out towards the Nab Tower. Yet apparently the discussion still continues.

Is not the Solent like a concertina? If this is so, there must be a limit to discharge to the Solent; it cannot be increased *ad infinitum*.

WEBBER Certainly, there is a limit to the discharge of effluent (other than fully treated) to the Solent if the environment is not to be adversely affected. But, at the same time, the extent of exchange with the English Channel and the capacity for self-purification of a large body of flowing water should not be minimised.

Looked at in the extreme, one can see that if the water in the central Solent and its tributary estuaries merely oscillated to and fro, without any exchange, then its constituents would be entirely fresh. But this is far from the case since the salinity at Calshot and elsewhere in the Solent proper is not very different to that in the open sea, so there must be a considerable degree of intermixture with the latter. There are enormous volumes of water flowing in and out of the system during each tidal period and there would still be a considerable exchange of water, even in the absence of any residual flow.

JURY (Newport, Isle of Wight). It is true that the public generally is very deeply concerned about the subject raised by our last speaker: the idea of contaminated polluted matter travelling backwards and forwards *ad infinitum* up and down the Solent. There is a very strong feeling that any main sewer outfalls or even outfalls discharging partly treated sewage, should be situated on the back of the Island where there is a strong tidal stream down the English Channel which will take effluent away completely.

WEBBER People often recoil instinctively when they hear about a new sewage outfall proposal. They tend to think of those far from satisfactory outfalls where the sewage slick is clearly visible from the shoreline and where there is an undoubted local pollution problem. Unfortunately, quite a number of outfalls around the Isle of Wight do fall within this category, although steps are generally in hand to improve the situtation.

The proposed outfall at Gurnard which will handle the sewage from Newport and Cowes is a case in point. The outfall is to be located in about 10 m water depth and, when fully operational, will discharge 0. 1 m^3/s. As compared with an average flow in the West Solent of about 20 000 m^3/s, this is a very small proportion indeed. There would be something like a hundredfold dilution in the rise to the surface, and, with a fairly rapid dispersion subsequently, it is doubtful whether there would be any visual indication of sewage, except perhaps in calm weather. In view of the very considerable demands currently being made on the public purse, it would seem a quite unnecessary luxury at the present time to pay an additional million pounds, or possible much more, for full treatment or for piping the effluent to the south side of the Island.

NEBRENSKY (University College of Swansea). It is surprising how many difficulties are encountered in application of even such a simple model. Would Mr. Webber comment on the degree of confidence he has in the salinity distribution used to evaluate mixing between the segments? Could he also give more details about the way in which the stations for salinity measurements were selected, and how does he cope with the vertical and transverse distribution of salinity as shown on some of his slides?

WEBBER The salinity data for the Itchen estuary were obtained from the analysis of samples taken at representative stations along the margins of the estuary during a number of separate tidal periods, whereas those for Southampton Water were obtained by means of a boat operating in the central deep-water channel during the time of high water slack.

In view of the very limited data, no attempt has so far been made to take account of non-uniform salinity distribution, except in the longitudinal direction. In fact, the estuary system on the whole is quite well mixed so that it would only appear to be of real consequence in respect of the upper reaches, where stratification is certainly evident.

However, it is relevant to point out that the salinity does vary with tidal state, tidal range and fresh water inflow, and the effects of these factors are currently being examined in relation to estuarine circulation and flushing.

There does not appear to be any field evidence to answer this question, but when tests were recently conducted in the Solent Model to observe the behaviour of effluent discharge from the mouth of Cadland Creek, on the western side of Southampton Water, it was noted that the rate of lateral dispersion was quite limited in relation to the longitudinal excursion.

JARMAN (C.E.G.B.). Warm water from Fawley or from Marchwood Power Stations issues largely as a jet and steadily disperses as it mixes with the adjacent water. The fine details of the patterns of movement have not been established. Despite this, it is usually possible to predict the general pattern of movement with some confidence. The question has been asked as to what would be the accuracy of results from these rather crude methods of working out dispersions over many tidal cycles. In Southampton Water the lack of detailed information on the vertical and lateral variations made it necessary to consider a rather simple model.[25] This was used to calculate the warming up of the estuary from the warm water put out from Marchwood Power Station and Fawley Oil refinery. At the same time, a few calculations were made assuming the salinity distribution was slightly different than assumed in the main calculation. It did not make a great deal of difference to the temperature rise predictions. We predicted a temperature rise of 0.7°C. Large errors in the salinities assumed in these calculations would probably have led to an error of about 10-15%, so these crude methods do give a reasonable answer. The computed temperature rises were similar to the results from thermographs taken over many years. It was quite pleasing to find out how close the results obtained with these very approximate theoretical methods were to the results of many years of monitoring.

WEBBER Mr. Jarman has made some valuable and constructive comments. His paper on the dispersion of heat from cooling water adds usefully to our knowledge of fluid behaviour in the estuary. It is gratifying to note that the simplified mathematical model, which assumes complete mixing, except in the longitudinal direction, does result in predictions of temperature which are quite near to those observed in nature.

REFERENCES

21. Raymont, J. E. G., 'Some Aspects of Pollution in Southampton Water', *Proc. Roy. Soc. Lond. B,* 1972, **180**, 451-468.
22. Woolley, P. R. G., 'An Investigation into Southampton Water and the Test Estuary, with particular reference to their Pollution Status', Report No. CE/10/73, Univ. of Southampton, Dept. of Civil Eng., 1973.
23. Bowden, K. F., 'The Flow of Water through the Straits of Dover', *Trans. Roy. Soc. Series A.* 1956, **248**, 517-551.
24. Key, D., 'The Stanswood Bay Oyster Fishery', Ministry of Agriculture, Fisheries and Food, Fisheries Leaflet No. 25, Aug. 1972.
25 Jarman, R. T. and de Turville, C.M., 'Dispersion of Heat in Southampton Water', *Proc. Inst. Civ. Eng.,* 1974, **57**, 129-142.

SEDIMENTARY PROCESSES WITHIN ESTUARIES AND TIDAL INLETS

E. J. Humby, B.Sc. (Eng)., C.Eng., M.I.C.E. and J. N. Dunn, B.Sc.

Department of Civil Engineering, Portsmouth Polytechnic

INTRODUCTION

1. The water circulation within estuaries is frequently used as a means of dispersing waste products arising from industrial processes and human and urban development. The waste products are often in a form which has its counterpart in the natural estuary either as a material in solution, or in particulate form submerged or floating at the water surface.

2. Some aspects of estuarine sedimentation are reviewed in order to describe the environment in which waste disposal occurs and to indicate areas where knowledge of sedimentary processes may assist those concerned with predicting the fate of unwanted material dispersed in estuaries.

3. Sediment transport in this context deals with the erosion, transport, sorting, and deposition of particulate matter by hydraulic forces, but classical sedimentation theory inadequately describes the process. Estuaries are areas of high biological productivity and the organisms within the area may modify the sediment properties.

4. The movement of material in solution, the drift of floating debris under the influence of wind and currents, and the effects of water movement on living organisms will be ignored except where of relevance to the mechanism of sediment transport.

5. The circulation within an estuary is caused by the combination of tidal flow, river discharge, and density currents formed by salinity and temperature differences. Meteorological disturbances and wave action exert a generally aperiodic influence on the currents and flow patterns within an estuary.

6. The natural sediments in an estuary may be of organic, or inorganic origin. In many estuaries this sediment is augmented by particulate waste.

7. The response of sediment to hydraulic forces is principally determined by the shape, size, and density of the individual particles which is typified by the settling velocity.

8. Particles such as sand grains retain their individual properties but clay particles form flocs which readily change in size and density as they migrate through the flow. Sorting of materials also occurs and although the properties of the individual grains may remain unaltered, the bulk characteristics of the sediment change.

9. Pollutants can be absorbed by the particulate matter and the actual amount of suspended material present can influence the ecology of an estuary.

10. Many of the waste products discharged into estuaries are hydraulically similar to natural sediments and therefore the study of the natural sediment distribution and circulation is of value when assessing the efficiency of a waste dispersal system.

TYPES OF ESTUARY AND TIDAL INLET

11. The concept of an estuary as the meeting of a river with the sea is formalised in Pritchard's definition of an estuary being "a semi-enclosed coastal body of water which has a free connection with the open sea and within which sea water is measurably diluted with fresh water derived from land drainage".

12. A number of coastal features having some of the attributes of estuaries are excluded by this definition, particularly areas where the influence of the inflowing rivers is no longer a dominant feature. These areas will be referred to as tidal inlets since this term seems appropriate when used to describe several harbours of the Solent.

13. The degree of mixing that takes place within an estuary is dependent upon the river flow, the tidal volume and the width and depth of the estuary. The least mixing occurs where the fresh water flow is of an appreciable volume compared with the influx from the sea. Stratification of the salt and fresh water may take place with the result that the fresh water flows out to sea over, what is in effect, a shoal of sea water. Complete stratification is unlikely, because some mixing between the salt and fresh water occurs at their interface and as a result of this mixing a landward flow of salt water along the bottom occurs to replace the salt water that mixes with the river flow.[30]

14. The implications of the effects of a vertical salinity gradient on the pattern of sediment transport is that there is upstream movement of sediment which accumulates at the end of the salf water intrusion. These effects will be greater in the case of the stratified estuary than in the well mixed estuary, where the salinity differences in the vertical plane are small and the extreme velocities of the stratified case are not present. The overall pattern of circulation is similar, with stronger bottom currents than would be the case in a uniform salinity tidal inlet. The effects of a varying river discharge on salinity gradients was considered by Simmons [35] who showed that as the river discharge increased at Savannah Harbour, the surface flow became predominantly downstream and at the bottom the percentage of total flow downstream diminished sharply from 50% with no river discharge, to a minimum of about 25%. Further increase in river discharge gradually increased the percentage of downstream flow to about 43%.

15. Model and prototype tests in the Thames have shown that the salinity distribution curve is displaced seaward by increased river flow and that the increased flow has little effect on the rate of mixing in the central part of the estuary. (12)

16. Investigation of siltation in both the Mersey and the Thames show that salinity differences between the surface and the bed of the order of 1 or 2 parts per thousand are sufficient to produce a net landward drift of the saline water close to the bed of the channel. (29)

17. In wide estuaries with a high tidal velocity and where the mixing is sufficiently strong to eliminate the vertical salinity difference, a longitudinal salinity gradient may be established causing a lateral circulation pattern.

18. Salinity gradients may occur in tidal inlets where evaporation increases the salinity above that of the open sea.

THE SOURCE OF ESTUARINE SEDIMENTS

Erosion products from the drainage catchment

19. One source of estuarine sediment is the catchment area of the rivers draining into the estuary. The amount of material entering the system is a function of the hydrology, geology and land usage of the area.

20. Most rivers transport organic debris from the river and the catchment area, which may be augmented by farming and sewage effluent. This represents the organic fraction of the river sediment load.

21. The supply of material from the upper reaches of the river to the estuary is usually intermittent. When the river is in flood general movement of sediment in suspension occurs with the rate of progression of the same order as that of the stream velocity.

22. Engineering structures such as weirs offer limited hindrance to the movement of sediment in suspension. As the flood subsides the material settles out of suspension and is deposited on the river bed where it may move as bed load or remain undisturbed until the next flood occurs. If the sediment moves along the bed, the rate of progress is diminished and engineering structures may now prevent its passage to the sea until the onset of the next flood.

23. The intermittency of transport is also dependent on particle size. Clays in suspension may pass along the length of the river without a halt, but large stones and cobbles may be moved only by exceptional floods.

24. Study of several East Yorkshire river catchments showed that within a relatively small region the factors affecting sediment listed above produced rates of erosion varying from 0.6 to 254 $m^3/km^2/year$.

In the case of the catchment area with the highest yield 88% of the yield was transported by flows which were only exceeded 5% of the time.[11]

25. Solid sediment yields for a number of British catchments have been published and the values given range from 16 to 82.2 $m^3/km^2/year$ [7]. The author has estimated from geomorphological evidence that the sediment yield of a particular Cornish river catchment is 8.3 $m^3/km^2/year$. Complete and uniform retention of this material within the estuary of the catchment represents a depth of deposit of 0.2 mm/year.

Material entering from the sea

26. The amount of suspended material present in coastal waters is relatively small compared with that contained by an estuary. Armstrong and Atkins [1] reported the annual variation of inorganic material in the English Channel to be 0.45 to 2.7 mg/l. The organic component of the suspended matter is variable but Hayes[10] states that it comprises 3% or more zooplankton, and at certain times up to 60% phytoplankton. Organic detritus can be as high as 90%.

27. The effect of salinity currents on the ingress of sediments to an estuary is now well known. Inglis and Allen [12] in their investigation of the River Thames showed that much of the material dredged from the shipping channels and dumped far down the estuary was in fact taken back upstream by currents at the estuary bed, rather than being dispersed seaward.

28. Investigation of the siltation of the River Mersey has also shown that material arriving at the mouth of the estuary has been transported upstream by salinity currents.[29]

29. In the absence of marked density currents material in suspension may enter a tidal inlet with the flood tide. Where the inlet is situated on an exposed coast line the amount of suspended material resulting from wave action on the adjacent beaches may be considerable. The tendency of this material to block the entrance is balanced by the flushing action of the tidal streams through the entrance. Bruun and Gerritson [5] have shown that a relationship exists between the tidal volume of the inlet and the area of the inlet channel and that this must be satisfied if the entrance is to be kept open.

Locally derived sediments

30. Within an estuary erosion and re-working of sediments by waves and currents occurs. In the absence of protection, erosion occurs at the shore line providing new material to the estuary, and existing deposits are re-worked and sorted. Evidence of this has been found in the Tay [4] and Langstone Harbour.

31. Shells and shell fragments are a common feature of estuarine sediments and can be derived from geological sources or from the animals found in the area. The quantities of shell present are therefore dependent on the random features of the geology and the ecology.

32. There is a variable quantity of organic matter in the upper layers of the sediment averaging 2.5% for nearshore sediments [36] but it may be much higher depending on the hydrography of the area. This is mainly composed of detrital material produced by organisms living in or above the sediment surface, but there are also large populations of micro organisms.

33. The effects of the discharge of organic waste material in the Mersey were investigated by O'Connor and Croft [25] who showed that organic waste combined with inorganic sediments and both were retained within the systems.

34. Organic material absorbed by the sediments in the United States estuaries may be as high as 10% of the total sediment load and becomes an integral part of the flocculated sediment layer [2].

MODIFICATION OF SEDIMENTARY PROCESSES BY BIOLOGICAL ACTIVITY

35. Living organisms or their remains can affect the properties of sediments both before and after deposition.

36. Large numbers of planktonic and benthic suspension feeding animals are found in estuarine areas. These pump a substantial volume of water and remove both organic and inorganic material from suspension. Some material is rejected and may be loosely bound by mucus to form pseudofaeces, and the rest is ingested. The undigested remains are voided as faecal pellets, which have different properties to those of their component particles. Lund [19] found that oysters increased the volume of material settled from suspension by a factor of eight. The particles are bound by mucus and the pellets are often very resistant. They may form up to 100% of a deposit, [21], and will contribute towards the mixing of fine and coarse material in the same deposit [6]. Verwey [39] has calculated the volume of material deposited in this way by cockles in the Waddensee as being $0.71 \times 10^6 m^3$/year, equivalent to a deposit of 0.12 mm/year.

37. The presence of mussel banks, marine plants such as *Zostera* and the macroalgae on the estuary bottom will alter the turbulent structure of the sediment surface and may create areas favourable for sedimentation.

38. Fager [9] found that the presence of a large population of the tubeworm *Owenia* tended to stabilise the sediment in an area of shifting sand. The growth of a mat of blue green algae binding individual sediment particles makes the surface less liable to erosion [23]. Local

scour around bivalve shells on a sand surface was investigated by Johnson [14] who found the scour action tended to bury the shells. Burial of shells, therefore, cannot be used as an indication of recent deposition.

39. The most significant influence on deposited sediments is that produced by the deposit feeders who extensively rework the upper layers of sediment. The nature of the sediment changes as it passes through the digestive tracts of these animals; some of the organic material is removed, particles undergo mechanical abrasion and certain substances become more soluble.

40. Many deposit feeders are burrowers, and their burrowing activities alter the bulk density and drainage of sediments and tends to produce a rough surface liable to erosion and preferential resuspension [32]. Vertical sorting of sediments also takes place as deposit feeders reject particles they are unable to ingest which then fall to the bottom of their burrows [33].

41. The degree of bioturbation of a sediment is reflected in the water content of the surface layers [32], but will have a seasonal variation as well as being dependent on the rate of sedimentation, and for this reason is particularly difficult to quantify.

SUSPENDED MATERIAL IN ESTUARIES

42. The concentration by weight of the suspended inorganic material in estuaries has been reported on numerous occasions. Concentrations ranging between 300 and 2000 mg/l were measured in the Thames Estuary [12] and Kendrick [15] measured concentrations ranging from below 50 mg/l in the seaward limits of the Thames Estuary to 50, 000 in the Mud Reaches. The higher value was thought to be due to the formation of fluid mud at the sampling site, and that a value of the order of 3000 mg /l was more representative of the material in suspension.

43. In the upper reaches of the Mersey the author found that concentrations of 2000 mg/l were common with occasional values in excess of this figure, whilst in the Solent estuaries concentrations varying from 20 to 200 mg/l have been measured.

44. The mean diameter of a typical Thames mud was given by Inglis and Allen [12] to be 35 μm with about 5% of the sample being less than 5 μm diameter. Some samples however contained 50% of material finer than 2 μm diameter. Samples obtained from the Mersey show that 40% of the material is of clay size. The sizes referred to above apply to the dispersed material but clay is usually present as flocs.

45. Tests made to determine the settling velocity of silt in suspension in the Thames showed that under conditions of turbulent flow salinity had little effect on the settling velocity of Thames silt and that during

spring tides the settling velocity increased linearly with concentration but on neap tides it varied approximately as the square of the concentration [15].

46. The concentration of material in suspension varies during the tidal cycle and also on a seasonal basis. The development of instruments to monitor sediments in suspension has facilitated the compilation of long term records. Analysis of records obtained in this way for the River Humber shows that there is a linear relationship between the concentration of sediment and the tidal range and that it appears to vary inversely with temperature [13].

47. A long term measurement of inorganic suspended material in the Menai Straits gave a similar relationship with temperature as found by Jackson [13] and possible biological causes were suggested [3]. This was not thought to be the case off the North Yorkshire coast [24].

48. The ecology of an estuary is dependent upon the turbidity of the water and will be affected by any change in the sedimentary regime.

MOVEMENT OF SEDIMENT BY FLUID FLOW

49. The movement of dense particles by fluid flow occurs when the lift and drag forces exerted by the fluid overcome the sliding or rolling resistance of the particles. If the lift force is sufficiently large then the particle may rise from the bed and either be redeposited or carried into suspension by the turbulence within the flow.

50. This simple concept is insufficient to describe adequately the sediment transport process as it actually occurs, since the properties of individual grains vary and separate classes of sediment exist. The structure of the fluid flow is complex and subject to internal fluctuations of velocity due to turbulence, and the fluid and its erodable sediment boundary interact with each other producing the complex bed forms familiar as ripple marks and sand waves.

51. When water is flowing over a bed of sand the onset of movement is when a few particles here and there start rolling or sliding along the stream bed. Increasing velocity of flow causes a general motion of the particles on the bed and sand ripples form and travel in the direction of the flow. Increasing velocity sweeps the ripples away and movement takes place over the entire surface of the bed and an increasing proportion of the sand is carried into suspension by turbulence. Some of the sand will fall to the bed again but an atmospheric type of distribution will prevail with the concentration of material decreasing from the bed towards the upper surface of the flow.

52. It is usual to classify two modes of sediment transport, the bed load and the suspended load. The first is the movement that occurs in the immediate vicinity of the bed, usually in the form of ripples and waves, and the second is the transport of material distributed

throughout the depth of the flow. The division between the two classifications is arbitrary but has practical value since the transport of suspended load occurs at the same order of velocity as the flow, whereas the bed load or "bed creep" as it is sometimes called, occurs at a much slower rate.

53. The types of sediment movement described above are common to rivers, tidal flow in estuaries, or sediment movement by waves. They are however modified by the sediment properties and in the particular case of clay, by cohesion, which gives rise to different bed forms than sand. A further property of clay is that it will flocculate when suspended in sea water and this is of great importance in estuaries.

FLOCCULATION OF COHESIVE SEDIMENTS

54. Particles of clay size have a high ratio of surface area to mass and therefore the surface forces acting on the particles can control the behaviour of the material when it is in suspension. Both attractive and repulsive forces are present and when the clay is suspended in sea-water the attractive forces predominate and if the particles are brought into contact a lasting bond is formed. The initial aggregation of the flocs is brought about by Brownian motion of the particles in suspension and further aggregation can occur due to fluid shear or by differential settling of the particles. The theory of particle interaction is described by Kruyt [17] and van Olphen [37], and Packham [26] has reviewed the current theory of coagulation.

55. Aggregation by fluid shear, orthokinesis, was shown by Reich and Vold [31] to be largely a reversible process. The rheology of clay flocs from San Fransisco Bay was studied by Krone [16] who advanced a theory of aggregation which predicts the volume, density and settling velocity changes which occur during this process.

56. Microscopic examination of the flocs show that they consist of a diffuse assembly of particles of ill defined shape. The mechanism of flocculation of clay particles at a salt water/fresh water interface has been investigated for a number of clay minerals. [34]

57. Generally the bonding of the particles forming the floc has been described in terms of the surface physics of the particles but flocculated sediments in estuaries exist in an environment of high biological activity and several workers have reported its effects on clay minerals. During an investigation of the behaviour of clay minerals in saline water Whitehouse et al [40] showed that the interaction of organic compounds with clay minerals could modifty their settling velocity in saline water. Carriker [6] and van Straten and Kuenen [38] have indicated the importance of organic binding in estuarine processes. During an investigation of sediment transport in waters off the East Frisian Islands, Luck [18] examined aggregates of mineral particles adhering to an organic slime. A photomicrograph of the aggregates

showed that their structure was similar to the flocs occurring in South Coast harbours. These flocs frequently have fine sand grains adhering to them. Nelson [22] reported that fresh water streams discharge both flocculated and dispersed erosion products [6]. The authors sampled the sediments suspended in the River Avon and found the flocs contained minerals typical of the catchment area.

58. The essential difference between the floc and the sand grain is that whereas the granule has virtually constant properties those of the floc are dependent on the flow regime when in suspension, and on consolidation processes when it forms part of the bed of the estuary. Krone [16] has shown that successive aggregation increases the volume and decreases the density of the floc, but that the net result is to increase the settling velocity.

SEDIMENT MOVEMENT IN ESTUARIES

59. The sediment process in estuaries is complex, being subject to the tides and salinity differences which are dependent on the seasonal variation in river flow, the ecology of the estuary, and wave action at the coast line.

60. Density currents set up by salinity differences can assist the ingress of sediment from the sea and retain material brought down by the rivers. This has been of great practical consequence as was shown in the case of both the Thames and the Mersey.

61. The range of particle size within an estuary extends from shingle to clay particles and so a wide spectrum of sediment transport modes are present varying from the suspension of clay flocs to bed movement of sand as ripples or waves.

62. Inshore mud flats are typical of both the estuary and the tidal inlet and Postma [27] has suggested an explanation for the accumulation and retention of material at the inshore margins of estuaries and inlets in terms of a scour lag and a settling lag. The scour lag is the additional velocity required to resuspend a particle in excess of that at which it would settle to the bed. The settling lag is the time taken for a particle to reach the bed after the tidal velocity has diminished to the point at which the particle can no longer be held in suspension. Postma has shown that these effects together with the tidal velocity pattern can bring about the inshore accumulation of fine sediment. Van Straaten and Kuenen [38] considered that biological processes also contributed to the inshore accumulation of material.

63. Another form of sediment movement found in estuaries is the flow of fluid mud at the bottom of channels. The presence of such a layer was reported by Inglis and Allen [12], and a particle size analysis of a fluid layer was made by Prentice et al [28], showing that only about a third of the material was clay the rest being sand or silt.

64. Migniot[20] has stressed the importance of flocculation, consolidation and rheology as properties governing the hydrodynamic behaviour of fine sediment.

65. The stability of sand banks in the presence of relatively high transport rates has been explained by a circulatory movement of sediment and many estuaries have distinct ebb and flow paths.[28]

66. The compilation of a sediment balance for an estuary is useful in indicating the relative importance of the various sedimentary factors operating within the area[8].

67. The volumetric stability of an estuary is usually assessed from survey records and change may be due to some or all of the following factors:-

(1) the net volume of the material entering or leaving the system
(2) the volumetric changes in any residual sediment arising from (1) due to flocculation before, and consolidation after deposition.
(3) the effects of biological processes on the volume of the sediments.
(4) the volume of organic matter produced within the system.
(5) the effects of density changes on material locally eroded within the system.
(6) the variation of volume relative to mean sea level due to subsidence or elevation of the general land mass.

An estimate of the volumes associated with these factors can be made, particularly (1) and (6). The other estimates such as (2) and (5) can be made with considerably less precision and for the biological factors only an order of magnitude for a particular site could be estimated at present.

68. It is not surprising that a unique equation predicting the transport of sediment in estuaries is not available at this time. Methods of study have evolved using field measurement, tidal computations, tracer studies and physical models of estuaries.

CONCLUSIONS

69. The study of sedimentary processes has revealed retentive mechanisms which accumulate fine material at the inshore margins, retain material brought down from the river catchment and facilitate the ingress of material from the sea. In the case of tidal inlets the retentive characteristics of water circulation due to salinity gradients are absent, but those due to the sedimentary process at the inshore margin are usually present.

70. The effects of the retentive character of an estuary are not confined to the particulate matter, since clays have been shown to absorb large quantities of organic and nutrient material which may result

in unacceptably high accumulations of these pollutants within the sediments of an estuary.

71. Each estuary and the problems produced by its development are probably unique. General principles may be used to indicate a possible pattern of sedimentation but these should be supplemented by field and laboratory data. This information will help to predict the dispersal and the eventual disposal of particulate waste.

REFERENCES

1. F. A. J. Armstrong & W. R. G. Atkins. The suspended matter of sea water. *J. mar. bio. Ass. U.K.*, 1950, **29**, 139-143.
2. R. A. Baltzer, J. J. Leendertse, J. B. Lockett, B. W. Wilson & F. D. Masch. Research needs on thermal and sedimentary pollution in tidal waters. *Proc. Amer. Soc. Civil Eng.*, 1970, **96**, HY7, Paper 7426, 1539-1548.
3. S. Buchan, G. D. Floodgate, & D. J. Crisp. Studies on the seasonal variation of the suspended matter in the Menai Straits. I. The inorganic fraction. *Limnol. Oceanogr.*, 1967, **12**, 419-431.
4. A. T. Buller & J. McManus. The formation of the turbidity maximum in the Tay estuary and its relationship to bottom sediment distribution. Preprints to Joint Geological Society—Estuarine and Brackish-Water Sciences Association meeting on 'the characteristics of estuarine sedimentation,' 1973.
5. P. Bruun & F. Gerritson. Stability of coastal inlets. North Holland Publishing Co., Amsterdam. 1960.
6. M. R. Carriker. Ecology of estuarine benthic invertebrates: a perspective. In Estuaries, ed. G. H. Lauff. American Association for the Advancement of Science Publication No. 83, 1967, pp. 442-487.
7. I. Douglas. Sediment yields from forested and agricultural lands. In The role of water in agriculture. J. A. Taylor (Ed) University College of Wales, Aberystwyth Memorandum No. 12, 1969, pp. 57-88.
8. K. R. Dyer. Sedimentation in estuaries. In The Estuarine Environment. R. S. K. Barnes & J. Green (Ed). Applied Science Publishers, London 1972, pp. 10-32.
9. E. W. Fager. Marine sediments: effect of a tube building polychaete. *Science*, 1964, **143**, (3604) pp. 356-359.
10. F. R. Hayes. The mud-water interface. *Oceanogr. Mar. Biol. Ann. Rev.* 1964, **2**, 121-145.
11. A. C. Imeson. Variations in sediment production from three East Yorkshire catchments. In The role of water in agriculture. J. A. Taylor (Ed). University College of Wales, Aberystwyth Memorandum No. 12, 1969, pp. 39-56.
12. Sir C. C. Inglis, & F. H. Allen. The regimen of the Thames estuary as affected by currents, salinities and river flow. *Proc. Inst. Civil Eng.*, 1957, **7**, 827-868.

13. W. H. Jackson. An investigation into silt in suspension in the River Humber. *Dock and Harbour Authority* August 1964, pp. 120-122.
14. R. G. Johnson. Experiments on the burial of shells. *J. Geol.*, 1957, **65**, 527-535.
15. M. P. Kendrick. Siltation problems in relation to the Thames Barrier. *Phil. Trans. R. Soc. Lond. A.*, 1972, **272**, 223-243.
16. R. B. Krone. A study of the rheological properties of esturial sediments. SERL report No. 63-8 University of California, Berkeley, 1963.
17. H. R. Kruyt. Colloid Science 1. Elsevier Publishing Company, London, 1952.
18. G. Luck. Observation of sediment motion by underwater television. Proc. 12th Coastal Eng. Conf., Washington D. C., 1970, pp. 687-708.
19. E. J. Lund. Self-silting by the oyster and its significance for sedimentation geology. *Publ. Inst. Mar. Sci. Texas.* 1951, **4**, 320-327.
20. C. Migniot. Study of the physical properties of various forms of very fine sediments and their behaviour under hydrodynamic action. *La Houille Blanche,* 1968, **23,** 591-620.
21. H. B. Moore, The muds of the Clyde Sea area, III Chemical and physical conditions: rate and nature of sedimentation; and fauna. *J. Mar. Biol. Ass. U.D.* 1931, **17,** 325-358.
22. B. W. Nelson. Transportation of colloidal sediment in the fresh-water-marine transition zone. In International Oceanographic Congress, preprints M. Sears, (Ed.) A.A.A.S., Washington D.C., 1959, pp. 640-641.
23. A. C. Neumann, C. D. Gebelein, & T. P. Scoffin. The composition, structure, and erodability of subtidal mats, Abaco, Bahamas. *J. Sed. Petrol.*, 1970, **40,** 274-297.
24. A. J. Newton & J. S. Gray. Seasonal variation of the suspended solid matter, off the coast of North Yorkshire. *J. mar. biol. Ass. U.K.*, 1972, **52,** 33-47.
25. B. A. O'Connor & J. E. Croft. Pollution in a tidal estuary. *Effluent and water treatment Journal,* 1967, **7,** 365-374.
26. R. F. Packham. The theory of the coagulation process 1. The stability of colloids. *Proceedings of the Society for Water treatment and examination,* 1962, **2,** pp. 50-63.
27. H. Postma. Sediment transport and sedimentation in the estuarine environment. In Estuaries, G. H. Lauff. (Ed.) A.A.A.S. Publication No. 83, 1967, pp. 158-179.
28. J. E. Prentice, I. R. Beg, C. Colleypriest, R. Kirby, P. J. C. Sutcliffe, M. R. Dobson, B. D'Olier, M. F. Elvines, T. I. Kilenyi, R. J. Mandrell & T. R. Phinn. Sediment transport in estuarine areas. *Nature,* Lond. 1968, **218**, 1207-1210.
29. W. A. Price & M. P. Kendrick. Field and model investigation into the reasons for siltation in the Mersey estuary. *Proc. Inst. Civil Eng.*, 1963, **24**, 473-517.
30. D. W. Pritchard. Estuarine circulation patterns. *Proc. Amer. Soc. Civil Eng.*, 1955, **81**, 717.

31. I. Reich & R. D. Vold. Flocculation-deflocculation in agitated suspensions. *J. Physic. Chem.*, 1959, **63**, 1497-1501.
32. D. C. Rhoads & D. K. Young. The influence of deposit feeding organisms on sediment stability and community trophic structure. *J. Mar. Res.*, 1970, **28**, pp. 150-178.
33. D. C. Rhoads and D. J. Stanley. Biogenic graded bedding. *J. Sedimentary Petrology*, 1965, **35**, 956-963.
34. W. Sakamoto. Study on the process of river suspension from flocculation to accumulation in estuary. *Bull. Ocean Res. Inst. Univ. Tokyo*, No. 5, pp. 1-46. 1972.
35. H. B. Simmons. Some effects of upland discharge on estuarine hydraulics. *Proc. Amer. Soc. Civil Eng.*, 1955, **81**, 792.
36. P. D. Trask. Organic content of recent marine sediments. In Recent Marine Sediments. P. D. Trask. (Ed.) 2nd edition 1954. Dover Publications, Inc. New York, 1939. pp. 428-453.
37. H. Van Olphen. An introduction to clay colliod chemistry. Interscience Publishers, 1963.
38. L. M. J. U. Van Straaten & Ph. H. Kuenen. Tidal action as a cause of clay accummulation. *J. Sed. Petrol.*, **28**, pp. 413.
39. J. Verwey. On the ecology of distribution of cockle and mussel in the Dutch Waddensea, their role in sedimentation and the source of their food supply, with a short review of the feeding behaviour of bivalve molluscs. *Arch. Neerl.*, 1952, **10**, pp. 171-239.
40. U. G. Whitehouse, L. M. Jeffrey, J. D. Debbrecht. Differential settling tendencies of clay minerals in saline water. Proc. 7th Conf. Clays Clay Mins., 1960, pp. 1-79.

DISCUSSION

HUMBY The Authors are respectively a civil engineer and a biologist whose common interest at present is the biological modification of sediments in estuaries. The field work connected with this interest has been mostly confined to Langstone Harbour which is to the east of Portsmouth Harbour and has an area of about 2000 ha. It is filled mainly from the Eastern Solent but has connections at high water with both Chichester and Portsmouth Harbours.

At low water the flooded area is reduced to about one eighth of that at high water. Low water uncovers an area of sand banks and mud flats which are sub-divided by the three or four main channels in which water stands to a depth of several meters.

The northern half of the harbour is typically mud flats and salt marsh, the mud consisting largely of clay, organic matter, silt and sand.

Near the harbour entrance the material of the sea bed is sand and gravel, resulting from the higher degree of exposure of this area to wind and waves.

Historical research into charts of the harbour shows that it was changed little during the past few centuries. At present there is only modest industrial development mainly associated with the sand and gravel industry. Recreational use of the harbour is intense and it is an area of great scientific interest. The harbour is used to disperse effluent from sewage works sited in the North East of the area.

The effluent discharged from the City of Portsmouth sewage outfall at the harbour mouth is flushed into the Eastern Solent by the ebbing tide from Langstone Harbour.

The sediments present within the harbour are typical of those associated with estuaries and are circulated within the harbour by the ebb and flow of the tide. The sea bed of the harbour was explored by the Authors on one occasion using a side-scan sonar device which revealed that sand waves and ripples extended over the bed and along the sides of the channels in the southern half of the harbour.

Bed samples show that considerable sorting of materials occurs in this area (Fig. 1) and in contrast to the samples from the mud flats in Fig. 2, most of the material below 0. 2 mm diameter has been removed.

A recent aerial photograph of part of the harbour is reproduced (Fig. 3), and this shows bed configurations for a wide range of particle size. The area at the seaward end of the photograph is known as Sword Sands and is traversed by sand waves indicating the movement of sand along the surface of the sea bed.

One can also see the complex pattern of small channels draining the upper levels of the mud banks and carrying mud in suspension.

As part of a general study of the harbour earlier aerial photographs of Sword Sands taken at low tide were examined and these showed that the wave pattern was present on each photograph and that the general configuration was little changed over a period of about 30 years. Since the area is accessible at low water an attempt was made to measure the movement, if any, of the waves across the sand bank. The waves were found to move but in an intermittent manner apparently associated with progression of the tides from neaps to springs. Since the measurements showed that material was in fact being transported northwards, the apparent stability of the area could only be explained if a circulating system was present. Detailed examination of the area showed that material probably moved northwards in the centre of the bank but southwards along the edges.

The extent of the circulation area is not known to the Authors. Also waves perpendicular to the general movement were seen but as yet are unexplained.

The pattern of movement at Sword Sands is an example of dynamic stability being achieved in the presence of high transport rates due to the existence of separate ebb and flow channels. The literature of fluid

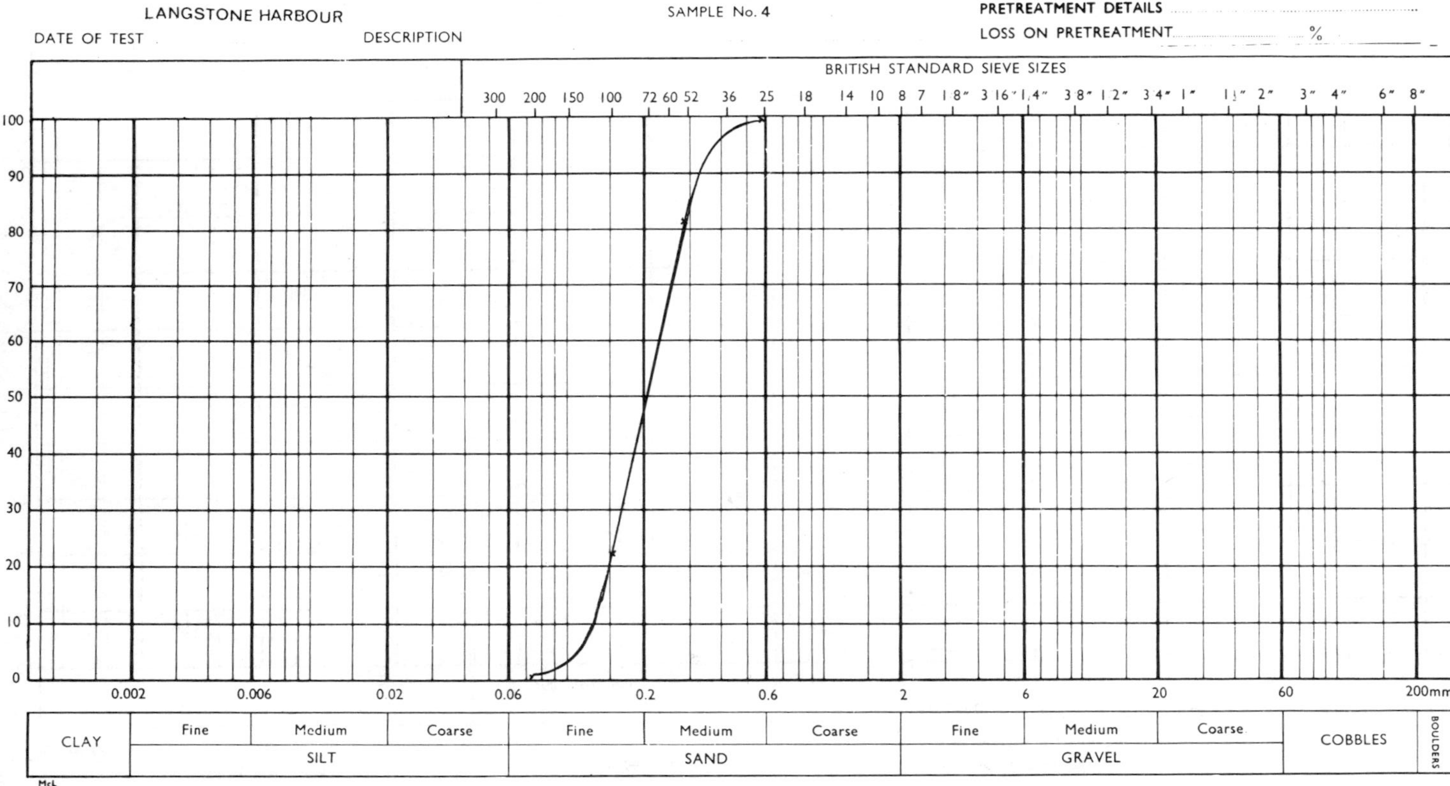

Fig. 1. Particle size analysis- sea bed sample

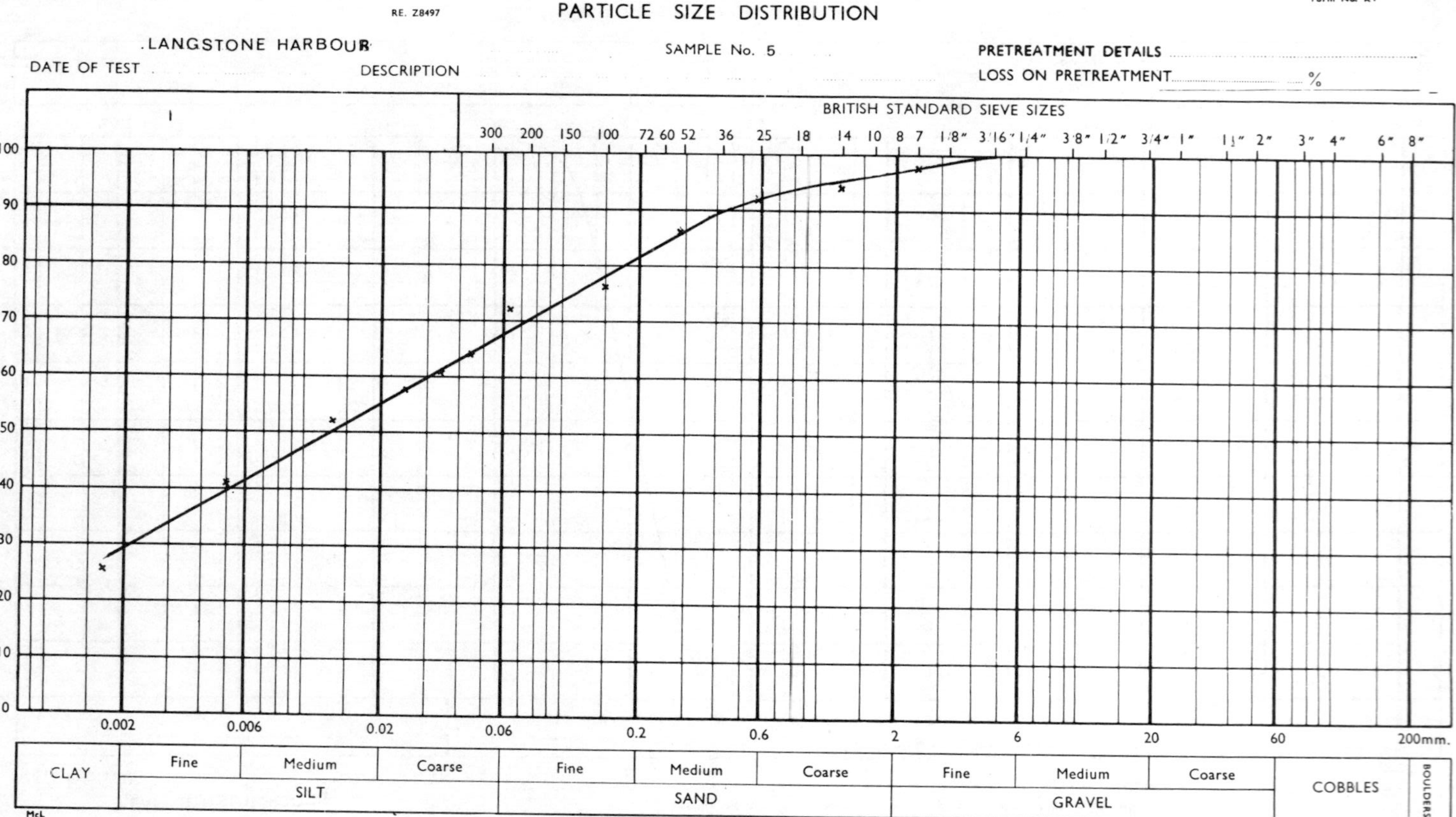

Fig. 2. Particle size analysis- mud bank sample

mechanics contains many examples of the production of different flow patterns when flow is reversed through comparatively simple boundaries and similar results obtain in the more complex boundaries of a harbour.

The material in suspension in the water of the harbour when sampled and subjected to microscopic examination has the appearance of the flocs shown in (Fig. 4).

The particle itself is about the size of medium sand and the largest inclusion is comparable to a particle of coarse silt. X-ray diffraction

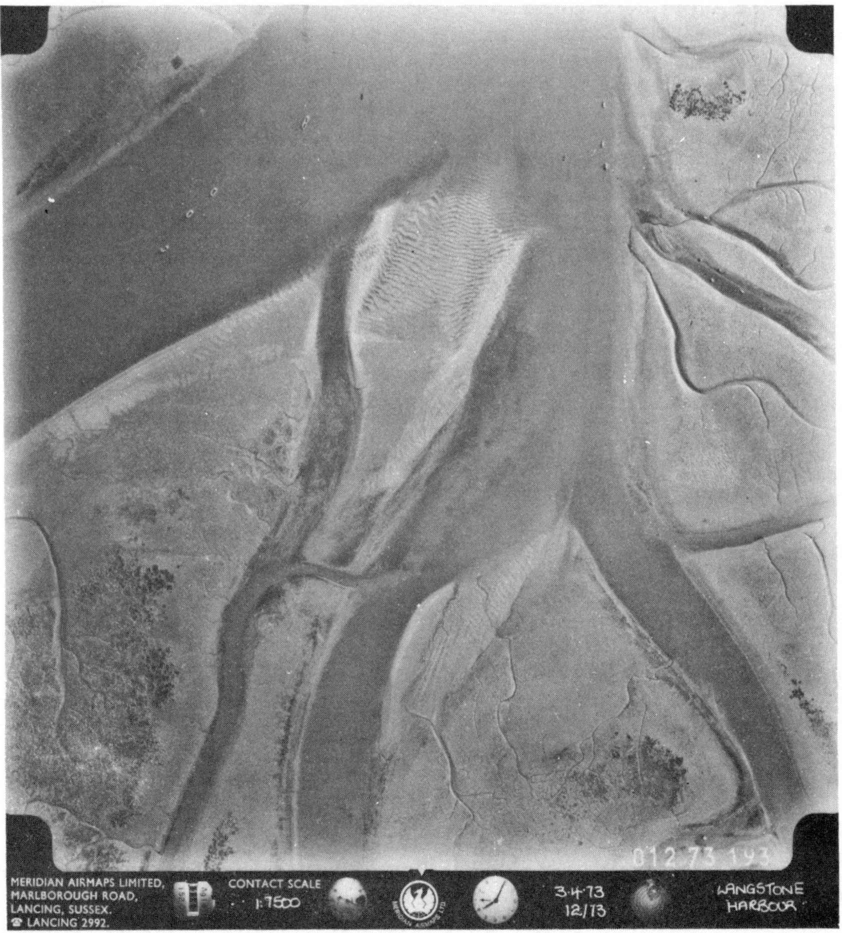

Fig. 3. Recent aerial photograph of part of Langstone harbour. Area covered approximately 1.5 km square

analysis of similar flocs indicate that they include clay minerals matching the local geological material.

Both the size of the flocs themselves and the inclusions suggest that the bonding of the floc is not entirely due to ionic forces but that some form of mucous binding occurs. This form of biological binding has been suggested by other workers and Luck[18] has produced similar photographs of sediments from the North Sea coast.

The flocs are fragile and sampling can destroy the *in situ* form of the floc. In suspension in the harbour they do not represent a stable form but can be successively disrupted and aggregated by the hydraulic forces present in the flow. The particles are difficult to define hydraulically, since the density, size and shape of the particles in suspension is dependent on the turbulent structure of the water flow.

Similar flocs were found in fresh water in the Rivers Avon and Arun and X-ray diffraction analysis showed that those flocs were composed in part, of the sediments of the respective catchment areas.

There is similarity in appearance between the sedimentary flocs and those derived from sewage effluents. The sedimentary behaviour of both must be similar.

The work described has been undertaken as an attempt to establish the background to sedimentary processes in a natural harbour and to recognise the interdependence of the sedimentary, hydraulic and biological processes in an estuarial type of environment.

In conclusion it can be said that a study of the sedimentary processes in estuaries can be helpful when deciding the effect of injecting particulate wastes into a natural environment. In areas outside of the Solent the pollutants are not necessarily sewage but may include mining residues and other industrial wastes.

The estuarial system is dynamic and apparent stability may conceal constant movement of material. Therefore stability may be destroyed both by addition to or depletion from the supply of material to the system.

When proposing measures which will modify the flow of material to a system care must be taken to assess the probable consequences and this caution must be equally exercised when causing a well established practice to cease. A chain of events may start which results in unexpected and not wholly beneficial consequences.

CHARLTON (University of Dundee, Tay Estuary Research Centre). The aerial photograph (Fig. 3) showing the sand bank with its dunes or ripples is extremely interesting. Similar effects have been observed on the Tay and they occur frequently in other places. It is typical of a situation where the flood currents take a different direction to the ebb currents over banks which subsequently become exposed. The photograph shows

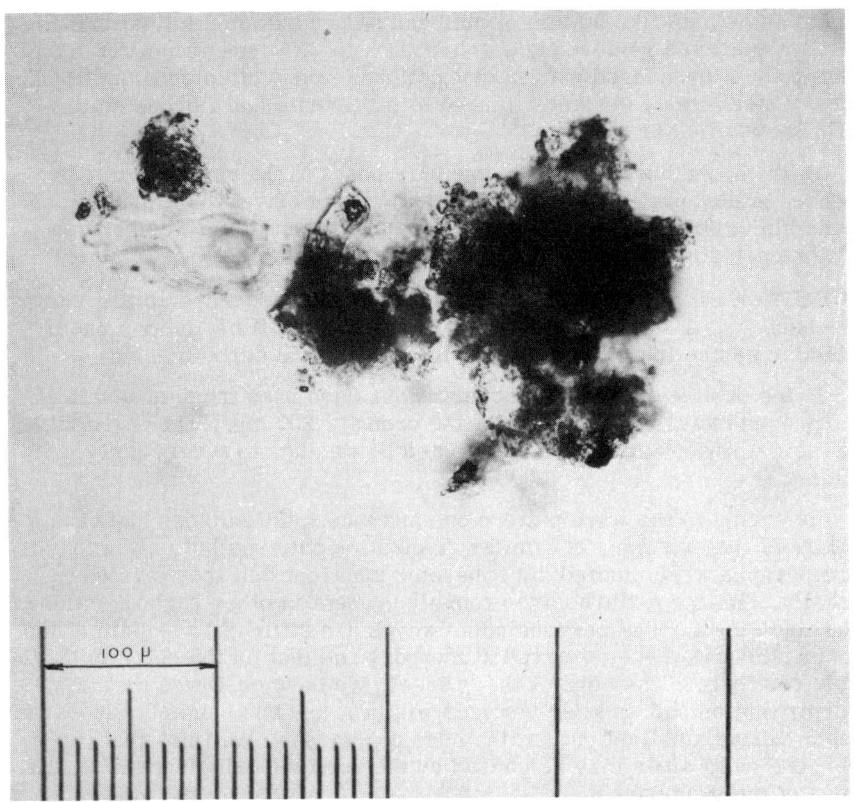

Fig. 4. Photomicrograph of suspended material from Langstone Harbour

typical flood dominant dunes, where the tide comes in over a wide front, and also ebb dominant dunes where the ebb tide follows the main channels and leaves the flood dunes as they were set up on the flood tide.

From studies of sub-surface dunes, it is possible to establish the direction of ebb- and flood-dominant channels. They can be quite large and very distinctive and it is possible to decide what the dominant flow is by looking at them. From the pollution point of view, we cannot assume that the flood and the ebb are likely to flow the same channels, they very rarely do.

It is interesting to note that in this type of estuary there is likely to be very little input either from the sea or from the land to the silt budget of the estuary. The amount of sediment circulating is almost a fixed amount, fluctuating over a range depending on weather conditions.

Up to something like 50 mm of mud can be deposited on a flat surface in a calm spell and removed within half an hour if waves get up. It is to this body of suspended matter that pollutants very often become attached. Heavy metals may become attached to particulate and fibrous matter, either adsorbed or absorbed.

Have the Authors made any measurements of the variation of suspended solids, particularly in the mouth of this estuary? It may be possible to observe these changes and relate then to the weather conditions pertaining at or just prior to that particular time.

HUMBY We have not previously monitored the suspended solids within the harbour. A start is being made at the present time, using a photoelectric device installed on one of the rafts in the harbour.

In the past, some isolated observations have been made of levels of suspended material. They are of the order of 200 mg/l. The turbidity of the water does appear to be effected by weather to a very large extent.

Referring to the wave pattern on sand banks, the Authors had been aware of the existence of similar circulation patterns but only when air photographs were studied did it become apparent that they existed locally. Similar patterns are probably present in other harbours along the south coast. The perpendicular waves are difficult to explain at the moment, but we have observed at Sword Sands that on the ebb tide large vortices form at the edge of the 'wedge'. We have observed that this vortex motion has smaller vortices within it and one can actually watch them cutting a helical path at the edge of the sand. We think that the perpendicular lines may be caused by the vortices being shed from the edge of the sand bank.

CHARLTON Generally speaking, a vortex will be generated by some obstacle or some change in bank formation. Very similar streaks to this have been observed extending for something like 500 m in water with a depth of flow of 2 to 3 m and a velocity of up to a metre per second.

OAKLEY (J. D. and D. M. Watson). Can anybody comment on how one can distinguish between naturally occurring sediments and those deriving from pollution sources?

HUMBY The Authors have attempted to classify flocs according to origin by visual means. Some 250 photomicrographs were examined and loosely classified as being of solely mineral origin, as containing faecal pellets or being sewage flocs, but the results were inconclusive.

BAKER (Field Studies Council, Oil Pollution Research Unit). The speaker has no experience of sewage accumulating in mud, but there is quite a lot of evidence of oil accumulating in mud. This takes place by the absorbtion of small oil droplets on to small particles of sediment, which sink to the bottom and gradually build up. These small oil droplets can result either from refinery effluent or from oil slicks treated by various detergents.

GILSON (Committee for Environmental Conservation). Have any carbon determinations been made on the flocs ? The photomicrographs showed that the materials were held together by organic material of some sort. These may derive from breakdown products of dead Spartina leaves, many of the animals that inhabit the area, the muddy parts of the area, and so on. There are abundant supplies of organic material in a place of this sort. The flocs are largely bonded by such materials, but the precise origin and history of them is a much more difficult story.

HUMBY It is interesting to hear this account of the origin of the material bonding the flocs together. It is difficult to examine these materials and identify their origin and the Authors' work was limited to visual and X-ray diffraction methods. The complex history of a floc makes the study of cohesive sediments much more difficult than that of the granular sediments.

PAGE (J. D. and D. M. Watson). In their study of the Mersey Estuary in 1938, W. P. R. L.[(41)] reported on the belief that sewage particles might cement natural sedimentary material. They could find no such correlation nor could they identify what organic material was of sewage origin as opposed to organic material from other sources. The Authors refer in Para. 43 to suspended sediment concentrations of 2000 mg/l in the Upper Mersey. Samples taken in the lower reaches of the Estuary between Birkenhead and Liverpool showed concentrations between 300 and 1000 mg/l. This is not due to sewage particles being discharged to this area but to the large movement of sedimentary material from Liverpool Bay.

During discussion, the Authors referred to sand and gravel abstraction within the Langstone Harbour. The study of the Mersey Estuary showed that in general only the top 60 mm of estuarine mud exert oxygen demands on water and therefore any such operations would result in some oxygen depletion.

An aspect of estuarine pollution as yet undiscussed is dredging work. The total quantity of material moved from the Mersey Estuary, from the dock estates, from the Manchester Ship Canal and from the approach channels, is about 8 to 10 million wet tonnes per annum. This material is highly pollutant when disturbed. It is not of sewage origin, but of natural organic origin, and it has values in BOD and COD terms of about 2500mg/l BOD and 30 000 to 40 000mg/l COD. The Mersey Docks and Harbour Company found that although the spoil ground is some 12-15km from mouth of the estuary, approximately 80% of the material returned to its source. This must be a quite considerable contribution to pollution loads of the estuary since the material which is removed from the Mersey exerts a pollution load at the point of deposition equivalent to that from almost half a million population, although the immediate dispersion of the material is not fully understood.

HUMBY The figure mentioned for the Mersey was for mineral matter in suspension. It was not intended as a measure of pollution, but was included to establish relative orders of magnitude of suspended material.

ALLEN (The Ecologist). In estuaries showing stable flow and sediment behaviour, would an ecological stability also be expected? If so, would not changes due to pollution be easier to see than in estuaries showing greater natural fluctuations. In other words, might it not be less difficult to establish pollution criteria for a relatively stable estuary like Langstone Harbour than for more unstable ones?

DUNN In theory it would be attractive, but in Langstone it has not been possible. The reasons for this are a lack of historical data, and also an extremely variable faunal distribution in apparently similar situations.

REFERENCE

41 Estuary of the River Mersey: The effect of the discharge of crude sewage into the estuary of the River Mersey on the amount and hardness of the deposit in the estuary, H.M.S.O., 1938.

PLANNING AND POLLUTION—A CASE STUDY OF SOME ECONOMIC AND SOCIAL PRESSURES ON COASTAL WATERS—SOUTH HAMPSHIRE

J. F. Barrow, B.A., Dip.T.P., M.R.T.P.I.

Manager, South Hampshire Plan Technical Unit

INTRODUCTION

1. Planning has been described as having at least two essential properties. It is *anticipatory decision making*, requiring a time lapse between making decisions and carrying them out; permitting reconsideration of decisions previously made. It also involves a *system* of decisions inter-acting one with another; a system so large and complex that it cannot all be done at once. It ought, therefore, to be broken down into independent sub-groups of decisions—that can be reviewed by considering the results of decisions made and evaluating the effects. Planning also takes place in a *dynamic* context in which the environment, for which decisions are taken, is constantly changing in ways that would still affect the very system being planned for, unless adjustments are made. Hence the growing importance which is being attached to monitoring plans and subsequent decisions continuously in order to examine their effects and to consider whether a review of the plan is required.

2. The emphasis on decisions and sub-groups of decisions in a dynamic context might be regarded as underlying the new forms of development plans now being introduced into the British planning scene—Structure Plans as they are termed—which are regarded first and foremost as decision documents concerned with groups of decisions of a strategic nature. They focus attention on the Local Planning Authorities' intention to initiate, control and encourage change in their areas. One of the reasons underlying their introduction in the Town and Country Planning Act of 1968 is that they should be prepared in a manner and through procedures which are flexible enough for the making of major decisions and related groups of decisions at a time when these are needed; thereby enabling the deferment of other decisions of a more local nature which are not immediately needed. The emphasis in Structure Plans is on their written nature and their time scale is normally of twenty years—with policies and broad proposals on strategic issues, including the general locations of employment growth, shopping development, housing areas, the role of the countryside, and transportation systems, with an indication of the regard that has been paid to the resources of finance and manpower in order to demonstrate the feasibility of the Plan. It is these issues—policies and general proposals of a strategic nature—which are considered by Central Government through submission of the Plan to the Secretary of State for the Environment, leaving the strategic proposals to be worked out in more detail for a shorter time scale, usually of about ten years, in the preparation of Local Plans adopted by the Planning Authorities themselves.

3. The re-organisation of local government places the responsibility for the preparation of land use and transportation policies and proposals and for putting them into effect through development control in the charge of different Authorities in the new local government system. In this arrangement the Structure Plan will be the responsibility of the new County and forms the basis of all Local Plan proposals (principally the responsibility of the new District Councils); and the policies of the Structure Plan underlie many policy decisions relating to the physical environment and the control of development.

4. Local Planning Authorities are thus charged with a statutory duty of preparing development plans. Such plans must be decision documents relating to those matters over which planning has control or influence, and the term "structure" is defined as to mean the social, economic and physical systems of the area so far as they are subject to planning control or influence. As decision documents of the Local Planning Authority they do not, and should not, pre-empt decisions of other bodies and organisations. However, they must be prepared with careful regard to, and co-operation with, those statutory organisations who are responsible for other services; and if development plans are to be realistic, advice and co-operation of those bodies responsible for pollution control in all its forms, is a pre-requisite. Perhaps, therefore, this conference will accept a prima facie case for the involvement of land-use planners in studies of estuarine pollution, for as this paper attempts to demonstrate, the policies and proposals for land-use development have a direct bearing on the subject. Equally, the need to avoid and remove pollution can operate as a constraint on land-use development.

5. In South Hampshire there have been two decades of economic growth and population increase which are likely to continue in the next twenty years. These are associated with increasingly demanding complex industrial and recreational pressures on the natural resources of land, air and water, and are occurring at the same time as public awareness of, and concern for, the conservation of physical resources and the consequences of further physical development.

6. Among the many problems of preparing development plans for future growth and change, one of the more difficult is to ensure that economic expansion, if desired, can continue in the face of social and potential demands for conservation and the improvement of the physical environment. In preparing any plan it is essential to identify the pressures for growth and change and their likely consequences; to examine these against the measures available and action necessary to ensure that no harmful effects on the environment occur or at least are minimised; and at the same time to explore whether existing pollution can be eliminated.

7. This case study examines how the subject of water pollution has been integrated into the preparation of the recently published draft South

Hampshire Structure Plan and views the type of policies and criteria for pollution control which such a plan could include.

BRIEF BACKGROUND

8. South Hampshire has featured in several regional studies of South East England as a growth area and is among five which were selected in the most recent regional strategy—the Strategic Plan for South East England.

9. That Strategy states:

"South Hampshire is potentially the core of a largely self-contained City region, with a substantial existing employment base and good employment growth prospects and a number of attributes favourable to future rapid and successful growth. Growth in this area is particularly desirable since local employment growth momentum is great enough to exert proportionately only relatively small demands on mobile employment, particularly the limited amount capable of moving over long distances within the region.

"There will, however, be need for heavy investment on the infra-structure, particularly on roads and for foul drainage. Much of this investment will be needed even at present growth rates. The impetus for growth appears to be sufficient to warrant an urgent decision on the required infra-structure so that the works, which will take some years to design and carry out, will not inhibit a progressive build up to a higher rate of growth should this continue to seem necessary through the 1980s and thereafter."

10. The Strategy for South East England provided an indication of the scale of growth as part of the regional distribution and estimated that the present population of some 880 000 could well increase to 1 million in 1981 and to 1.210 million by 1991, with further growth thereafter.

11. One of the purposes underlying the regional strategy for South East England is to provide a basis for the preparation of Structure Plans by the Local Planning Authorities. In the case of South Hampshire the three Local Planning Authorities—Hampshire County Council and the City Councils of Portsmouth and Southampton—had accepted that further large-scale growth was occurring and was likely to continue for many years ahead; and that a new development plan was necessary for the whole of the area.

12. The organisation to steer the preparation of this new development plan—the South Hampshire Plan Advisory Committee—was set up in 1968. The responsibility of this Committee (consisting of senior members of the three Authorities) has been to recommend a plan to the three Local Planning Authorities. The responsibility of the Chief Planning Officers and Engineers, meeting as a Directing Team, has been to provide technical guidance to the teams engaged full-time on plan preparation (Fig. 1).

The technical work began early in 1969 and has recently culminated in the production of a draft Structure Plan published for consultation and public representation in accordance with the Town and Country Planning Act, 1972. During that period a considerable number of studies were completed, among which a significantly important number were concerned with main drainage and pollution, especially those related to Southampton Water and the Solent.

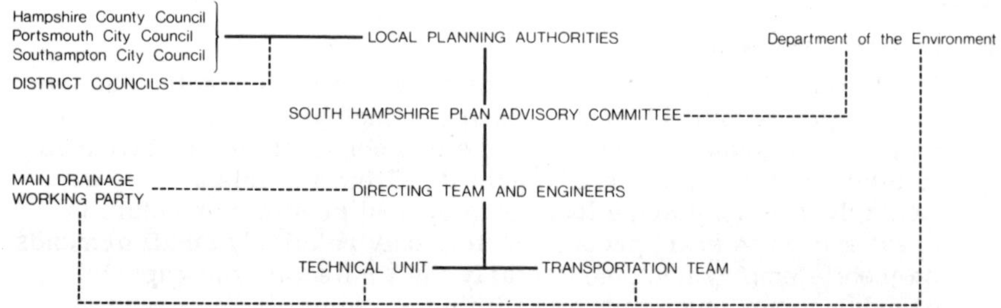

Fig. 1 Organisation for the preparation of the draft South Hampshire Plan.

POLLUTION AND STRUCTURE PLAN STUDIES

13. An outline of the technical process adopted for Structure Plan preparation is shown in Fig. 2, which shows a simplified step-by-step approach which was actually cyclical and on-going. The systematic surveys and studies included the plan area forecasts of employment and population expansion, housing need, recreational demands, shopping requirements, etc., (Box 2); investigation of problems and opportunities over a wide spectrum of activities, from recreation to shopping (Box 3); and how such problems might change in future.

14. The studies and forecasts gave rise to indications of the size of future growth by 1981 (1 014 000) and by 1991 (1 160 000) which allowing for margins of error in forecasting were broadly equatable with the population indications given in the Strategic Plan for South East England. Forecasts were made of the strategic land needs of industry, offices, shopping and recreation, in addition to housing, as a basis for the preparation of alternative strategies (Box 6). The studies of the main problems and opportunities were paralleled by work on devising broad strategic aims pointing the main direction in which planning action needed to progress (Box 4). These aims were progressively supported and refined by objectives and criteria to give greater precision to the technical work. (Box 5)

15. The eight broad aims assisted in the preparation of alternative 'strategies' to accommodate the forecast growth (Box 6). These were tested against objectives and criteria to reveal their advantages and

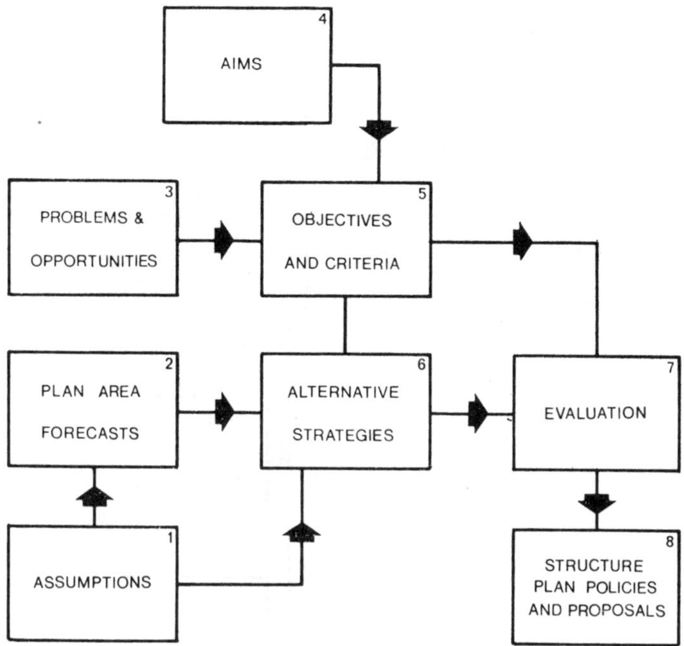

Fig. 2 The structure plan process.

disadvantages so as to provide a basis for a preferred strategy (Box 7). The chosen strategy was then refined to form a Structure Plan of broad proposals supported by policies, criteria and standards. (Box 8)

16. Some of the basic studies have a particular bearing on pollution control,particularly those involving forecasts of the likely economic growth of South Hampshire. *Employment*, as a measure of economic growth and change (embracing both commerce and industry) had been identified and accepted as a key factor governing the future development of South Hampshire. The forecast of employment growth was regarded as a prime determinant of the future increase in population and thus of projections for housing and other needs which are likely to make growing demands for urban land (Fig. 3). The forecasts of employment revealed major growth prospects largely in light industry and electronics, offices, finance, insurance; sectors of the economy with generally low demands for water and no obvious problems of effluent quality.

17. Forecasts of *population increase* not only established likely totals by 1981 and 1991, but of socio-economic and age characteristics; enabling more precise consideration to be given to the demand for facilities such as recreation.

18. The studies of *recreation* concentrated on strategic uses likely

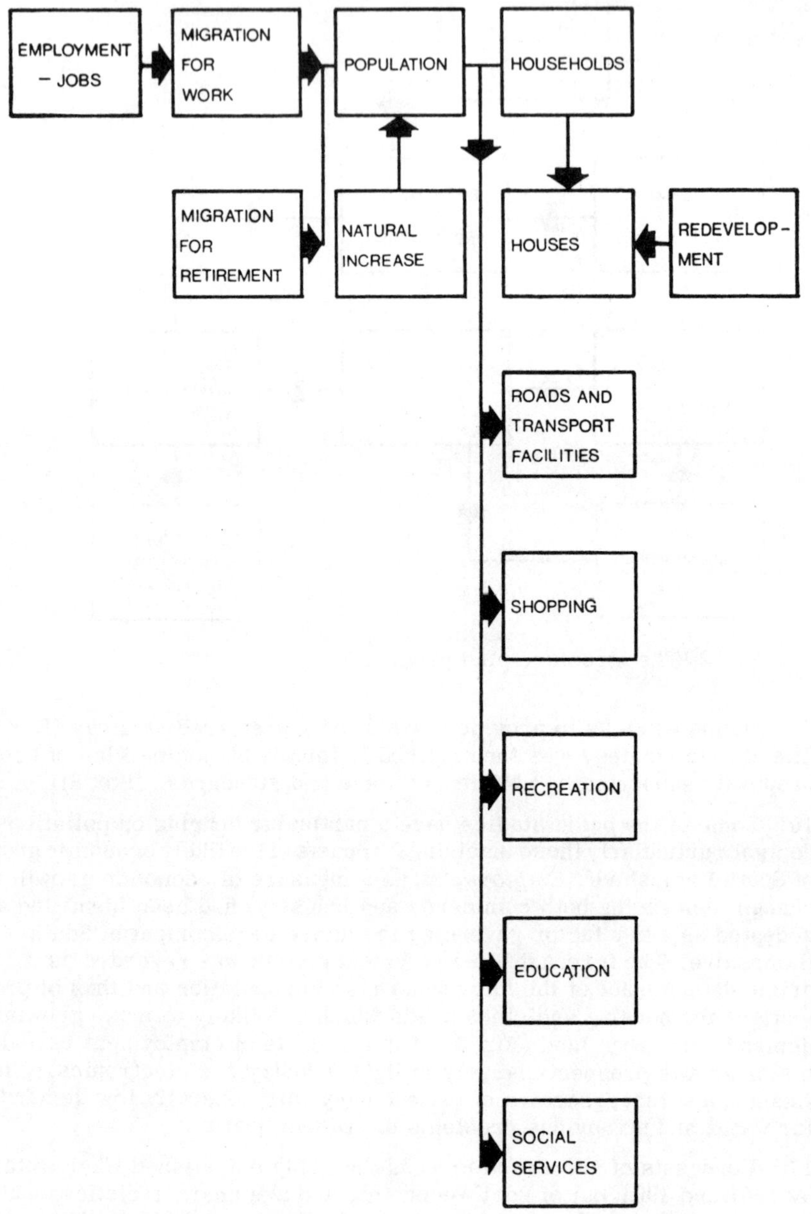

Fig. 3 Linkage of activity forecasts.

to place major demands on land and water. The studies and forecasts highlighted the role, functions and importance of the rivers and estuaries and the need to safeguard the coastline if the likely demands for fishing, sailing and day trips to the countryside and coast were to be met in the face of other conflicting demands for land reclamation for docks, industry, etc.

ESTIMATED PARTICIPATION RATES AND LEVELS OF DEMAND IN SOUTH HAMPSHIRE 1966, 1981 AND 1991

	Participation Rates* (per 1000)			Level of Demand		
Activity	1966	1981	1991	1966	1981	1991
Day trips to the Countryside	68.5	95.2	113.0	56 000	97 000	131 000
Day trips to the Coast	99.2	117.4	129.5	81 000	119 000	150 000
Sailing and Boating	10.2	16.4	20.5	7 000	13 000	18 000
Golf	4.3	7.1	9.0	3 000	6 000	8 000
Fishing (Inland)	3.1	4.3	5.1	2 000	3 500	4 500

* The rates for Sailing, Golf and Fishing exclude the under-12s.

19. Studies of *natural history* of South Hampshire revealed those areas especially sensitive and ecologically important; and for the coastline those sections which are vulnerable to damage and despoilation through increased public use of all kinds and from pollution in many forms.

20. The examination of *water supply and forecasts of water demands*, together with studies of *main drainage*, also brought into prominence possible conflicts between increased abstraction and amenity and fishing interests; the problems likely to be occasioned by increased effluent into chalk streams and minor water courses; and the need to safeguard sites for reservoirs.

21. Joint studies with the C.E.G.B. (continuing from wider regional studies) demonstrated the need for one and possibly two further power stations in South Hampshire requiring to be sited near large bodies of water, among other engineering criteria. Consideration of some seventeen alternative siting possibilities resulted in the reduction to a short list and the selection of the most probable satisfactory location from the engineering and land-use aspects.

22. Finally, in this summary of some of the studies having pollution and estuarial associations, were the wide ranging and detailed consideration given to *main drainage*. This included an examination of existing

problems of capacity, effluent discharge and pollution; forecasts of further effluent flow based on population and water use predictions; alternative possibilities for new systems which would gradually remove present problems and eliminate effluent discharge into many of the rivers and remove local pollution such as in the upper reaches of Portsmouth Harbour.

23. All these studies were carried out in close collaboration with, and in some cases jointly by, specialist organisations such as the Nature Conservancy, Ministry of Agriculture, C.E.G.B., Drainage Authorities and others; and included contributions from organisations in South Hampshire as part of an extensive programme of participation.

THE DRAINAGE STUDIES

24. The possibility of continued large-scale growth in South Hampshire had been examined in the mid-1960s by a study specially commissioned by the then Minister of Housing and Local Government, the Hampshire County Council and the City Councils of Portsmouth and Southampton. The South Hampshire Study(1) demonstrated the feasibility of continued growth and postulated a flexible strategy which sought to accommodate higher and lower ranges of population increase. Among the many aspects which were highlighted for further critical examination, was that of foul drainage. The early work in the preparation of the South Hampshire Structure Plan confirmed the importance of examining comprehensively and systematically present and possible future systems in the whole of the plan area. A Working Party was set up in 1968 consisting of the Engineers from the fourteen Drainage Authorities working in association with the Hampshire River Authority and the Isle of Wight River and Water Authority, the Hampshire County Council and the South Hampshire Plan Technical Unit. This Working Party was asked to consider and advise on the drainage implications of proposals for major development during the preparation of the draft Structure Plan. They carried out a considerable volume of work assessing the existing sewage and sewage disposal systems, their capacities and the commitments to extensions already made by the Drainage Authorities. They examined the volume, nature and quality of receiving waters and their capacities for receiving further effluent, together with the likely constraints on additional discharge arising from demands for other water uses, especially domestic supply and for fishing.

25. The Working Party recommended consultants who were commissioned by the South Hampshire Plan Advisory Committee to investigate more closely alternative drainage systems, to assess acceptability of discharging treated or untreated sewage into the Solent and to advise on particular specialist aspects, particularly location of points of discharge. A list of the consultants and a brief outline of their studies and summary of their results is contained in Appendix I.

26. Throughout the technical examination by the Working Party and its advisers, close liaison was maintained with the teams preparing the draft Structure Plan and at key stages in that work as strategies were prepared, tested and refined. The following early conclusions were reached by the Working Party:

(1) There was no significant installed or proposed spare capacity in either sewer networks or disposal works.
(2) Some systems were severely overloaded and urgently required major improvement, even to cope with population increase already provided for in approved development plans.
(3) Existing systems could not serve urban growth beyond that already provided for in the development plans and were generally incapable of extension.

27. Much detailed work was also done on the problems of effluent disposal and outfall location and the pollution load which could be accepted by the various bodies of water in the plan area. The conclusions reached, based on the assumption that the rivers should be pure enough to support trout and salmon and suitable for public water supply and that the estuaries, harbours and sea should not be polluted, were:

(a) In many cases the quality of present effluents left much to be desired.
(b) If additional effluent was to be dealt with then, preferably, none should be discharged to the smaller streams or the estuaries of the Itchen or Hamble.
(c) Treated effluent would be accepted in the Test at Romsey and further upstream and in the upper reaches of the Itchen, in Langstone Harbour and the Test estuary, but in all cases the amount and effects of increased flows should be carefully studied.
(d) Sewage from properly sited outfalls could be discharged to Southampton Water in limited amounts and the Solent could probably accept large quantities. But it was recommended that any system using sea outfalls should include adequate provisions for the prevention of pollution, including land reservation for possible treatment works, which need not necessarily be located on the coast.
(e) The disposal of sewage sludge caused some problems at present and, dependent upon the methods of sewage treatment adopted in the future, could require further study.

STRUCTURE PLAN PREPARATION AND DRAINAGE ASPECTS

28. Reference has already been made to the technical process of Structure Plan preparation (Fig. 2), with postulation of a large number of alternatives and their reduction by evaluation and testing to a preferred strategy which best satisfied the aims of the plan and the more refined objectives and criteria. A major part of this process was the reduction,

by testing, of a large number of theoretical possibilities to a set of four feasible and realistic alternatives. These 'Four Possibilities' (Fig. 4) were devised with careful regard to drainage constraints on physical development which had been identified in the basic studies. In particular, the preliminary conclusions from the drainage studies, limited the timing and location of areas otherwise thought possible for future development.

29. During the evaluation of alternatives, the assistance of the Working Party already described was invaluable in assessing the likely capital costs of drainage, and as part of a learning process of evaluation, modifications were made to the developing ideas for the location and timing of urban growth. These lessons provided a major input to the preparation of the preferred strategy (Fig. 5) in which the timing of development was particularly constrained by the administrative and procedural difficulties of implementing a major new sewerage system for central South Hampshire to serve the 'growth areas' of the Structure Plan.

SETTING UP OF SOUTH HAMPSHIRE MAIN DRAINAGE BOARD

30. Whilst work was continuing on the preparation of the draft Structure Plan during 1970, it became obvious that the administrative and technical problems associated with providing trunk sewerage systems for further large-scale development would have to be undertaken either by co-operation between the various Authorities in the construction of new systems or alternatively by the formation of a joint Board. The Drainage Authorities in central South Hampshire and the Hampshire County Council decided to adopt the latter means of ensuring that technical studies were further advanced and the joint Board could construct, operate and maintain all such sewers, pumping stations and sea outfall sewers and all such works of sewage disposal as may be necessary for effectively receiving and disposing of the sewage from the joint Board's area.

31. The studies which were undertaken direct through the Working Party and with the assistance of consultants were thus extended by the Main Drainage Board whose decision to treat sewage fully before discharge to the Solent is a fundamental part of the main drainage scheme; and one which has been influenced by the recreation, amenity and ecological importance of the rivers, estuaries and coastline.

32. Where general *proposals* for future development are included in a Structure Plan, close collaboration in its preparation is essential so as to ensure that proposals are not made which are likely to prove unrealistic in the provision of main drainage; do not place any unacceptable financial demands on the Drainage Authorities; nor add to the risks of pollution. The collaborative working with the Drainage Authorities in South Hampshire ensured that these general intentions were translated into objectives which were essential for the preparation and evaluation of strategies and in the refinement of the preferred strategy to form

Fig. 4 Possibility A, B, C, D.

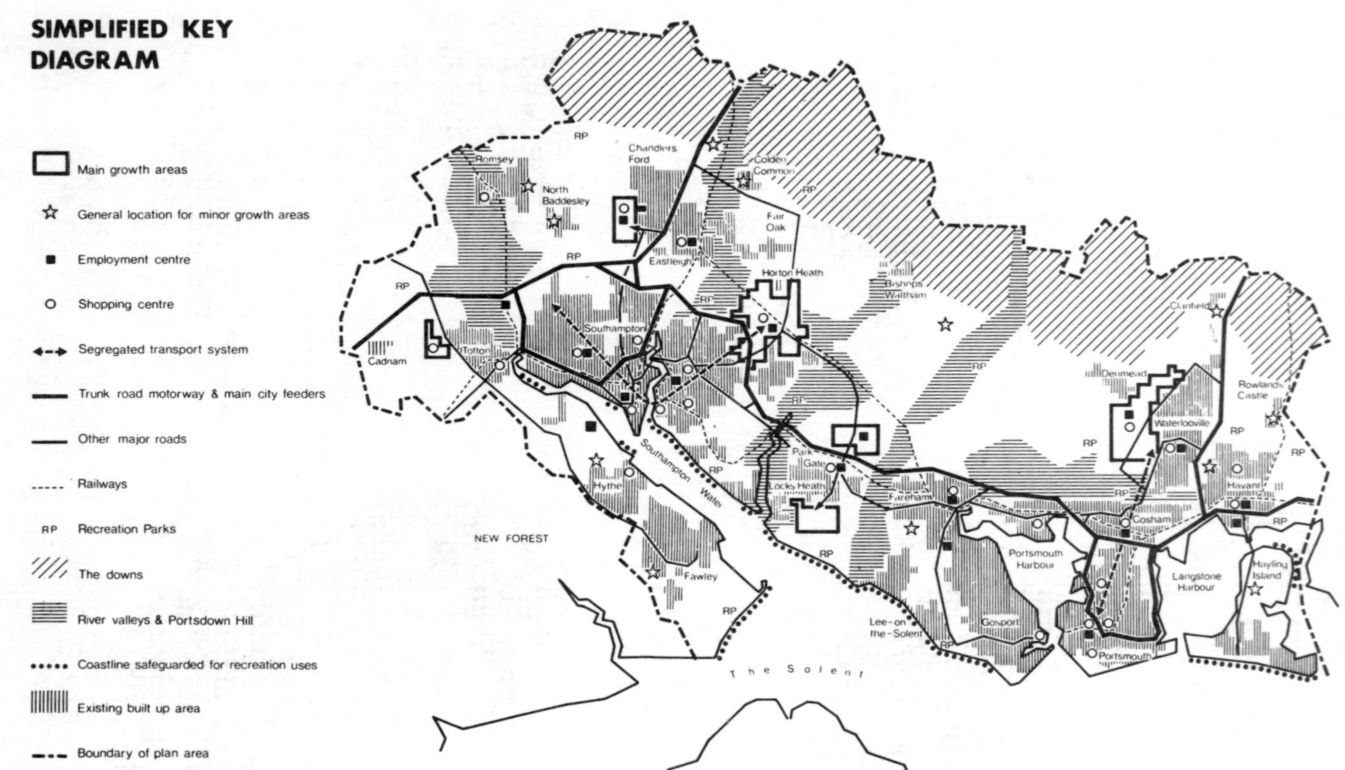

Fig. 5 South Hampshire Structure Plan.

the draft Structure Plan proposals for urban development. They were, with other objectives, instrumental in the proposals in the draft plan:

(1) to safeguard an area for the water pollution control works;
(2) to emphasise the recreational, amenity and ecological importance of the rivers and the safeguarding of major stretches of undeveloped coastline;
(3) to locate and phase proposals so that land for development was not released until main drainage was available.

33. One of the purposes of *policies* which form a major part of the Structure Plan as a written document is to put into effect the strategic proposals of the plan. Whilst general proposals of the plan have been derived from a reconciliation between conflicting objectives, the policies are intended to provide a framework for future decisions; particularly in the control of development and for the guidance of developers of all kinds.

34. Since the Structure Plan can only be concerned with aspects of physical development which are within the control or influence of the Local Planning Authority, the control of water pollution cannot be the subject of direct policy and action by the Local Planning Authority. In cases where there is a risk of water pollution, reference must be made to the responsible drainage and pollution control Authorities. One of the underlying intentions of policies in a Structure Plan must be to ensure that such Authorities are involved at an early stage so that drainage considerations and measures for the control and elimination of pollution are adequately incorporated in decisions about development.

35. With these considerations in mind, policies have been drafted which recognise the relationship of the Local Planning Authorities (as the body concerned with decisions on land-use development) with the other statutory bodies directly responsible for drainage and pollution control. For example, for foul and surface water drainage these policies are:

(1) Development requiring foul or surface water drainage will only be permitted where such facilities can be made available.
(2) No development will be permitted which would hinder the efforts of the Drainage Authorities to avoid harmful pollution of the rivers, watercourses, estuaries, harbours and seas of the plan area.
(3) Planning powers will be used to assist the Drainage Authorities in exercising their functions including, where necessary, the provision of water pollution control works on suitable sites, the safeguarding of foul and surface water drainage sewers and channels and other appropriate matters.

36. The criteria to be applied in order to put these policies into effect will require continuing close liaison with the Drainage Authorities involved. They are:

7.14

Policy 1

The Policy is to ensure that development only takes place where adequate provision for foul and surface water drainage can be made available at reasonable cost, and that neither pollution nor technical or administrative difficulties in the provision and maintenance of main drainage systems are created by piecemeal development and the use of local disposal systems, cesspits and septic tanks.

(1) The Local Planning Authorities will ensure that no development takes place:
  (i) which cannot be served by foul and surface water main drainage at reasonable cost;
  (ii) which will create other problems in terms of maintenance of the disposal or surface water system, or proliferation in the use of local disposal systems, cesspits and septic tanks;

until adequate drainage facilities, capable of catering for likely future demand can be made available.

Policy 2

The Policy is to ensure that no harmful pollution is caused to the rivers, watercourses, estuaries, harbours or sea by domestic, industrial or agricultural discharges.

(1) No development will be permitted which would overload existing or proposed facilities such as to cause harmful pollution.
(2) No industrial or other uses will be permitted which would discharge harmful wastes or effluents.

Policy 3

(1) In considering sites for water pollution control works, the Local Planning Authorities will have particular regard to:
  (i) the supporting criteria of the relevant Built Environment and Countryside Policies;
  (ii) the ability of the proposed works to cater for future extension of growth if necessary;
  (iii) minimising the visual effects of such works by requiring that all engineering works are incorporated into a landscape scheme of high standard;
  (iv) minimising the effects of smell or other nuisance;
  (v) minimising the costs of development, commensurate with environmental considerations.
(2) No development will be permitted which would:
  (i) conflict with potential water pollution control work sites;
  (ii) interfere with major foul and surface water drainage sewers.

37. These criteria are intended to be used in development control as a check list to ascertain the circumstances in which the specialist advice of the Drainage Authorities will also enable advance warning of concern

under particular criteria or in specified areas to be built into the development control system. This combination should ensure early and adequate consultation and action to avoid pollution. The manner in which the policies and criteria are framed should ensure that they are equally applicable and practical in the present and re-organised local government systems.

38. Altogether the approach adopted amounts to a comprehensive arrangement whereby the Structure Plan has been developed to take full account of pollution aspects in making strategic proposals for development and change. It also provides a continuing framework for the early identification and control of pollution in the development control process. However, Local Planning Authorities are not the executive authorities responsible for the application of the various statutory powers embracing pollution control, and the Structure Plan cannot propose direct action to eliminate existing problems (although it may draw attention to them) or exercise more than an advisory and advance warning service in the future.

39. The statutory bodies responsible have wide powers, frequently amended and updated to reflect technological progress and changing public attitudes. They also employ highly skilled technical officers to advise on pollution matters.

40. At the present time, and for the immediate future, it is clearly more satisfactory for the Local Planning Authority to be able to call on these bodies for technical advice, advise them of proposed development and rely on them as independent arbiters of pollution aspects rather than to include the evaluation of pollution potential as part of the already complex task of assessing applications for planning permission. Yet with greater technical knowledge of pollution, especially its cumulative effects, can we rule out the eventual possibility that applications for development will be refused even if it is demonstrated that ecological harm or environmental damage is the only objection to the proposal? To do so, will require greater technical and scientific evidence of the direct and indirect effects of pollution on ecological systems and on the human environment generally; the assessment of whether changes to these are directly attributable to those effects; what alternative means are available to prevent, reduce or remove harmful changes; and the monitoring of the systems and the changes occurring in them over time. In addition, there would have to be the political will to introduce the necessary statutory powers and the specialist expertise made available to monitor, advise, recommend and provide evidence. In the meanwhile the following improvements to present arrangements are desirable:

(1) Clearer definition of which bodies to consult on particular aspects of pollution, in the preparation of development plans and in the control of development. The Department of the Environment should be advising all Local Planning Authorities immediately after local government re-organisation since many of them

will have greater planning responsibilities than before.

(2) As some Local Authorities are often unaware of changes in industrial processes which cause new or additional pollution risks, technical advice is needed, especially where such Authorities are favouring or encouraging development and have not hitherto exercised control over trade effluent. Clarification by the Department of the Environment is needed on the desirability, acceptability and legality of conditions relating to pollution when planning permissions are granted for development if land-use planning is to make any further contributions to the avoidance and elimination of pollution risks.

CONCLUSIONS

41. This case study has attempted to demonstrate that in the preparation of development plans of a strategic nature, early and effective collaboration with bodies responsible for drainage, water supply and pollution control is essential if account is to be taken of the increasing demands and pressures on the physical environment. Such collaboration must include the best technical advice on the effects of greater human activity; the likely implications of the location and phasing of development; and the probable costs to Drainage and Water Authorities of possible alternative strategies for urban growth.

42. Such technical advice is needed for the formulation of objectives, alternative strategies and their evaluation and in the refinement of a preferred strategy. The policies for land-use development in a Structure Plan should be directed towards assisting those statutory bodies who have responsibility for pollution control and the provision of main drainage and water services. The relationship of structure planning to pollution control, as interpreted in South Hampshire, appears to provide a satsfactory administrative framework for Authorities to make their own decisions, but in a co-operative and collaborative manner.

43. Finally, one of the emphases in structure planning is the monitoring of assumptions and other information upon which the plan has been based. It will be necessary for the Structure Plan's policies to be monitored for their effectiveness in achieving the Plan's objectives, which themselves will change as the problems of the area alter over time. The Local Planning Authorities can and will develop the systems which will enable many forecasts of growth and change to be monitored, especially those relating to employment and population growth. Attention needs increasingly to be turned to how and by whom other aspects of human activity and its effects can be monitored, leading to a review of objectives and policies relating to the physical environment.

REFERENCE

1. Colin Buchanan and Partners, South Hampshire Study. H.M.S.O., London, 1966

APPENDIX I

Consultant	Subject	Result
John Taylor and Sons, Consultant Civil Engineers	To examine and recommend on alternative schemes for the sewerage of the central part of South Hampshire and to provide estimates of capital and running costs.	Outline plans of possible alternative schemes, together with some indication of their relative costs.
University of Southampton, Civil Engineering Department	The use of the Solent Hydraulic Model to ascertain the possible effect of future major outfalls in various locations.	The positioning of outfalls off Lee-on-Solent or off Browndown would give reasonable dispersion of the sewage, but as the model was limited in extent, outfalls from Gilkicker could not be tested.
Local Government Operational Research Unit	To develop a computer model to design and cost sewerage networks.	Although it proved possible to produce a computer model capable of performing limited design and comparison of alternatives, it was found to be unreliable when used to produce sewer costs.
Dr. B. A. Southgate, C.B.E.	To make a preliminary assessment of the acceptability of discharging treated or untreated sewage into the Solent, and its effect on the ecology.	Recommended that partial treatment should be given prior to discharge, and that further investigations should be carried out to ascertain the position of, and method of diffusion at, the proposed outfall. In addition, Dr. Soutgate recommended that the sludge produced should be treated on land if found to be practicable.

Consultant	Subject	Result
J.D. and D. M. Watson, Consultant Civil Engineers	To recommend on the nature, cost, and time-scale of the studies which would be necessary to determine suitable points for outfalls in the Solent and Spithead.	A proposed programme of hydrological and associated studies of the Solent and Spithead.
J.D. and D. M. Watson, Consultant Civil Engineers	To confirm the acceptability of disposal of partially treated sewage and to determine the most suitable point of discharge.	Recommended that the most suitable point for the discharge of partially treated sewage was Horse Sand Fort some three miles south east of Gilkicker point. They also recommended that any discharge near Gilkicker point should be fully treated.

DISCUSSION

WRIGHT (Hampshire River Authority). About four years ago the Author was responsible for setting up the South Hampshire Main Drainage Working Party to review the drainage of the area which was to become the subject of the South Hampshire Plan and also to devise sewerage and sewage disposal systems to serve the very large increase in the population of the area which was expected to take place before the end of the century. The Working Party provided an excellent opportunity for consideration to be given at the planning stage to one of the most critical aspects of development and the Author was to be congratulated on his foresight.

Neither the anticipated future population or where the development would take place were known when the Working Party was set up. Initially therefore it attempted to assess the capacity of the various receiving waters to assimilate effluents, bearing in mind the uses that are made of them, such as fishing, sailing, canoeing, water skiing, and as a source of drinking water and also the uses which might be made of them in the future, because recreational uses are increasing enormously.

It was eventually decided that the most satisfactory way of protecting the waters in South Hampshire was for the sewage from the area between Southampton and Portsmouth (but excluding the two cities) to be taken by means of a new long trunk sewer to the coast near Gosport and discharged into the Solent. Having taken this decision a series of detailed investigations were commissioned.

The tidal model at Southampton University was used in a preliminary exercise to assess the suitable outfall points. Later, Messrs. J. D. and D. M. Watson were appointed to carry out a detailed study including a biological and chemical survey of the Solent and also very extensive hydrographic surveys which were later used to develop a mathematical model of the area. No evidence of pollution was obtained from this work.

The Consultants conclusion was that the probable volume of settled sewage from the area could safely be discharged from a long outfall situated at Horse Sand Fort (that is roughly in the middle of the Solent off Portsmouth). The effect on the beaches of discharging settled sewage at Horse Sand Fort was the same as that of discharging a Royal Commission effluent from an outfall roughly at the point selected by the South Hampshire Main Drainage Board off Browndown Point. It would however take five hours for the sewage effluent to reach the beach from the Horse Sand Fort outfall but only two hours in the case of the shorter Browndown Point outfall.

One of the problems associated with the disposal system advocated by the Working Party is the widespread belief that pollution cannot occur if the sewage is purified to an arbitrary and often meaningless standard and therefore that so called 'fully purified effluent' can be discharged to any accessible body of water. Conversely it is generally believed that an effluent which fails to comply with the arbitrary standard inevitably causes pollution. It can be shown from the River Authority's records that this belief is not in all cases in accordance with the facts. For example, the level of dissolved oxygen in both the Itchen and Hamble Estuaries is decreased and the concentration of sewage pollutant is significantly increased by discharges of purified sewage effluent. Coliform counts in the Itchen Estuary are frequently in excess of 10^6/100 ml (and sometimes in excess of 2×10^6/100 ml) and coliform concentrations vary between several thousand and a few million per 100 ml are found along the northern shores of Langstone Harbour as a result of a discharge of fully treated sewage works effluent. The Solent, which receives sizable discharges of macerated sewage, is, however, fully saturated with dissolved oxygen, concentration of sewage pollutants are extremely low (e.g. the concentrations of ammoniacal nitrogen does not exceed 0.05 mg/l), coliform counts are generally less than those found in a river upstream of all significant discharges and even in the slick from the Gosport outfall they do not exceed 2000/100 ml. For comparison a Hampshire river providing a 10-fold dilution to a fully treated sewage works effluent of Royal Commission standard contains between 20×10^4 and 5×10^5 coliform bacteria per 100 ml. It would therefore appear that measurable (sometimes harmful) concentrations of sewage pollutants are praiseworthy if they are derived from sewage from which the bulk of the oxgyen demand (but little else) has been removed, but that the lower (often immeasurably small) concentrations derived from macerated sewage discharged to the Solent are unacceptable and reprehensible.

The South Hampshire Main Drainage Board has decided to treat the sewage from their area to Royal Commission standard and to discharge effluent to the Solent 1000 metres off Browndown Point. This outfall will cost some £8 million. If the public believe, as they appear to do, that if it's purified it's safe, discharge it to the nearest river, creek or estuary or on to a bathing beach, there will be a tendency to abandon the outfall to save money. If this occurs there will be a widespread reduction in the quality of receiving waters and possibly serious pollution in South Hampshire. It would be interesting to know if, in the Author's opinion, the public are more concerned about the quality of receiving waters (which is the *raison d'etre* of pollution prevention) or about securing the purification of all effluents—in other words what do the public want?

Amenities, which have been discussed during the conference, are of course most important. Unfortunately estuarine waters can contain a lot of silt scoured from the bed. This can cause serious discolouration and even foaming. Before using amenity criteria it is therefore necessary to know the cause of what is being objected to. It could for example be natural and nothing to do whith discharges of effluents. Certainly the colour in Ashlett Creek which was referred to by Drummond is not due to the discharge of sewage effluent.

Following the decision to build a new main trunk sewer, building development is going ahead. Flows are increasing at all existing sewage works and injury to water resources is likely to occur between now and when the new main trunk sewer comes into being, in fact one incident has already occurred. It is impossible to stop development completely, yet development can only result in a general deterioration, temporary albeit, in river conditions.

Another problem which will be generated by the development which takes place before the completion of the trunk sewer is that of a temporary sewage treatment works. What protection can the planners provide to ensure that privately owned temporary works are closed down at the earliest opportunity? Also will they be in a position to give information on the complete land use downstream of the temporary works so that a detailed assessment can be made of the effect of the discharge on the purposes for which the receiving water is used.

Finally, the Author had shown a plan of the relationship between effluents and river flows. This does not refer to current conditions and it would appear likely that it refers to a forecast alternative to the proposed scheme of the Main Drainage Board.

BARROW What do people want? They want maximum improved standards at minimum costs. In the future, at public enquiry, the objectives of schemes will be under cross-examination. The public will have to say what actually they do want and at what cost.

Can we stop development altogether? South Hampshire is a 'growth' area; accepted as such in the Strategic Plan for S.E. England and endorsed by the Government and local planning authorities. But because of sewerage problems there have been embargoes on development in various parts of the area—for example in the Fareham area. Yet pressures are mounting from council members concerned about the slowing down of housing development; from central government (White Paper on the 'Next Steps in Housing'). There perforce is an *a priori* assumption that housing will be permitted in growth areas like South Hampshire. To limit development, there must be clear evidence that such developments should not be allowed for specific reasons, e.g. that it is in an area of outstanding natural beauty, or likely to give rise to ribbon development. Difficulties of drainage or road access are expected to be overcome. The demand for houses will grow, notwithstanding a fall in birth rate, and if development is delayed because of the absence of main drainage, temporary sewage treatment plants will be proposed by developers and others. Proposals have and will be made that they are provided either by the developers or, if the R.W.A. agrees, by some contribution by the developer to a temporary system which is then phased out and related into a new major system of the kind now being prepared for South Hampshire.

In an era when housing is a social and economic issue, the authorities will be under very great political pressure and it is doubtful whether this can or should be resisted. Unless evidence can be accumulated to demonstrate that the effect of allowing temporary systems will be to add significantly and irredeemably to pollution, advice against them may well be ignored by politicians who find themselves under pressure to release further housing land.

We must also recognise that from 1st April, 1974, there will be two major responsibilities for planning in local government. The new Counties will be responsible for strategic (structure plans); the new Districts for local planning and the control of development. In view of this it will be essential to establish working links between all the new authorities if development proposals are to be closely related to the provision of main drainage.

WAKEFIELD (Coastal Anti-Pollution League Ltd.) Mr. Wright seems to be in somewhat of a dilemma. But tracing the problem of pollution over the last 15 years I know at first hand that there was considerable pollution on the shores of the Solent particularly from the Lee on Solent and Gosport outfalls. The people at that time asked their council to stop the pollution. There was a great deal of argument, lasting at least 5 years. The people was assured, on every count, that there would never be any further pollution if an outfall 1500 m long was put out from

Browndown, near Gosport. They were assured that there would be maceration and screening, and that there would no longer be any visible evidence whatsoever. Photographs of the outfall were taken by Hunting Surveys Ltd. They had no difficulty in locating the outfall. They could see it at all times. I have been out there once, in a boat to look for it, and I lighted on it straight away. Whether it is better to put an outfall out at Horse Sand Fort with untreated sewage or to put a shorter outfall in and discharge treated sewage is a technical question. People want a clean environment and at the moment they do not see it materialising. It is to be hoped that the present plans will succeed to some extent in producing this clean environment.

BARROW The co-operation of technical experts on this problem provides the best hope of arriving at agreed solutions. Land use planners have to take advice, but test that advice from time to time. For example, members and officers of a Committee responsible for land use planning should probe and cross-examine the responsible authority such as the River Authority to satisfy themselves that all aspects and possibilities have been explored before a firm decision is reached.

In the long term we can properly expect to have cleaner environment, even with the growth prospects of South Hampshire, as long as capital and other resources are available, perhaps of the order of £20 million. The introduction of a new system should be coordinated with the phasing out of existing systems within the area.

But in the short term, one cannot be so optimistic. The preparation and design of a new system has been completed, but the procedures for implementing it— public enquiry, compulsory purchase orders, and so on, together with the uncertainty of capital resource availability could combine to delay the provision of a new system. Yet without it, in the near future, there will be gradual, but perceptible, deterioration in the physical environment.

DOWNING (Water Pollution Research Laboratory). Cleaner environment is a planning objective. What one is concerned with at this conference is the criteria by which the cleaner environment is defined. Waste water must be put somewhere, and this will affect water quality. What quality of water is it that allows enjoyment of the various benefits of the environment?

CHARLESWORTH (Association of Rural District Councils). Many of us have suffered from development plans which did not have regard to the basic services of the town. Despite the enlightened approach in South Hampshire, there will be an unavoidable hiatus between the setting up of the Water Authority who will be responsible for all foul and surface water sewerage and water supply, and the provision of the drainage facilities which will be needed to implement the development plan.

The first priority of the Regional Water Authority, through the Department of the Environment, is going to be water supply, the second priority, much lower, is going to be foul sewerage and nowhere in the race will be surface water sewerage. As the plan develops, in view of present land prices, there will be high density development with incalculable alterations to the surface water run off pattern in South Hampshire. This is probably more important than arguing about the standard of treatment, whether full or partial treatment, and where one should discharge into the Solent. The cost of surface water sewerage systems is far greater than the cost of related foul sewerage systems.

There is over-emphasis on the treatment of foul effluents. Surface water balancing and the control of flooding is going to be increasingly important in the South Hampshire area.

WRIGHT (Hampshire River Authority). Except for one small area there are no particular surface water drainage problems in the area.

CAMP (J. D. and D. M. Watson). Storm water overflows have to be designed carefully, preferably screened with mechanically raked screens. So far as the South Hampshire Plan scheme is concerned, the design is for a particular peak rate of flow through the outfall at Horse Sand Fort, and a shorter but screened outfall of storm water into the Solent in the approximate position where the Plan now proposed discharges the fully treated effluent.

Taking a point raised by Mr. Wakefield, our recommendation to the South Hampshire Plan Advisory Committee on the outfall at Horse Sand Fort recommended partial treatment comprising screening, removal of screenings, and a brief sedimentation period so that in the effluent the *E. Coli* concentration was reduced to 10% of that of crude sewage. Crude sewage may have an *E. Coli* count of around 2×10^6/100 ml so that 10% of that is 2×10^6/100 ml. The drinking water standard is something like 3/100 ml so that a Royal Commission standard effluent is still a great way from drinking water standard. What does the public wish to pay for?

Going back to the recommendations for the Horse Sand Fort outfall, the aim was to find out what would be the conditions on the coast under certain conditions of wind and tide. Conditions considered were, a 24km/h wind and a 48km/h wind blowing from the North and from the South East. The level of coliform concentration at the coast under those wind conditions was estimated and a diffuser system designed so that this level was achieved. It is unlikely that there will be people on the beach when the wind is in excess of 50km/h, at which times the coliform concentration may exceed the proposed levels.

ARNOLD (Welsh Office). There is a danger of transferring pollution in the estuary to the land. Sewage treatment on land produces a sludge disposal problem, which is not yet adequately solved. A recent new sewage treatment works in Wales has improved conditions in the river, but has created a local smell problem.

THE POLLUTION OF PORTS

D. R. Houghton, B.Sc., M.I.Biol.

Head of Biological Division, Central Dockyard Laboratory

INTRODUCTION

1. It must be recognised from the outset that there are two major problems which have to be faced when talking about pollution. First, there is the very real difficulty of defining precisely what is meant and secondly there is a dearth of specific information regarding the subject. Even in those areas where considerable effort and money has been expended, such as pollution by oil, the position still remains that there are more problems than answers. Another real danger is that the easily recognised forms of pollution, which are not necessarily the most important ecologically, tend to be investigated with greater thoroughness than the more serious problems arising from less obvious sources. This paper deals specifically with the dangers of pollution of Ports and associated problems with an emphasis on the part played by the use of antifouling devices for ships and other structures. It has to be acknowledged straight away that as many ports are situated in estuaries aspects of pollution common to these areas need to be borne in mind.

2. Many definitions of pollution have been attempted[1,2,3,4,5,6] none of which is entirely satisfactory, any more than the one given here. For the purpose of this paper pollution will be taken to mean the introduction into the environment, by man or his agency, of materials which would not normally be found in that situation, or would not without his intervention be present in the same concentration.

THE ESTUARINE ENVIRONMENT

3. There are many ways and many rates at which an estuary may become polluted and as a consequence no two are ever the same. The effect may be on the nutrients that are present, the Biological Oxygen Demand, or on the quantity of particulate matter or directly on the living organisms. There are also other forms of pollution which may be present such as inhibitors, poisons and heat. Even aerial pollution may have a profound effect when it is washed down by the rain through the surface drainage into the rivers and hence the estuaries. Perhaps most important of all is the pollution due to agriculture which often has a direct run off into the rivers.

4. Even in the absence of pollution conditions in estuaries can vary very considerably from one year to the next. This is due to the fact that the distribution, seasonality and abundance of epifaunal species are dependent on such factors as salinity, temperature, turbidity and water movement.

It is therefore extremely difficult to assign the precise role of any pollutant in the estuarine condition unless controlled experiments are carried out to pinpoint the effect. They should also include studies to determine whether one pollutant is lowering the threshold response to another or to determine any synergistic effect.

5. Some ports are partially or practically enclosed and in this situation the forms of pollution may be more limited but those present are likely to reach far higher limits than in the estuarine condition. It has been calculated[7] that in enclosed basins with many ships alongside, a copper concentration of 1 ppm might be realised which would prove fatal to many forms of marine life. With the advent of the larger tankers and container ships there is a definite move towards constructing ports at exposed sites along the coast. This tendency may well have a beneficial effect upon our estuaries.

OIL POLLUTION

6. Oil pollution is one form which it is not proposed to discuss at any great length with regard to ports. It must be said, however, that there is an increasing risk for ports to be subjected to accidental spillage or spillage due to collision. Usually recovery to a single spillage is good but some evidence suggests that repeated oilings of salt marsh vegetation has a deleterious effect, e.g. from eight to twelve coverings results in considerable change taking place.[8] Recovery under these conditions is slow and where large bare areas of mud have been produced considerable change can take place due to erosion. From the biological point of view the most serious effects of oil pollution are experienced by sea birds although reports of adverse effect upon molluscs, barnacles and species of red algae etc., are well known. The age of the oil also has an important effect in so far as fresh oil contains the lighter fractions which disappear on ageing.[6] Another aspect of oil pollution is that it has been noted that persistent biocides such as D.D.T. become concentrated in oil slicks.[9]

ACCIDENTAL SPILLAGE OF TOXINS

7. One of the best examples of the adverse effect of pesticides comes from the Yacht Marina at Birdham Pool, Chichester Harbour.[10] In the late 1960s phorate was being sprayed from aircraft on to fields near the pool which was carried by the wind on to the water. The grey mullet previously introduced to scavenge the refuse from the yachts had flourished successfully up to that time. The result was that they died off in large numbers as well as other fish in the area. A spillage of insecticides in the Moroccan Port of Mohammedia was reported to have

been responsible for the death of large numbers of fish and seabirds.[11] Another accidental discharge of pesticides occurred at La Coruna, Spain, in 1971.[12]

COMPOUNDS USED IN ANTIFOULING PAINTS

8. With the severe penalties exacted by the fouling of ships, viz., extra fuel costs, loss of speed, loss of earnings and the cost of remedial action, it is not surprising that ship owners are anxious to have the bottoms of their ships protected with efficient antifouling agents. As far as the very large tankers and container ships are concerned it is highly desirable that they be kept at sea for as long as possible. It is not only the enormous cost involved in docking but the fact that there is a shortage of docks and those available are situated in geographically distant parts. This serves to reinforce the need for really effective antifouling devices which will last for long periods of time. Methods of underwater painting would be useful but they may well increase the risk of pollution.

9. The most commonly used antifoulant is copper (I) oxide although many other compounds have been tried including others of the same metal. The latter are less effective and are not therefore in general use. Mercury compounds such as mercury (II) oxide and (I) chloride are even more effective than copper (I) oxide but because of cost they were usually added to copper paints as 'boosters'. They are no longer used for antifouling purposes because they first of all represent a health hazard in the preparation of the outer bottom for repainting and secondly they have been implicated as environmental pollutants. Since some forms of bacteria are capable of converting the inorganic compounds into methyl mercury, the risk to man is increased through food chains. Other inorganic metallic compounds including zinc oxide have been found to be considerably less effective than copper (I) oxide. Arsenic being an accumulative poison has not gained general acceptance as an antifouling agent and in any case it is not very effective in its inorganic forms. Some organic arsenicals have nevertheless been tried. It is worth recording that 10, 10', oxybisphenoxarsine proved to be an effective antifouling agent. The Dow Chemical Company, however, carried out a number of tests on algae and found that the compound could be concentrated as much as 1 000 fold by these organisms. Because of the implications of the environmental hazard they withdrew from the market a potentially good antifouling compound.[13]

10. The bis tributyl tin compounds were found in laboratory tests to be about two orders of magnitude more toxic than copper (I) oxide. They are also degradable which led to their initial use in agriculture as fungicides.[14] They have not been used to the extent which at one time appeared possible because of their inability to control the brown alga *Ectocarpus*. They have been used to 'boost' copper paints.

11. The triphenyl lead and tributyl lead acetates have been found to be reasonably effective as antifouling agents but since lead is no longer acceptable in the environment their use is being curtailed.

12. Because of the wide range of its effectiveness copper (I) oxide is the usual toxicant included in antifouling paints. Unfortunately, it loses its toxicity when the paint is dried out and has to be 'boosted' with other compounds for some purposes. This is particularly so where there is a large difference in depth between the ballast and load lines.

13. To remain effective copper has to leach out from the antifouling paint at approximately 10 $\mu g/cm^2/day$. At this rate 0.1 g/m^2 of copper is discharged into the water per day and a ship of 10 000 m^2 outer bottom would be releasing about 1 kg/day.[7] Under normal conditions in an estuary or port, with a good exchange of water, these quantities of copper would not present a hazard as an environmental pollutant. It is stated[6] that man can take as much as 100 mg of copper/day in food without any danger whatsoever. It is not usual for copper to become an environmental pollutant, except in mining areas, and even here adaptation has occurred in some species to withstand the high concentrations.

14. The possibility does exist, however, that copper could be regarded as an environmental pollutant in enclosed docks with a high density of shipping and practically no exchange of water.[7] If ten ships of 10 000 m^2 underwater area were enclosed in such a dock for 100 days, 1 tonne of copper would have accumulated. In a dock of about 1 000 000 m^3 the concentration would be about 1 ppm. This concentration of copper is effective in killing off the larvae and spores of sedentary organisms but would not constitute a direct hazard to a large number of higher organisms and man. It would of course upset the food chains giving rise to an indirect effect upon higher forms. The chances are that such an enclosed basin would have other pollutants carried into it of a more serious nature and would be considerably polluted on that account alone. It is also of significance that when copper is allowed to accumulate in the water adjacent to an antifouling paint relatively low concentrations have a stifling effect upon the leaching of further copper, even at concentrations as low as 0.1 ppm.

15. Another way in which copper could constitute a hazard is in the disposal of arisings from docking. It has been estimated[7] that 10-20% of the composition supplied for a ship falls to the dock bottom and that a well coated ships bottom has approximately 1 kg of copper per 40 m^2 left in the paint which is blasted off and falls to the bottom of the dock. Arisings are often dumped at sea in deep water where over a long period of time they may accumulate on the sea bed. The remains which are not removed from the dock will contaminate the water when the dock is filled. There is no direct evidence on this point but the marine growths in the vicinity of the discharge pipes, in some docks at least, indicate that there is no serious pollution from this source.

16. Sewage sludge dumping in the Firth of Clyde has been carried out for many years. The sludge has been contaminated with a variety of heavy metals. Copper together with lead, tin and zinc have a strong association with organic matter. The level of copper, however, in the whelk (*Buccinum*) from the vicinity is lower than it is in the Irish Sea.[15] The range of copper in sediments is from about 20 ppm in low copper areas to more than 4 000 ppm in heavily polluted estuaries.[16] Nevertheless, it is worthy of note that in the United States there are local ordinances, in Newport News, Virginia, for instance, that prohibit the disposal of paint scrapings in the harbour. In Holland[17] where the level of copper in coastal waters can fluctuate between 2-15 μg/l, it has been shown that there is virtually a linear relationship between the living organisms and the surrounding water. In some instances the margin between safe limits for copper sensitive species and the prevailing copper content appears to be small.

17. Before leaving antifouling paints it should be mentioned that the chlorinated diphenyl compounds have been used as plasticisers in many of these compositions. It is encouraging, from an environmental point of view, that Monsanto Chemicals have withdrawn all its range of chlorinated di- and ter-phenyls because of their suspected involvement in environmental pollution. PCB compounds have apparently been responsible for the high incidence of deformities[18] in, as well as the death of, a large number of sea birds.

CONCLUSION

18. Copper-containing antifouling paints do not appear to constitute an environmental pollution problem.

19. It will be necessary to ensure that compounds used to 'boost' copper based paints do not give rise to an environmental hazard.

20. It will also be necessary to ensure that any new compounds developed as antifoulants do not give rise to a pollution problem.

21. The indications, at present, are that on the whole industry is maintaining a watchful eye on the pollution problem and certainly in two instances have acted very responsibly.

REFERENCES

1. Cronin L. E. and Flemer D. A., Pollution & Marine Ecology. Interscience Publishers, 1967, 171-183
2. McKee J. E., Pollution and Marine Ecology. Interscience Publishers, 1967, 259-266
3. Hurley D. E., *Mar. Poll. Bull.*, 1970, **1** (9), 133-134
4. Lee N., *Mar. Poll. Bull.*, 1971, **2** (10), 151-153

5. Duke T.W., *Mar. Poll. Bull.*, 1972, **3** (NS) (8), 126
6. Yapp W.B., Production, Pollution, Protection. The Wykeham Science Series, 1972, p. 77
7. Gay P.J., Lloyds List, 14th June, 1972
8. Baker J.M., *Environmental Pollution*, 1973, **4** (3), 223-230
9. Walsh, G.E., *J. Wash. Acad. Sci.*, 1972, **62** (2), 122-139
10. Report in *Mar. Poll. Bull.*, 1970, **1** (NS) (8), 117
11. Report in *Mar. Poll. Bull.*, 1972, **3** (7), 101
12. Report in *Mar. Poll. Bull.*, 1971, **2** (1), 5
13. Zimmerman R.L., *Journal of Paint Technology*, 1973, **45** (580), 58-61
14. Kerk G.J.M. van der and Luitjen J.G.A., *J. Appl. Chem.*, 1954, **4**, 314-319
15. Mackay D.W., Halcrow W. and Thornton I., *Mar. Poll. Bull.*, 1972, **3** (1), 7-10
16. Bryan G.W. and Hummerston L.G., *J. mar. biol. Ass. U.K.*, 1971, **51**, 845-863
17. Koper in het Nederlandse milieu—*TNO nieuws* September, 1972
18. Report in *Mar. Poll. Bull.*, 1972, **3** (1), 4

DISCUSSION

SHILSTON (Sir William Halcrow & Partners). Pollution takes many forms and this conference has confined discussion largely to one type—that resulting from 'infected' water. Typical are the problems arising from stagnation or the inability to flush out an enclosed basin or a wet dock system. The marine growths in the inner part of Poole Harbour are probably the result of an unbalanced ecosystem; the phrase is different but the outcome is the same.

Surely, though, there are other types of pollution which are equally apt for this conference to consider. The thick layer of fine coal dust on the water surface of the Old Harbour at Hartlepool when the coal staithes were in use was apparently accepted as inevitable.

Many industrial processes are notorious for producing vast tonnages of waste material in the form of mud or sludge which is discharged into estuaries. Traces of such elements as iron, aluminium and titanium are not present in sufficient quantity, according to the industrial concerns, to justify their extraction. An example recently noted was the dumping from a British factory of 2 000 tonnes per day of titanium-rich sludge.

In addition there are the pollutants derived from the emission of noise, dust and fumes. The majority of estuaries are the site of industrial zones, and they will be even more so with the enlargement of the bulk carriers importing raw materials. The Author's example of the aircraft spraying insecticide was an excellent example of an accident, but this also should be very much to the fore in such a conference as we have here.

We must work to the standards available at the present time. This conference provides an opportunity for standards to be updated.

HOUGHTON There are indeed a lot of problems that would have to be tackled in dockyards in particular, and dust and noise are well to the fore in these. Also there are the industrial effluents from the workshops and so on which accumulate in the docks, an aspect which we have not actually mentioned. All these factors are important when considering the pollution of docks. However, this conference is dealing with the pollution of waters in estuaries.

HAMILTON (Institute of Marine Environmental Research). The Author states that mercury compounds are no longer used in antifouling paints. Is this a U.K. policy, or is it accepted internationally?

When using a plankton tow net one is sure to pick up paints in disseminated amounts. This is quite a serious problem in terms of analysis for elements such as Cu, Zn, Hg and Pb.

HOUGHTON Mercury is banned internationally. There were particularly bad experienced in Japan and Sweden. In the former there were a large number of deaths due to the presence of methyl mercury taken up through fish. Paint flakes certainly can contribute to pollution, but the solubility of copper is uncertain. A concentration of 1 mg/l can be reached, but it would be unlikely in large bodies of water.

LITTLE (National Yacht Harbour Association and Brighton Marina Co.) Could the Author give any indication as to whether the solution of antifoulants and cleaning and painting operations from small boats in yacht harbours could give rise to local pollution? Are the same compounds used as on large ships? The majority of big yacht harbours will hold 1 000 boats.

HOUGHTON At a guess 1 000 small boats may well be equivalent to one large vessel. The antifoulants which are supplied to yachts are usually not of the same high toxic content as those that are put on to ships which are going across the oceans of the world. They are mainly copper-based, some of them are just straightforward copper paints, but many are now boosted with tributyl tin. We have no evidence to suggest that tributyl tin is an environmental pollutant. It is de-graded in the environment.

The same criteria hold here as hold with the large docks. If there is a good interchange of water in the marina then there is no problem. Taking Birdham Pool (in Chichester harbour), for instance, where there are a lot of yachts, there is no evidence that the anti-fouling is giving trouble to the organisms that are growing there.

RAYMER (Southampton Corporation) When anti-foulants are spent they are scraped off. Why is this done? Presumably vessels go into a dry dock and the material is collected together and dumped out into the sea. Is the copper content so low that it is not reclaimable?

HOUGHTON It is not economic to reclaim the copper. Not all docks dump at sea, but some do. The evidence is that dumping does not seem to give rise to any major problems in the immediate environment on the sea bed. It tends to form a sediment rather than a solid block, and this will change the species that will be present in that place.

DODD (Berridge International Consultants) The cost of collecting and processing copper from these arisings would be totally uneconomic compared with the normal processes of obtaining copper.

Leaching of the copper from the anti-fouling coating to which the Author referrred is a progressive process from the outer surface inwards. When the copper has been leached from the paint matrix to a certain depth, then the leaching rate falls off to a point where the anti-fouling action ceases. At this point the whole paint layer has to be scraped off and new paint put on. The paint that is scraped off may well have 50% of its original copper still there, but it is quite uneconomical to try to recover it.

The arisings may contain about 15% cuprous oxide, but there would be considerable difficulty in collecting them from the bottom of a 'dry' dock, which invariably contains some water, and much preparatory work would be needed before recovery.

PARSELL (Coastal Ecology Research Station) Is there any information on the quantity of plasticisers being discharged to estuaries (Para 17)? They are also widely used in the plastics industry.

HOUGHTON I am not aware that plasticisers cause any trouble in estuaries. In the running of the laboratory at Eastney, only one case of such a problem has ever arisen. This was many years ago, and it was found that a lead plasticiser had been used.

The manufacture of PCB has now ceased, so that this particular substance should not be a problem in future.

DOWNING (Water Pollution Research Laboratory) Copper-based anti-foulants must have been in use for a long time. One would have thought that problems would have appeared already if they were going to appear at all.

HOUGHTON These compounds have been used by the Navy for 120 years. There is, despite this, no detectable increase in copper content in Portsmouth Harbour—where trouble would be expected. It is hoped to start an ecological survey in this area paying particular attention to copper. This will include a survey of the mud in the harbour, which has not previously been surveyed.

It is correct to say that copper binds protein.

In congested harbour and dock areas which have little exchange with the water outside, it is unlikely that there will be any demand on the area for recreational amenity. Some relaxation of standards should be possible in such a situation. Similarly, in such a situation there is unlikely to be a problem with fish, as they will be absent.

Two reports on grossly polluted harbours point to sewage as the pollutant, and do not refer to anti-foulants.

WAKEFIELD (Coastal Anti-Pollution League Ltd.) There are many estuaries where a multitude of outfalls, both of sewage and of industrial wastes, discharge in a haphazard fashion, and include discharges into the dock area. For example, there are 27 outfalls into Medina Estuary at Cowes, Isle of Wight. Should not these be collected by a ring sewer, and brought to one point for monitoring, treatment and discharge, to improve control and to clean up the harbour?

HOUGHTON This is an excellent suggestion. Ports were, and in some cases still are, very dirty areas. It seems to be generally accepted that old ropes and old rags can be thrown over the side; but this is all a form of pollution. Uncontrolled discharge down gullyways and drains has been the accepted practice.

CHAMBERS (Mason, Pittendrigh and Partners, Consulting Engineers) On the question of provision of interceptor sewers and the provision of adequate treatment facilities, it is very much more difficult than people probably realise. Sewage arising from domestic sources can be treated by biochemical processes. Trade wastes of a chemical nature are usually best treated at source, rather than in admixture. When the volume and quality of chemical wastes to be treated are unknown the problems involved in the design and operation of a works can be enormous. Along the 25 or 30 km of the tidal reaches of a heavily industrialised river such as the Tyne, there are literally hundreds of outfalls from all kinds of industrial sources. The problem cannot be resolved simply by putting in a large collector sewer along either side of the river leading to a works at the mouth and hoping that conventional treatment will result in a satisfactory effluent.

Another problem which needs to be seriously considered is the pollution caused from more common metals, and particularly from the effect of the rusting of iron and steel. Presumably the oxidized iron sinks to the bottom of the river where it mixes with the silts and organic mud which has accumulated. In severe cases this presumably destroys all plant life and invertebrates on the bed of the river with disasterous results to the ecology.

Some time ago we were called on to look at some steel sheet piles in the river Tyne about 10 km from the sea. These were rusting away very rapidly and an estimate of their useful life was 40 years compared with 110 years in the maker's literature. This accelerated rate of corrosion gives some idea of the high degree of pollution in the Tyne Estuary.

THURSTON (I.C.I.) (written contribution) ICI Paints Division currently employs a combination of tributyl tin compounds (usually considered bio-degradable) and cuprous oxide (to date thought safe) but these still do not meet the requirements of an anti-fouling paint with a life in excess of two years. The largest single item in the research effort in the marine coatings field is for better anti-fouling compositions. All new organo-metallic compounds have been discounted and the new toxins screened must be both of low mammalian toxicity and of low persistence. Very few of the many dozens of compounds tested annually meet the requirements. Two purely organic compounds, now on ships' trial, are thought to be suitable on the grounds of water pollution and hazard but it is not yet known whether performance is adequate.

BACTERIOLOGICAL, BIOLOGICAL AND CHEMICAL PARAMETERS EMPLOYED IN THE FORTH ESTUARY

R. W. Covill, Ph.D., D.P.A., Dip.Bact., C.Eng., F.I.Chem.E., F.I.P.H.E., M.I.Biol., M.Inst.W.P.C. (Dip), A.M.B.I.M.

Chief Technical Officer, Lothians River Purification Board

INTRODUCTION

1. In accordance with the provisions of the Rivers (Prevention of Pollution) (Scotland) Act, 1951, the Secretary of State for Scotland was empowered to establish River Purification Boards. The Lothians River Purification Board Establishment Order was made on the 27th February, 1953, and came into operation on the 6th March, 1953, and the staffing of the Board was initiated with the appointment of the Chief Technical Officer-River Inspector in June 1955. Whilst at the commencement of the Board's activities their powers were defined by the '1951' Act which only enabled it to impose conditions relating to any new discharge taking place after 1951, nevertheless, it was apparent that gross pollution existed in the estuary resulting from pre-1951 discharges of domestic sewage and industrial wastes.

2. In 1958 the Department of Health for Scotland served on the Board a copy of a Draft of the Firth of Forth (Prevention of Pollution) (Tidal Waters) Order, 1958, prepared by the Secretary of State for Scotland, applying the provisions of the Rivers (Prevention of Pollution) (Scotland) Act, 1951 to the tidal waters of the Firth of Forth. It was appreciated immediately by the Board that there would be opposition to the proposed Order which would result in a Public Inquiry, and it was, therefore, agreed that a limited survey would be made by the Board covering an area of approximately half of the estuary between a line drawn due North of the mouth of the River Almond and one drawn due North of Gullane Point. The known sources of pollution in 1958, were reported by Covill (1958 and 1971-2). [1 to 4]

(1) Discharges of crude sewage and industrial effluent from the City of Edinburgh = 241×10^3 m^3/day (53 mgd).
(2) Discharges of crude sewage and industrial effluent from the Midlothian County Council's Esk Valley regional trunk sewer = 68×10^3 m^3/day (15 mgd).
(3) Discharges of crude sewage and industrial effluent from Musselburgh, Prestonpans, Tranent, Port Seton, Cockenzie, Longniddry, Aberlady and Gullane = 19×10^3 m^3/day (4.25 mgd).
(4) Pit, mine and washery waters, drainage from bings from the following National Coal Board Establishments—

    Preston Links Colliery;
    Preston Grange Colliery;
    Bankton and Glencairn Collieries
    = 15×10^3 m^3/day (3.35 mgd).

9.2

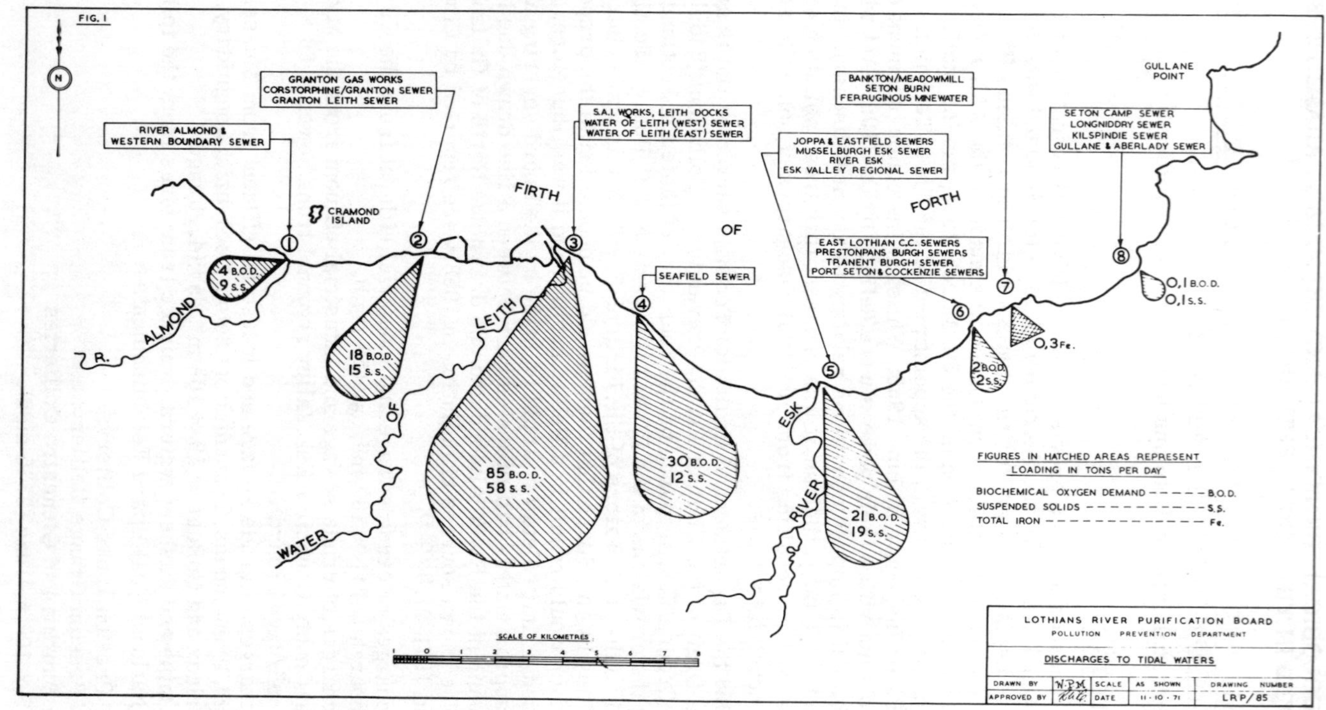

FIG. 1
N
RIVER ALMOND & WESTERN BOUNDARY SEWER
GRANTON GAS WORKS
CORSTORPHINE/GRANTON SEWER
GRANTON LEITH SEWER
S.A.I. WORKS, LEITH DOCKS
WATER OF LEITH (WEST) SEWER
WATER OF LEITH (EAST) SEWER
SEAFIELD SEWER
JOPPA & EASTFIELD SEWERS
MUSSELBURGH ESK SEWER
RIVER ESK
ESK VALLEY REGIONAL SEWER
EAST LOTHIAN C.C. SEWERS
PRESTONPANS BURGH SEWERS
TRANENT BURGH SEWER
PORT SETON & COCKENZIE SEWERS
BANKTON/MEADOWMILL
SETON BURN
FERRUGINOUS MINEWATER
SETON CAMP SEWER
LONGNIDDRY SEWER
KILSPINDIE SEWER
GULLANE & ABERLADY SEWER
GULLANE POINT
CRAMOND ISLAND
R. ALMOND
WATER OF LEITH
ESK RIVER
FIRTH OF FORTH
4 B.O.D. 9 S.S.
18 B.O.D. 15 S.S.
85 B.O.D. 58 S.S.
30 B.O.D. 12 S.S.
21 B.O.D. 19 S.S.
2 B.O.D. 2 S.S.
0,3 Fe.
0,1 B.O.D. 0,1 S.S.
FIGURES IN HATCHED AREAS REPRESENT LOADING IN TONS PER DAY
BIOCHEMICAL OXYGEN DEMAND — B.O.D.
SUSPENDED SOLIDS — S.S.
TOTAL IRON — Fe.
SCALE OF KILOMETRES
1 0 1 2 3 4 5 6 7 8
LOTHIANS RIVER PURIFICATION BOARD
POLLUTION PREVENTION DEPARTMENT
DISCHARGES TO TIDAL WATERS
DRAWN BY W.P.M.
APPROVED BY
SCALE AS SHOWN
DATE 11·10·71
DRAWING NUMBER LRP/85

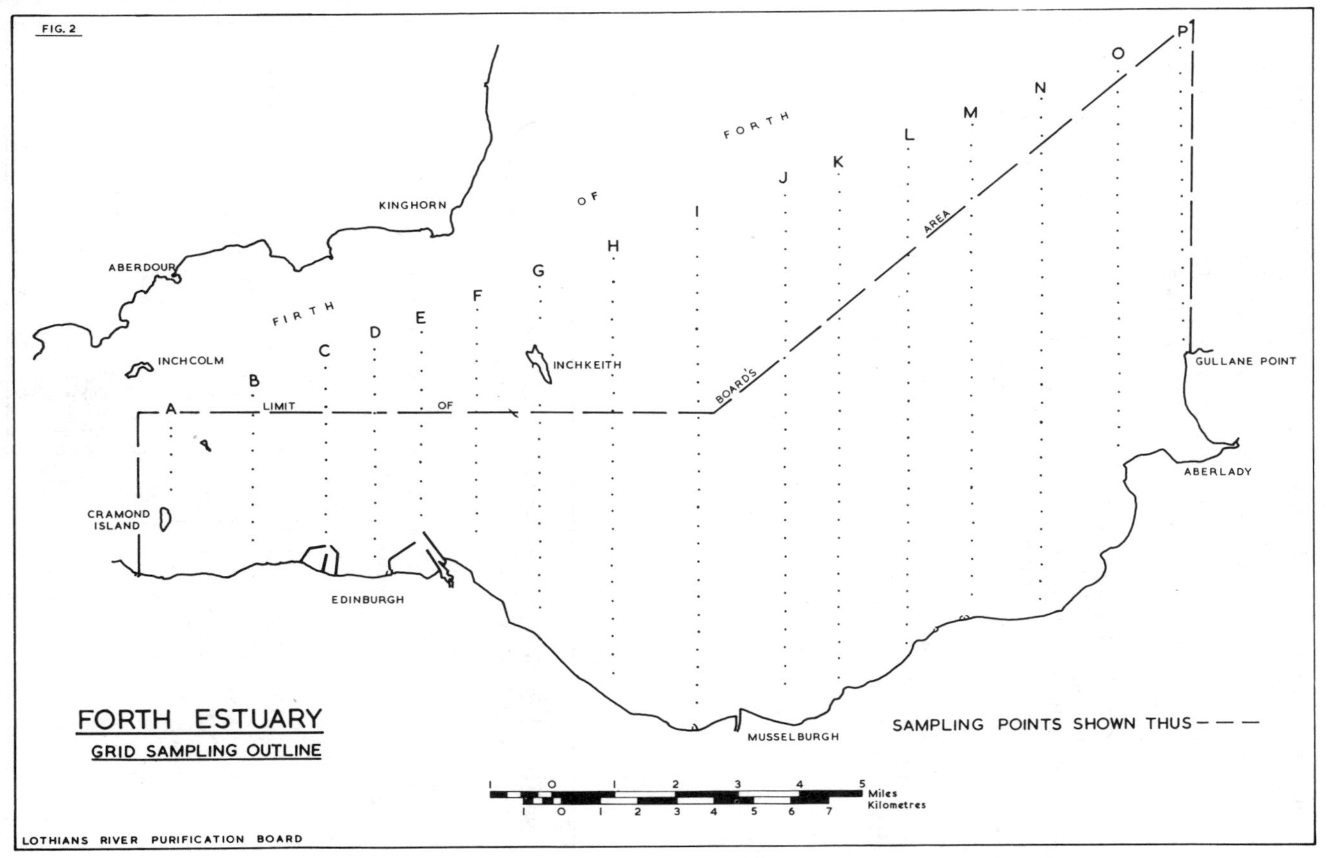
FIG. 2
FORTH ESTUARY
GRID SAMPLING OUTLINE
SAMPLING POINTS SHOWN THUS
LOTHIANS RIVER PURIFICATION BOARD
GULLANE POINT
ABERLADY
MUSSELBURGH
EDINBURGH
CRAMOND ISLAND
INCHCOLM
ABERDOUR
KINGHORN
INCHKEITH
FIRTH OF FORTH
LIMIT OF BOARD'S AREA
A
B
C
D
E
F
G
H
I
J
K
L
M
N
O
P
Miles
Kilometres

(5) Industrial Discharges from the Gas Board, Scottish Agricultural Industries, South of Scotland Electricity Board, Inveresk Paper Mills and Fowler's Brewery = 20×10^3 m^3/day (4.5 mgd).

3. The effect of the above discharges in terms of the tonnage of suspended solids and biochemical oxygen demand loadings per day is illustrated in Figure 1.

4. The object of the Survey was to ascertain the effect of such discharges on the estuarial waters, with particular reference to salinity concentration, distribution of the dissolved oxygen and the bacterial concentrations in terms of coliforms and *E. coli*. Biological assays at that time were confined to a general appreciation of the biota found in the samples.

5. A sampling grid was established, (Fig. 2) and 4000 samples for Bacteriological, Biological and Chemical examinations were taken during the period August-September, 1958. Table 1 classifies the samples.

Table 1

Bacteriological	Biological	Chemical
540	1220	2240

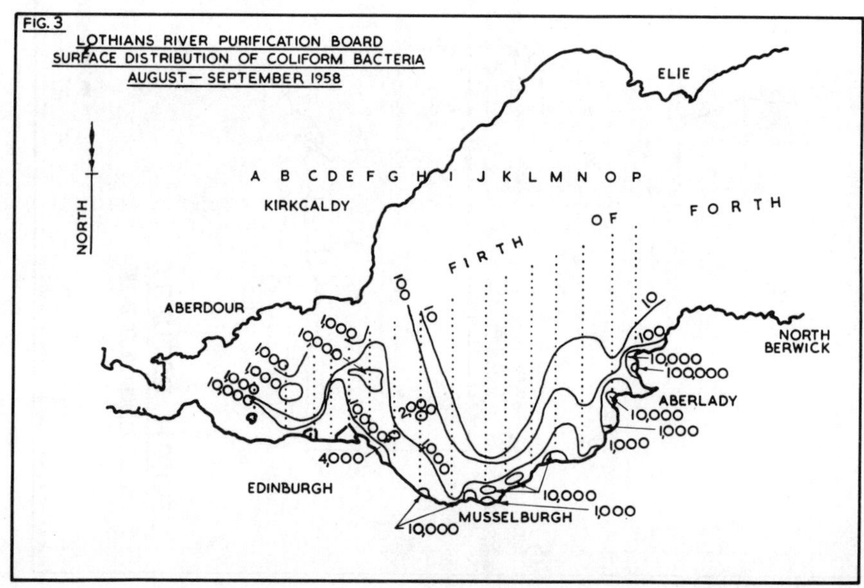

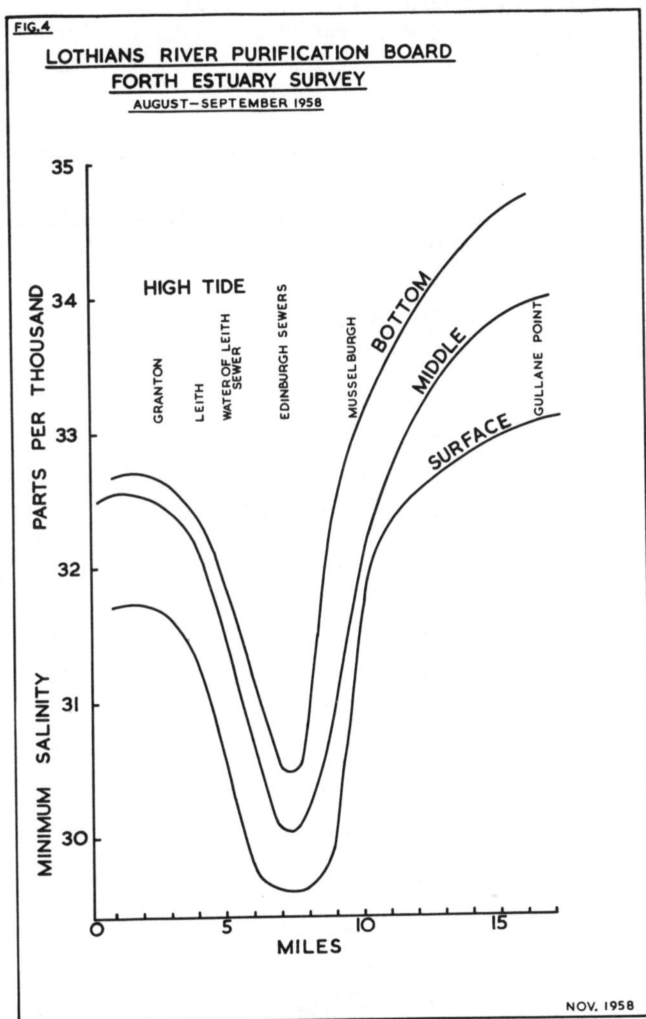

6. Data obtained from the survey may be summarised as follows:-

(1) Bacteriological

Bacteriological examinations in respect of the probable numbers of coliforms and *Escherichia coli.* were carried out on all samples. Figure 3 gives the surface distribution of coliforms and it was correct to say that *E. coli.* were found to be present in 100 ml in 97% of the samples. Figure 3 emphasises that, at that time, there was not only gross pollution in the Edinburgh area, but also in parts of the east coastal regions of the estuary.

(2) Biological

(i) Samples collected in the vicinity of Gullane Point contained typical marine flora and fauna.

(ii) Samples collected in the Cramond-Musselburgh area substantiated the heavy pollution in the entire area.

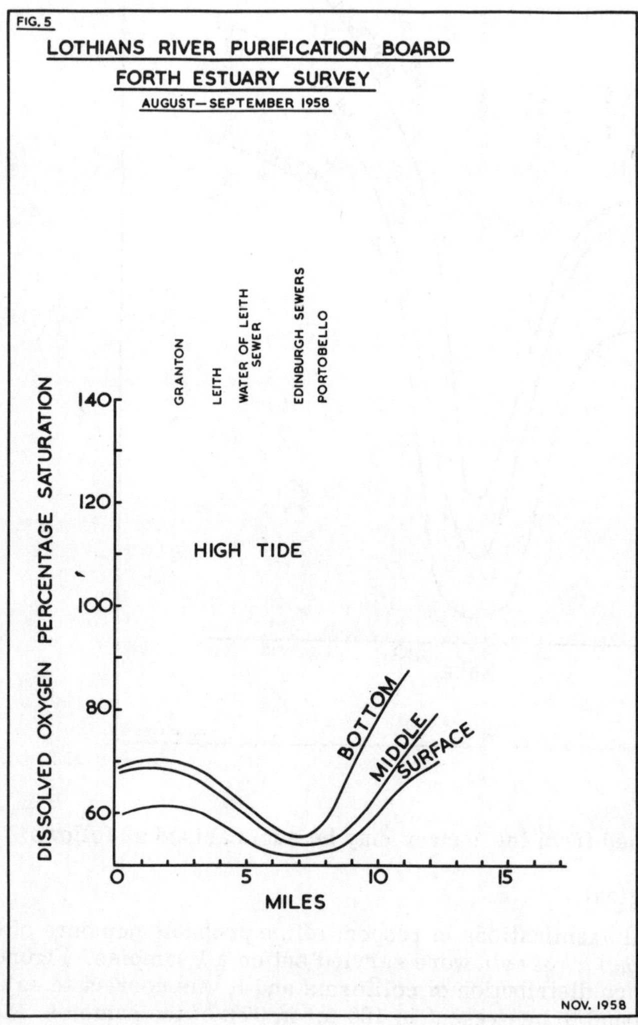

(3) Chemical

The salinity and dissolved oxygen sag curves, (Figs.4 and 5) stressed the concentration of pollution in the Granton, Leith, Seafield and Musselburgh areas. The lowest dissolved oxygen percentage saturation value at that time, was 50.

(4) Shore Survey

In conjunction with the 'grid survey' the shore was inspected for the entire length and samples were taken for bacteriological, biological and chemical examinations. The bacteriological results were interesting. The probable Coliform counts varied from as low as 150 coliforms/100 ml of sample to 21×10^6 coliforms/100 ml of sample. This wide variation is accounted for by the position of the points at which the samples were taken relative to a discharge of sewage polluted liquor to the beach. However, in view of the public health hazard in sea water so grossly polluted the necessity for the adequate purification of sewage before discharge to tidal waters was emphasised.

(5) General

The weather during the survey was fine for practically the entire period—25th August to 19th September 1958—as indicated by the following Table.

Table 2 Rainfall—Edinburgh Area

25th August, 1958-19th September, 1958

The values have been averaged from readings taken at the following meteorological stations:-

(i) Davidsons Mains;
(ii) Royal Botanical Gardens;
(iii) Portobello.

	Date	Inches of Rain
August	25th	Nil
	26th	0.36
	27th	0.18
	28th	0.02
	29th	Nil
	30th	Nil
	31st	Nil
September	1st	Nil
	2nd	Nil
	3rd	Nil
	4th	Nil
	5th	0.01

Table 2 Rainfall—Edinburgh Area

25th August, 1958-19th September, 1958 (cont:)

	Date	Inches of Rain
September	6th	0.28
	7th	0.40
	8th	0.03
	9th	Nil
	10th	Nil
	11th	Nil
	12th	Nil
	13th	Nil
	14th	0.03
	15th	Nil
	16th	Nil
	17th	Nil
	18th	0.19
	19th	0.04

Total throughout period = 1.54 inches

On two occasions only, could the sea be described as 'rough' for the 17m, 18-tonne diesel launch 'Jason' employed on the survey.

(6) Survey 1958—Conclusions

The results stressed that under the conditions pertaining during the survey—

(i) The pollution of the inshore waters, beaches and mussel beds resulting from the discharge of crude sewage was emphasised from the bacteriological, biological and chemical data produced. The main areas of pollution being Granton, Leith, Seafield, Portobello, Musselburgh and Aberlady Bay.

(ii) The dissolved oxygen concentration in the estuarial waters was considerably reduced by the discharge of crude sewages and industrial wastes. Fifty to fifty-five per cent saturation values were evident in the Edinburgh-Musselburgh area.

(iii) The pollution of the shoreline by discharge of pit and washery waters, together with the erosion of bings was pronounced in the Joppa, Fisherrow, Musselburgh, Prestonpans and Cockenzie regions.

(iv) Industrial discharges in the Edinburgh area and also to the River Esk aggravated pollution in the corresponding coastal areas.

(v) Further important sources of pollution particularly of a bacteriological nature are associated with the operation of storm sewage overflows. The overflows, especially in those areas in which the sewerage system has been in use for many years, and this is common in the area under consideration, frequently come into operation, since the sewers are often almost surcharged even in periods of dry weather and hence the slightest rainfall results in a discharge to the river or direct to the estuary. The quality of such discharges can, following a period of dry weather, by many times stronger than an average crude sewage and so the bacterial content can be proportionately greater than usual. Another form of bacterial pollution to an estuary, via a river, can result from the discharge of what are considered to be satisfactory sewage effluents from the chemical standpoint.

(vi) The above circumstances indicated the urgent need for the Board to control the tidal waters in order that the pollution might be reduced by—

(a) The primary treatment of all crude sewage and the treatment of the resultant sludge;
(b) The reconstruction or addition to major sewerage systems in order to give the required capacities for the increased drainage populations and so ensure that the storm sewage overflows operate only in accordance with specified conditions;
(c) The satisfactory treatment of industrial effluents;
(d) The adequate treatment of pit and washery wastes.

SURVEYS—1966 TO DATE

7. Since the 1958 Survey, it will be understood that the volume of discharges to the estuary has increased, and in order to ascertain the effect on the tidal waters, work has continued, and bacteriological, biological and chemical surveys have been made relative to the inshore waters and the intertidal zone on the shore. This work has been supplemented by hydrographical and hydrological surveys in the area.

Sampling Procedure

8. (1) Grid Samples

Following on the results of the 1958 Survey, from which it was evident that the greatest pollution, as anticipated, was in the inshore waters, it was decided to concentrate the sampling on the six inshore stations on each line of the grid, and also to give particular attention to those grid lines on or adjacent to the major sewage outfalls, namely Cramond, Caroline Park, Granton, Leith, Seafield, Joppa, Eastfield, Musselburgh and the Esk Valley Sewer. The survey vessel was, therefore, used to take sub-surface, middle and bottom depth samples.

(2) Sewage Slick Samples

Sub-surface samples were also taken along the line of the major sewage slicks in order to ascertain the 'fall off' in coliform counts, resulting from the mixing of the sewage with the estuary water.

(3) Shore Samples

These samples were taken in line with various grid stations in waters very close to the shore and seawards to wadeable depths only.

(4) Beach-Core Samples

Samples have been taken in the more polluted areas and examined for coliforms and salmonellae.

(5) Mussels *(Mytilus edulis)*

These have been sampled along the shore from Cramond to Gullane Point and examined for the presence of salmonellae.

Sampling Methods

9. (1) Sub-Surface Samples

Samples taken on the Grid or Sewage Slick were all grab ones using a sterilised medicine flat sample bottle.

(2) Middle and Bottom Depth Samples

The following equipment has been used—

- (i) Knudsen Reversing Bottle—2000 ml fitted with calibrated reversing thermometers.
- (ii) Nansen-Peterson Insulated Water Bottle.
- (iii) Ruttner Sampler.

All samples were transferred to a box partially filled with ice and delivered to the laboratory refrigerator as soon as practicable and in any case within 6 hours of sampling.

(3) Inshore Samples

For this purpose the sampler waded out to a depth of approximately one metre, then took samples from approximately 150 mm below the surface. The cap and paper bonnet of the sterilised sample bottle were removed without touching the mouth, the bottle submerged and filled with a steady forward sweep of the arm. The cap was quickly replaced, the sample labelled and stored in an ice-box.

(4) Beach-Core Samples

Further to the routine sampling of tidal waters for bacteriological and chemical examination, it was decided to commence a programme of core sampling in August 1970, at the same stations from which the inter-tidal zone water samples were taken. Between Cramond and Gullane Point, twenty stations were utilised for this purpose and these corresponded approximately to a landward extension of the lines used in defining the estuary grid sampling stations. Ten samples were taken each week for

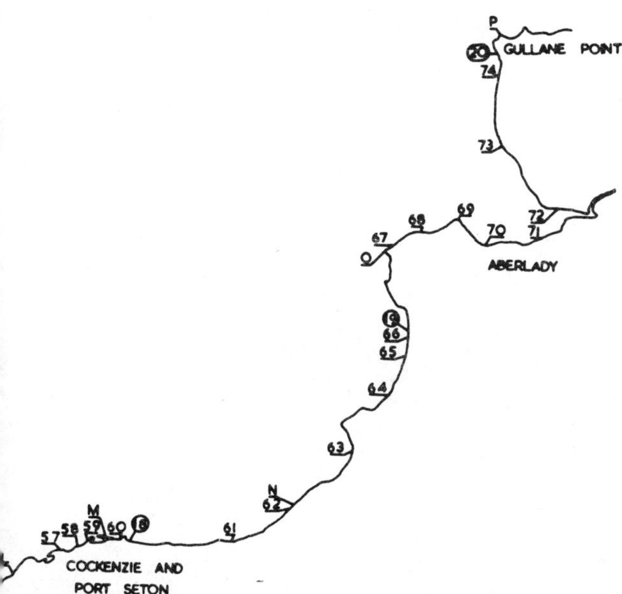
P
20
GULLANE POINT
74
73
72
71
70
69
68
67
O
ABERLADY
19
66
65
64
63
N
62
61
M
57
58
59
60
18
COCKENZIE AND
PORT SETON
TONPANS

bacteriological and chemical examination, and inshore water samples were obtained at the same time. Since it was intended to examine the samples for the presence of heavy metals, it was decided to intensify the number of sampling stations at Leith Docks and Seafield Bay where the main effect of the crude sewage discharges from three major outfalls was experienced.

The cores were obtained by using a 300 mm long stainless steel tube with a diameter of 30 mm. The tube was inserted into the sand and after withdrawal the core was released into a wide-mouthed sterilised glass jar. Some cores contained a high mud content whilst at other stations there was little mud or sand, and the cores were mainly of small stones, broken shell and coal particles. At the outset it was decided that the samples should be taken at half tide, one at each station. After some months of sampling it was observed that there was a degree of scatter in the results being obtained, particularly in respect of the heavy metal content. In order to investigate this aspect, the sampling method was changed. All ten samples were taken at the same station covering an area of two square yards for nine samples and taking the tenth sample ten yards northwards from the centre.

This method of sampling was pursued for several months and following consideration of the results of examination it appeared that the single sample method was representative of the area being examined and this was therefore reinstated, thus enabling ten stations to be sampled each week. Bacteriological samples were examined for Coliforms, *E. coli.*, *Clostridium perfringens* and Salmonellae employing standard procedures.

Sampling Stations

10. The accompanying Figure 6 shows the location of the sampling stations referred to in this section. The graphs (Figs. 7 to 37) include the results of three different sampling programmes in which different numbers of sampling stations were employed. The three different programmes were designated as follows:-

(1) Stations 1-74
(2) Stations A-O
(3) Stations 1-20

11. In order to relate the findings of all three surveys to a location common to each programme, the alphabetical key is used since the letters are, in fact, landward extensions of the estuary survey grid.

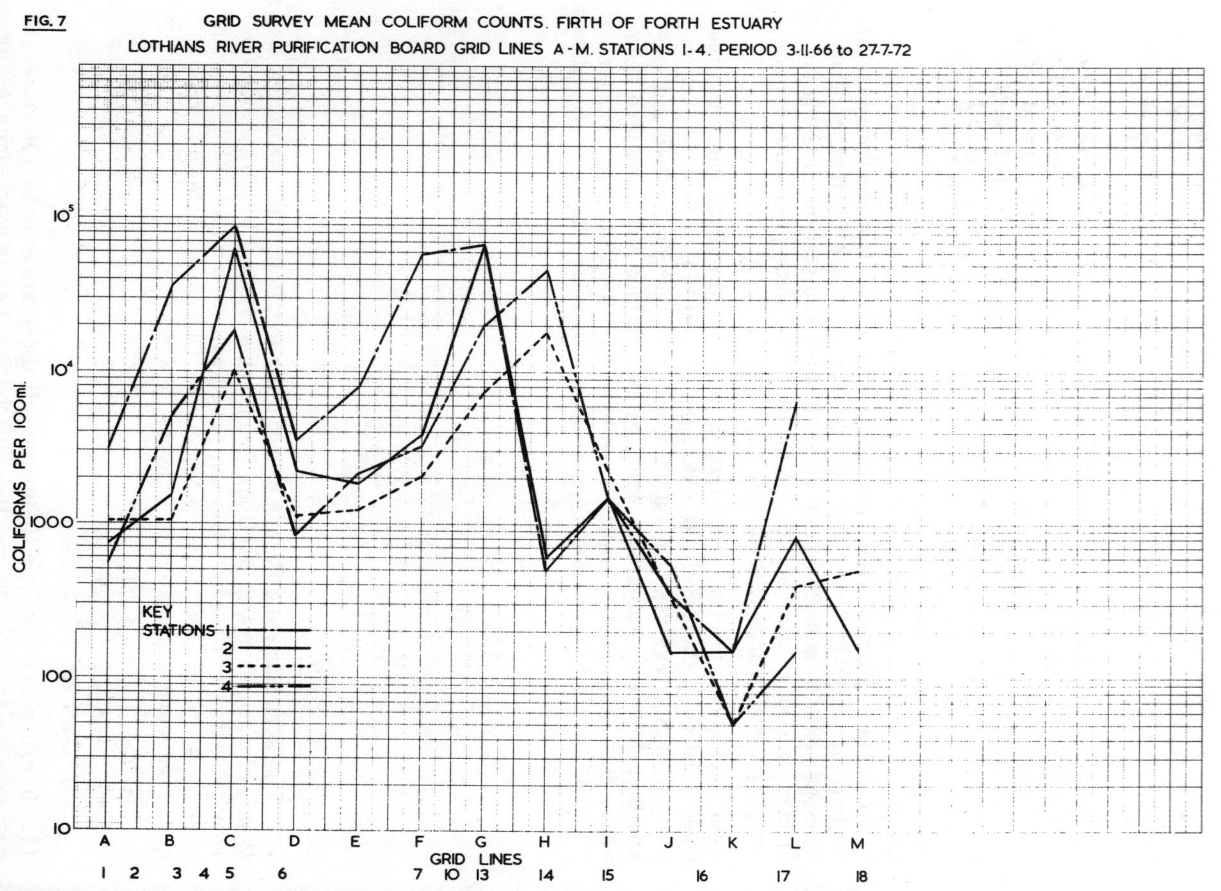
FIG. 7
GRID SURVEY MEAN COLIFORM COUNTS. FIRTH OF FORTH ESTUARY
LOTHIANS RIVER PURIFICATION BOARD GRID LINES A-M. STATIONS 1-4. PERIOD 3-11-66 to 27-7-72
COLIFORMS PER 100ml.
10^5
10^4
1000
100
10
A B C D E F G H I J K L M
1 2 3 4 5 6 7 10 13 14 15 16 17 18
GRID LINES
KEY
STATIONS 1
2
3
4

FIG. 8

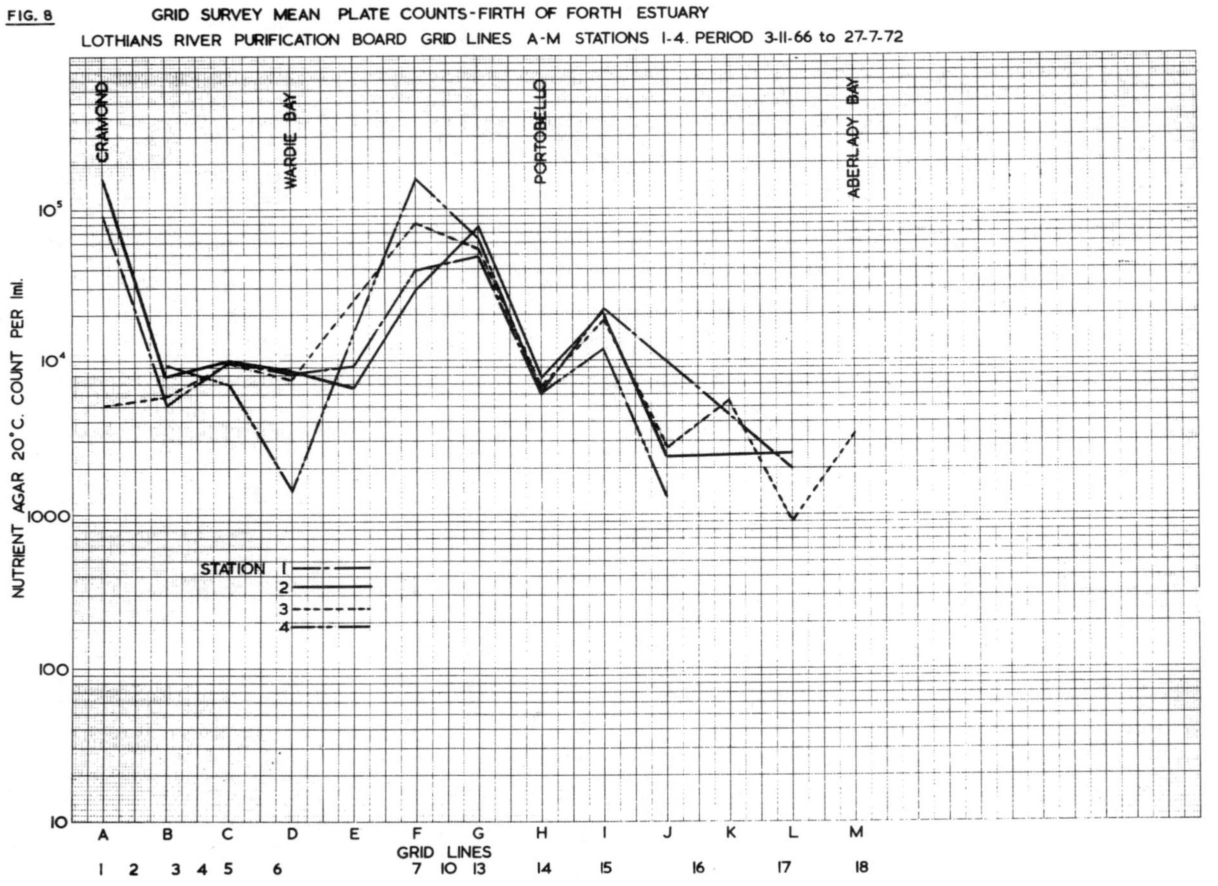
GRID SURVEY MEAN PLATE COUNTS-FIRTH OF FORTH ESTUARY
LOTHIANS RIVER PURIFICATION BOARD GRID LINES A-M STATIONS 1-4. PERIOD 3-11-66 to 27-7-72
CRAMOND
WARDIE BAY
PORTOBELLO
ABERLADY BAY
NUTRIENT AGAR 20°C. COUNT PER 1ml.
10⁵
10⁴
1000
100
10
STATION 1
2
3
4
A B C D E F G H I J K L M
GRID LINES
1 2 3 4 5 6 7 10 13 14 15 16 17 18

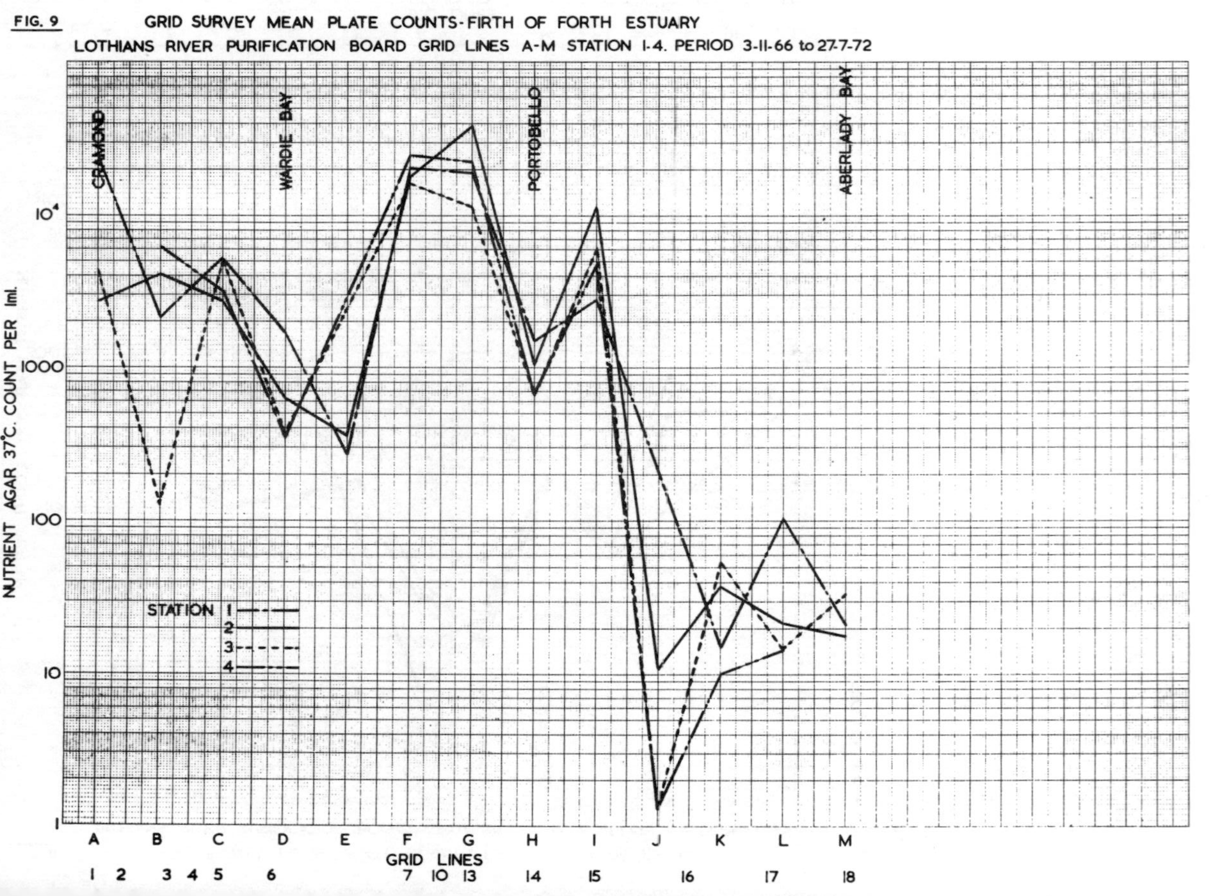
FIG. 9
GRID SURVEY MEAN PLATE COUNTS-FIRTH OF FORTH ESTUARY
LOTHIANS RIVER PURIFICATION BOARD GRID LINES A-M STATION 1-4. PERIOD 3-11-66 to 27-7-72
CRAMOND
WARDIE BAY
PORTOBELLO
ABERLADY BAY
NUTRIENT AGAR 37°C. COUNT PER 1ml.
10^4
1000
100
10
1
STATION 1
2
3
4
A B C D E F G H I J K L M
GRID LINES
1 2 3 4 5 6 7 10 13 14 15 16 17 18

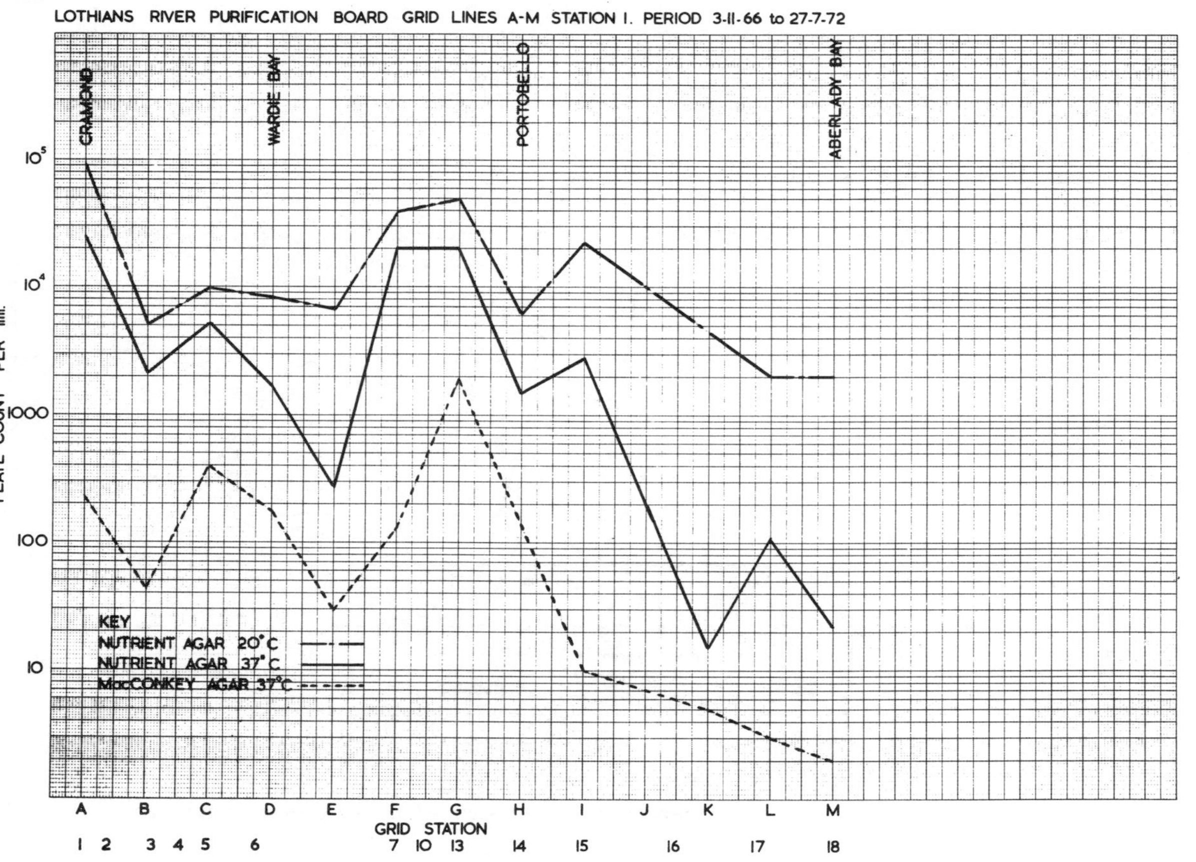
FIG. 10
GRID SURVEY MEAN PLATE COUNTS - FIRTH OF FORTH ESTUARY
LOTHIANS RIVER PURIFICATION BOARD GRID LINES A-M STATION I. PERIOD 3-11-66 to 27-7-72
CRAMOND
WARDIE BAY
PORTOBELLO
ABERLADY BAY
PLATE COUNT PER 1ml.
10^5
10^4
1000
100
10
KEY
NUTRIENT AGAR 20°C
NUTRIENT AGAR 37°C
MacCONKEY AGAR 37°C
A B C D E F G H I J K L M
GRID STATION
1 2 3 4 5 6 7 10 13 14 15 16 17 18

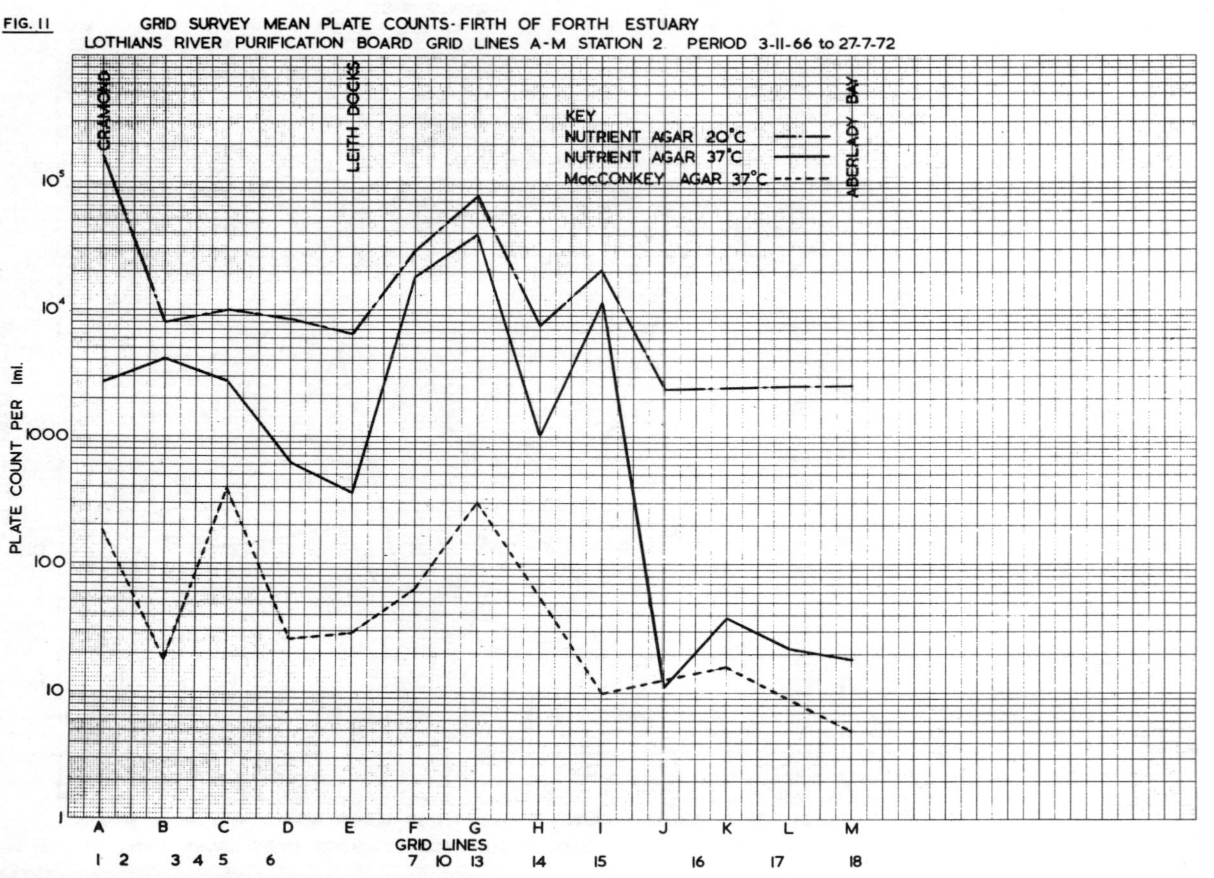
GRID SURVEY MEAN PLATE COUNTS - FIRTH OF FORTH ESTUARY
LOTHIANS RIVER PURIFICATION BOARD GRID LINES A-M STATION 2 PERIOD 3-11-66 to 27-7-72
CRAMOND
LEITH DOCKS
ABERLADY BAY
KEY
NUTRIENT AGAR 20°C
NUTRIENT AGAR 37°C
MacCONKEY AGAR 37°C
PLATE COUNT PER 1ml.
10^5
10^4
1000
100
10
1
A B C D E F G H I J K L M
GRID LINES
1 2 3 4 5 6 7 10 13 14 15 16 17 18

FIG. 11

FIG. 12

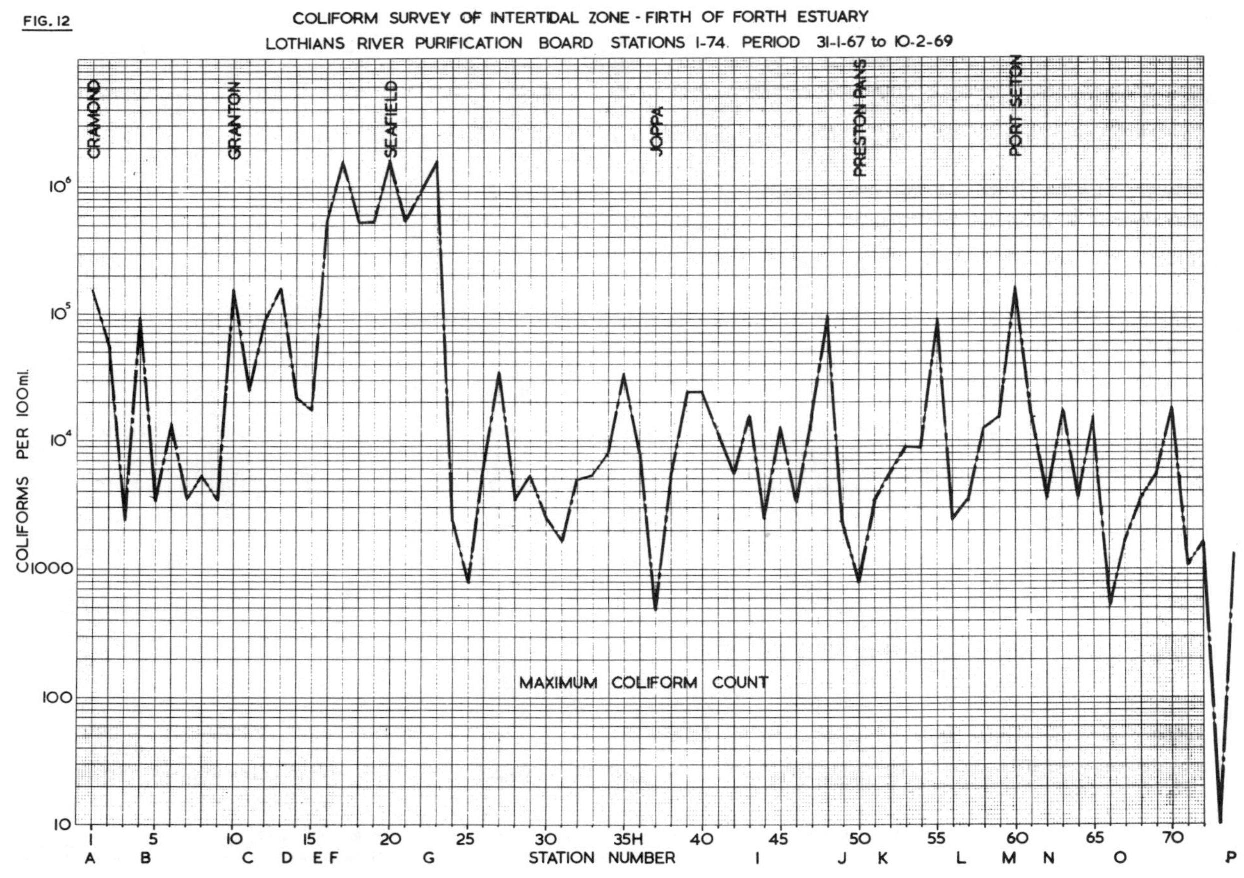
COLIFORM SURVEY OF INTERTIDAL ZONE - FIRTH OF FORTH ESTUARY
LOTHIANS RIVER PURIFICATION BOARD STATIONS 1-74. PERIOD 31-1-67 to 10-2-69
CRAMOND
GRANTON
SEAFIELD
JOPPA
PRESTON PANS
PORT SETON
COLIFORMS PER 100ml.
10^6
10^5
10^4
1000
100
10
MAXIMUM COLIFORM COUNT
1 5 10 15 20 25 30 35H 40 45 50 55 60 65 70
A B C D E F G STATION NUMBER I J K L M N O P

FIG. 13

COLIFORM SURVEY OF INTERTIDAL ZONE - FIRTH OF FORTH ESTUARY
LOTHIANS RIVER PURIFICATION BOARD STATIONS 1-74. PERIOD 31-1-67 to 10-2-69

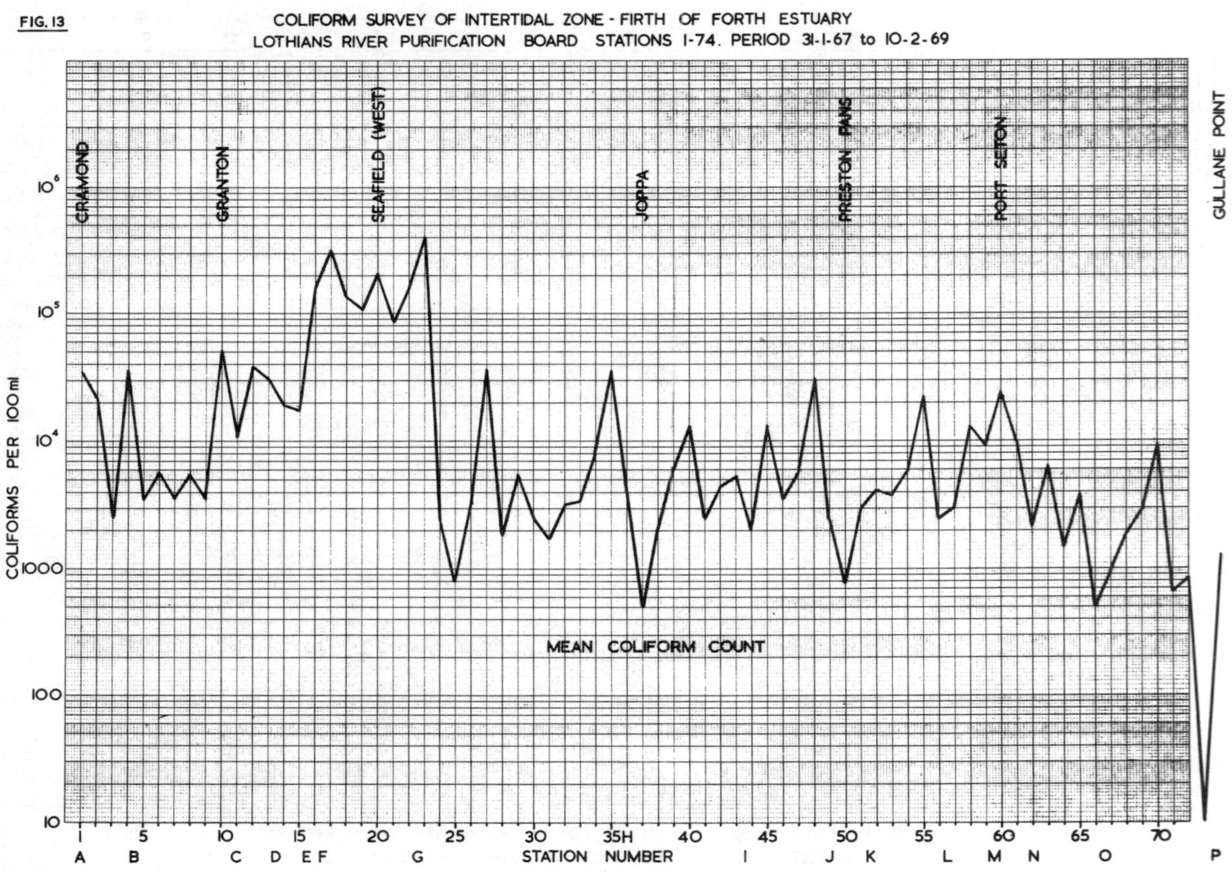

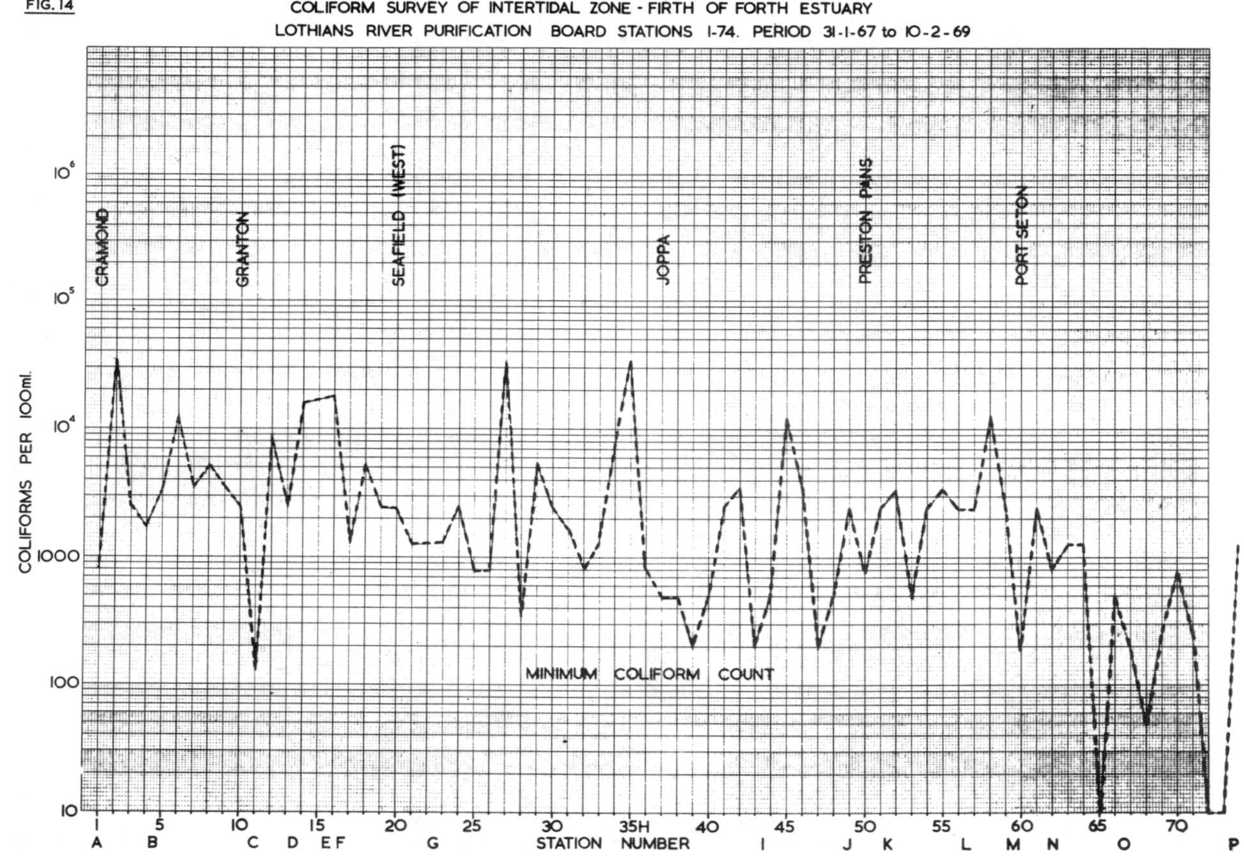

FIG. 14
COLIFORM SURVEY OF INTERTIDAL ZONE - FIRTH OF FORTH ESTUARY
LOTHIANS RIVER PURIFICATION BOARD STATIONS 1-74. PERIOD 31-1-67 to 10-2-69
CRAMOND
GRANTON
SEAFIELD (WEST)
JOPPA
PRESTON PANS
PORT SETON
COLIFORMS PER 100ml.
10^6
10^5
10^4
1000
100
10
MINIMUM COLIFORM COUNT
1 5 10 15 20 25 30 35H 40 45 50 55 60 65 70
A B C D E F G I J K L M N O P
STATION NUMBER

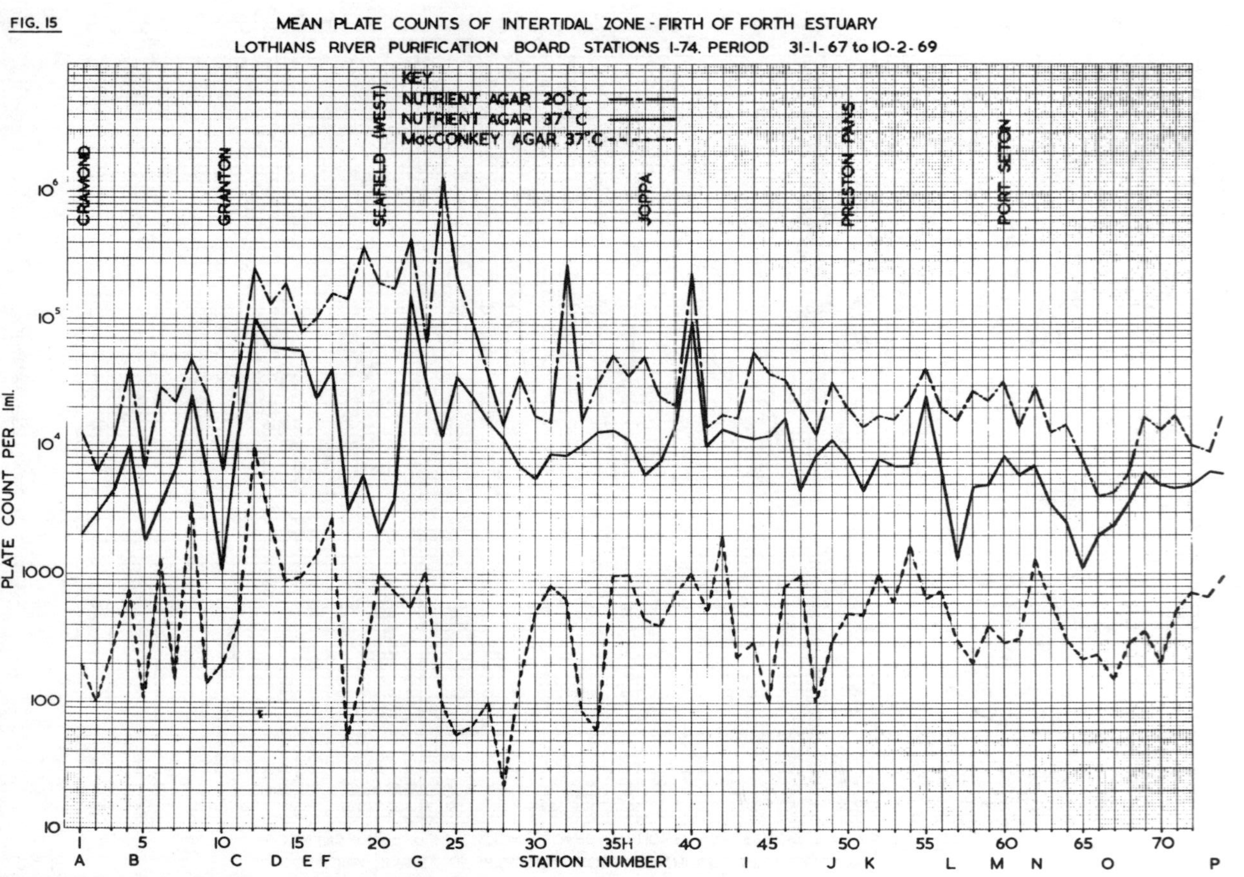
FIG. 15
MEAN PLATE COUNTS OF INTERTIDAL ZONE - FIRTH OF FORTH ESTUARY
LOTHIANS RIVER PURIFICATION BOARD STATIONS 1-74. PERIOD 31-1-67 to 10-2-69
KEY
NUTRIENT AGAR 20°C
NUTRIENT AGAR 37°C
MacCONKEY AGAR 37°C
CRAMOND
GRANTON
SEAFIELD (WEST)
JOPPA
PRESTON PANS
PORT SETON
PLATE COUNT PER 1ml.
10^6
10^5
10^4
1000
100
10
1 5 10 15 20 25 30 35H 40 45 50 55 60 65 70
A B C D E F G I J K L M N O P
STATION NUMBER

COLIFORM SURVEY OF INTERTIDAL ZONE EXTENSIONS OF GRID LINES - FIRTH OF FORTH ESTUARY

FIG. 16 LOTHIANS RIVER PURIFICATION BOARD STATIONS A-O. PERIOD 14-7-70 to 19-1-71

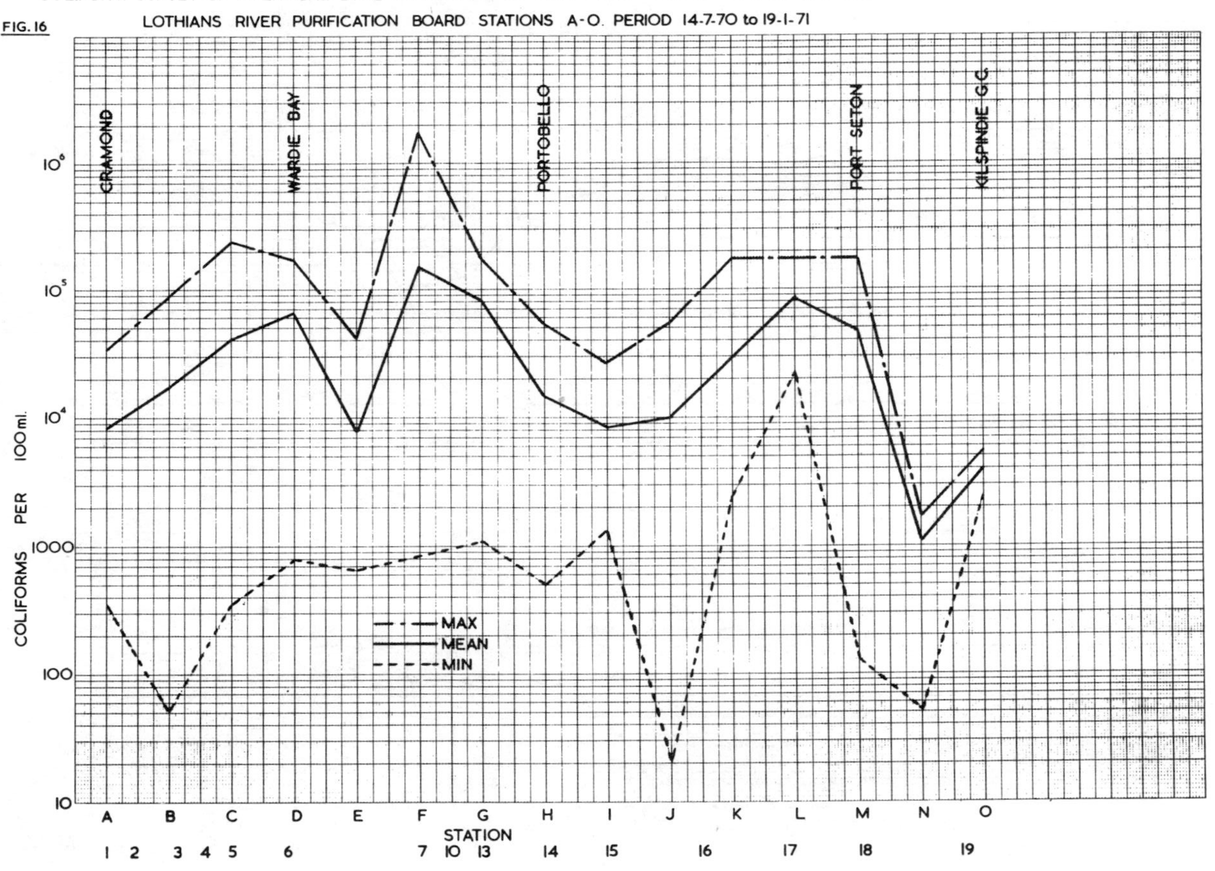

FIG. 17 MEAN PLATE COUNT OF INTERTIDAL ZONE EXTENSIONS OF GRID LINES - FIRTH OF FORTH ESTUARY
LOTHIANS RIVER PURIFICATION BOARD STATIONS A-O. PERIOD 14-7-70 to 19-1-71

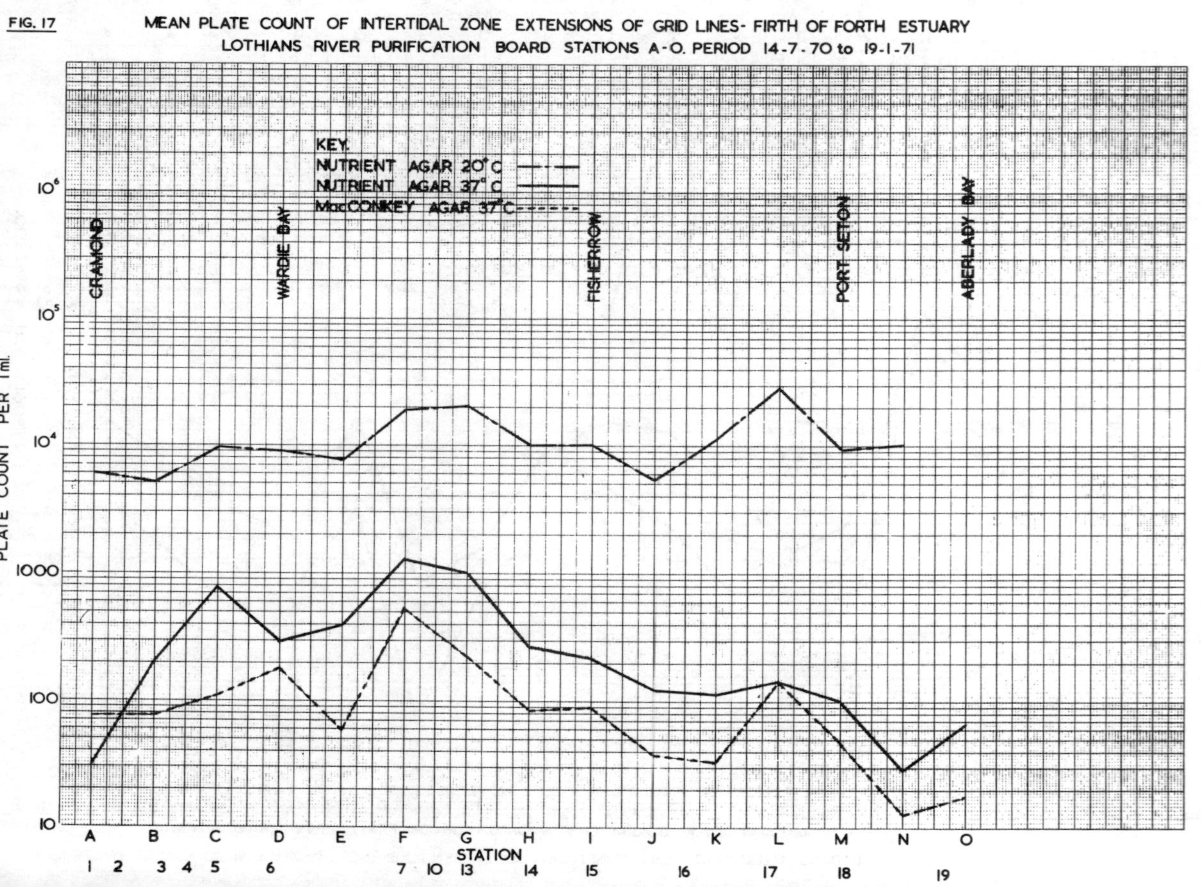

FIG. 18

COLIFORM SURVEY OF INTERTIDAL ZONE - FIRTH OF FORTH ESTUARY
LOTHIANS RIVER PURIFICATION BOARD STATIONS 1-20. PERIOD 10-1-72 to 10-7-72.

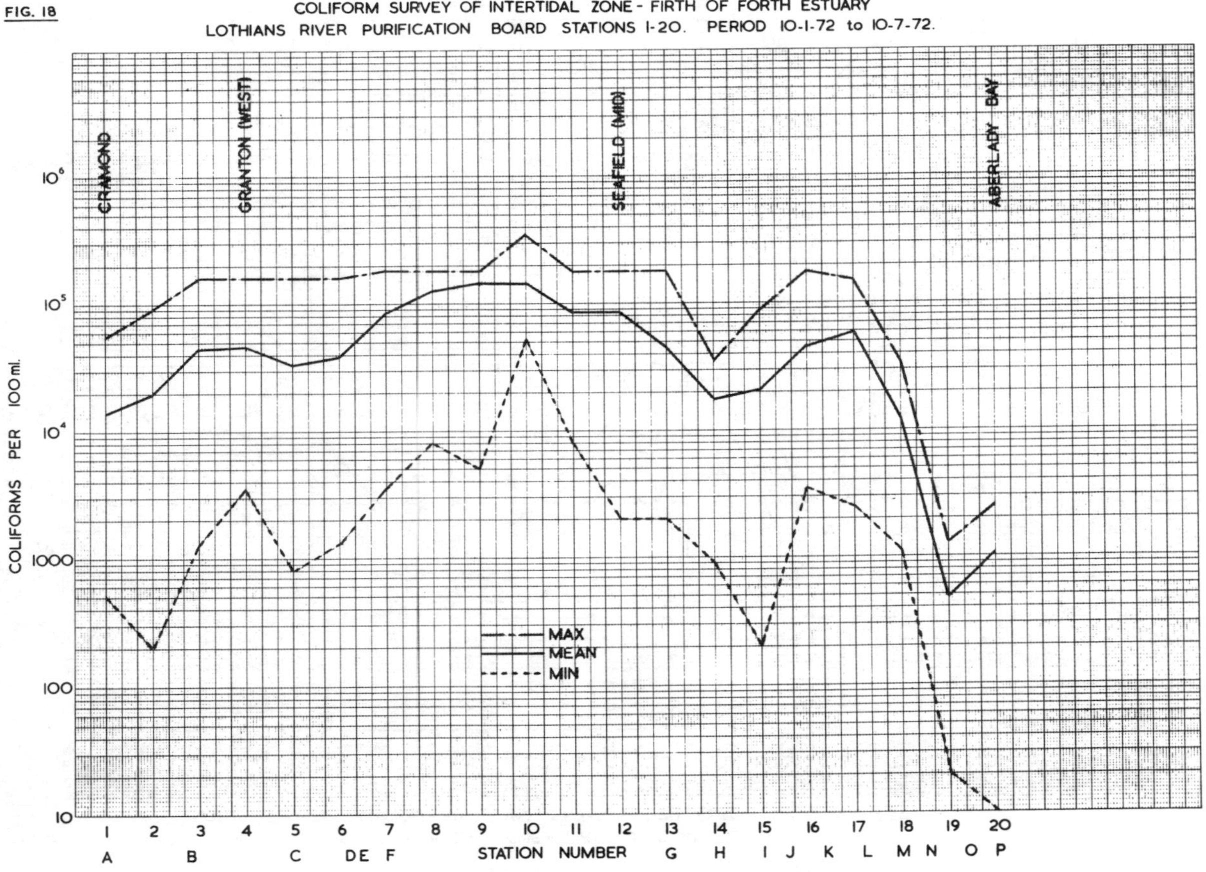

FIG. 19 E.COLI SURVEY OF INTERTIDAL ZONE - FIRTH OF FORTH ESTUARY
LOTHIANS RIVER PURIFICATION BOARD STATIONS 1-20. PERIOD 10-1-72 to 10-7-72

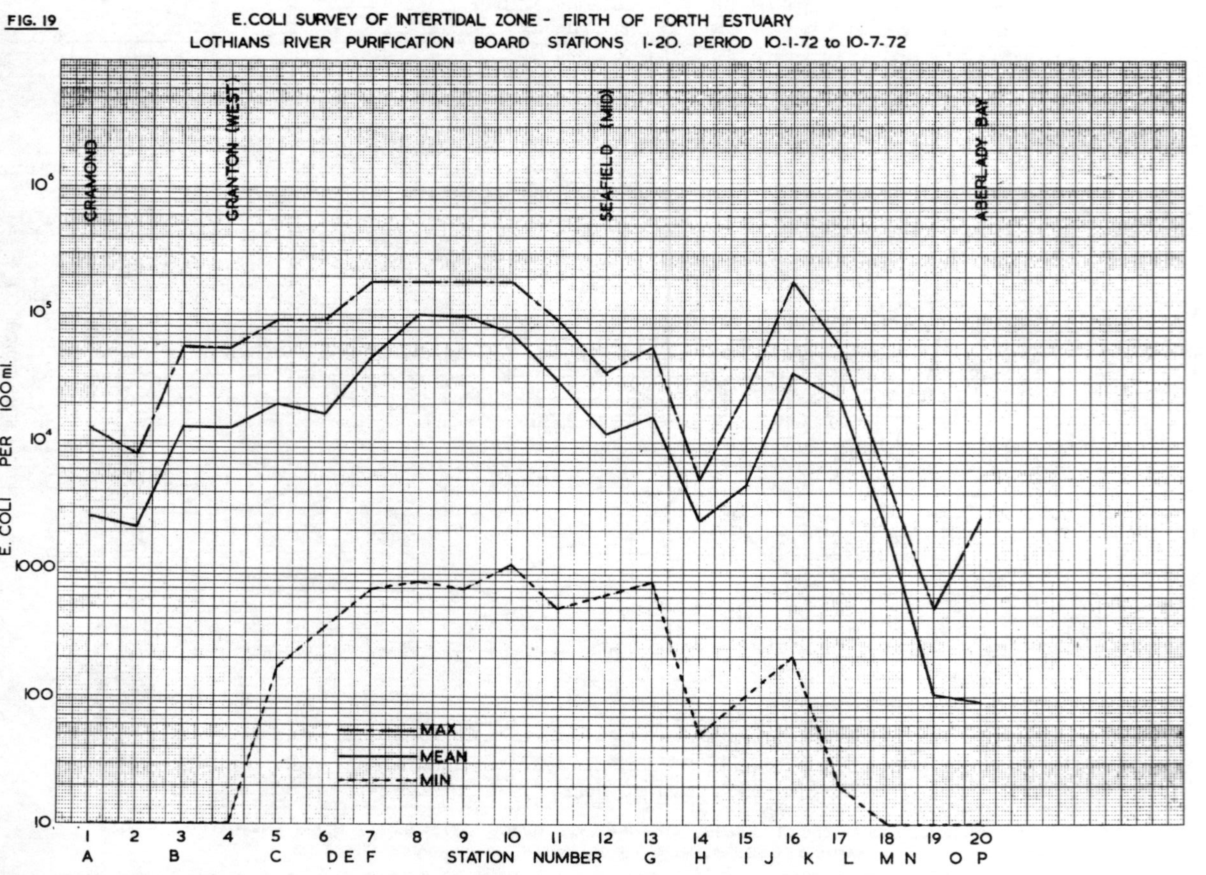

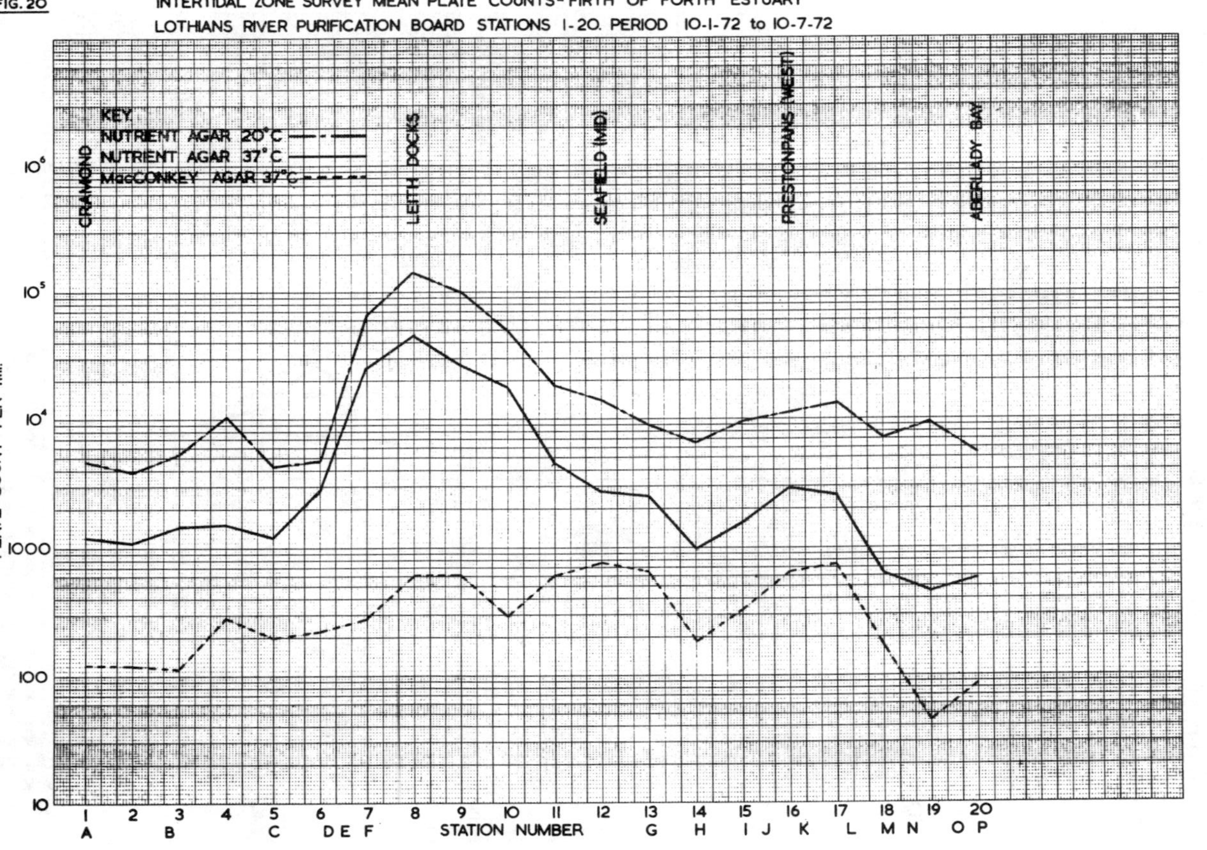
FIG. 20
INTERTIDAL ZONE SURVEY MEAN PLATE COUNTS - FIRTH OF FORTH ESTUARY
LOTHIANS RIVER PURIFICATION BOARD STATIONS 1-20. PERIOD 10-1-72 to 10-7-72
KEY
NUTRIENT AGAR 20°C
NUTRIENT AGAR 37°C
MacCONKEY AGAR 37°C
CRAMOND
LEITH DOCKS
SEAFIELD (MID)
PRESTONPANS (WEST)
ABERLADY BAY
PLATE COUNT PER 1ml.
10⁶
10⁵
10⁴
1000
100
10
1 2 3 4 5 6 7 8 9 10 11 12 13 14 15 16 17 18 19 20
A B C D E F G H I J K L M N O P
STATION NUMBER

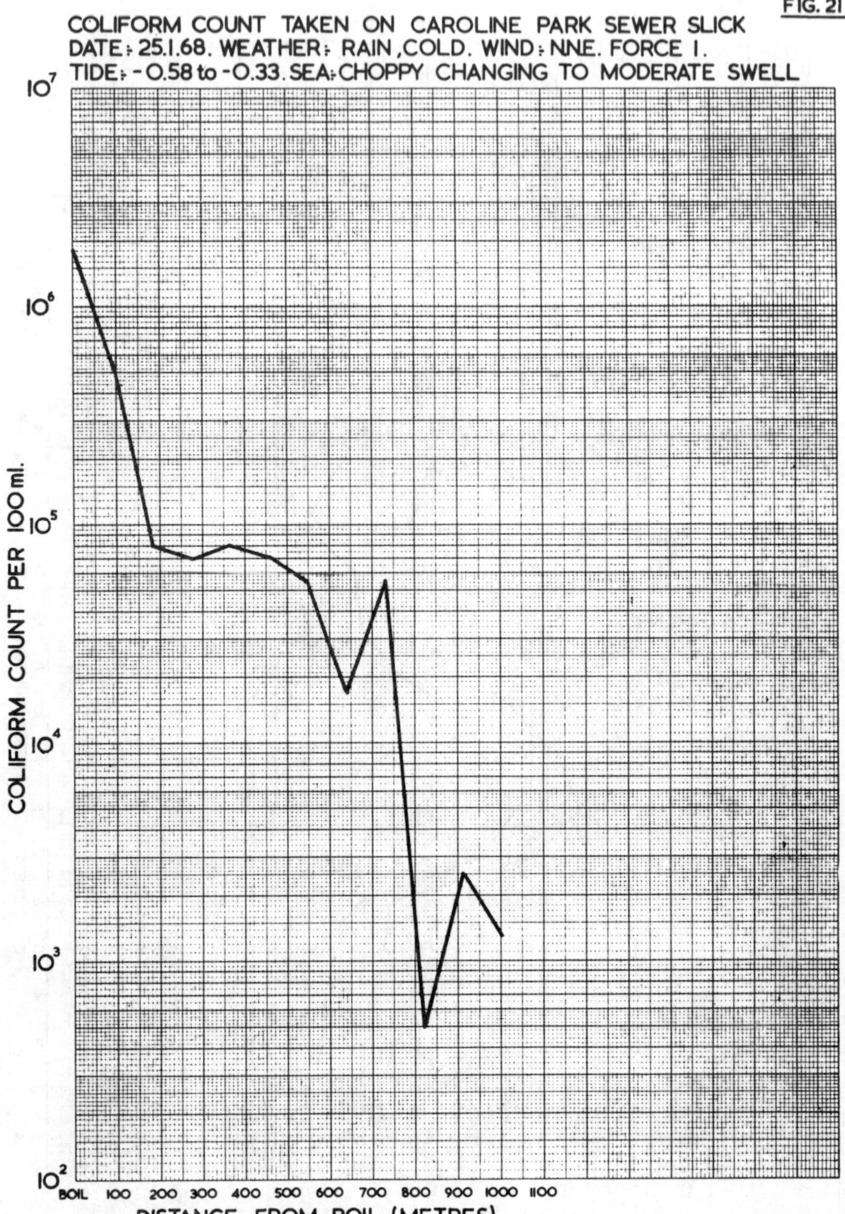
FIG. 21
COLIFORM COUNT TAKEN ON CAROLINE PARK SEWER SLICK
DATE:- 25.1.68. WEATHER:- RAIN, COLD. WIND:- N.N.E. FORCE 1.
TIDE:- -0.58 to -0.33. SEA:- CHOPPY CHANGING TO MODERATE SWELL
COLIFORM COUNT PER 100 ml.
10^7
10^6
10^5
10^4
10^3
10^2
BOIL
100
200
300
400
500
600
700
800
900
1000
1100
DISTANCE FROM BOIL (METRES)

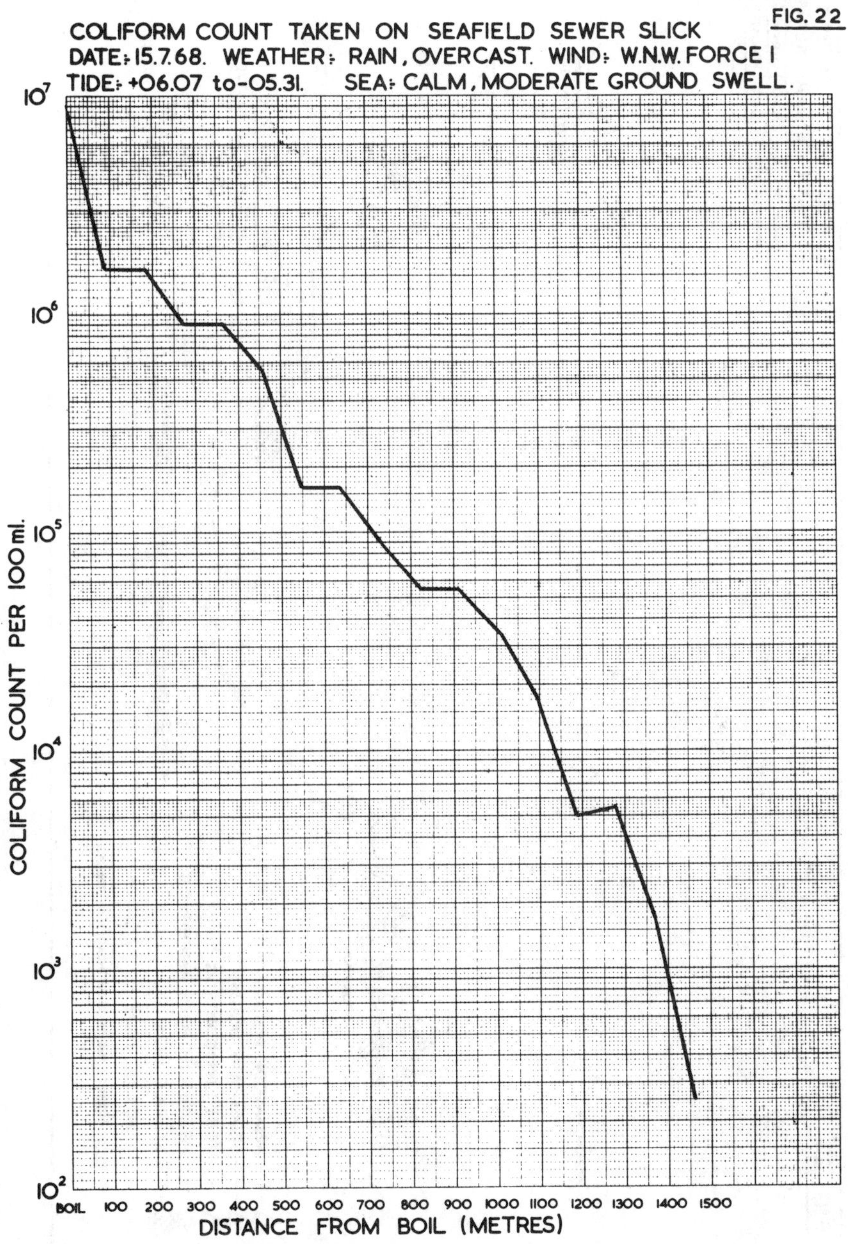
FIG. 22
COLIFORM COUNT TAKEN ON SEAFIELD SEWER SLICK
DATE: 15.7.68. WEATHER: RAIN, OVERCAST. WIND: W.N.W. FORCE 1
TIDE: +06.07 to -05.31. SEA: CALM, MODERATE GROUND SWELL.
10^7
10^6
10^5
10^4
10^3
10^2
COLIFORM COUNT PER 100 ml.
BOIL 100 200 300 400 500 600 700 800 900 1000 1100 1200 1300 1400 1500
DISTANCE FROM BOIL (METRES)

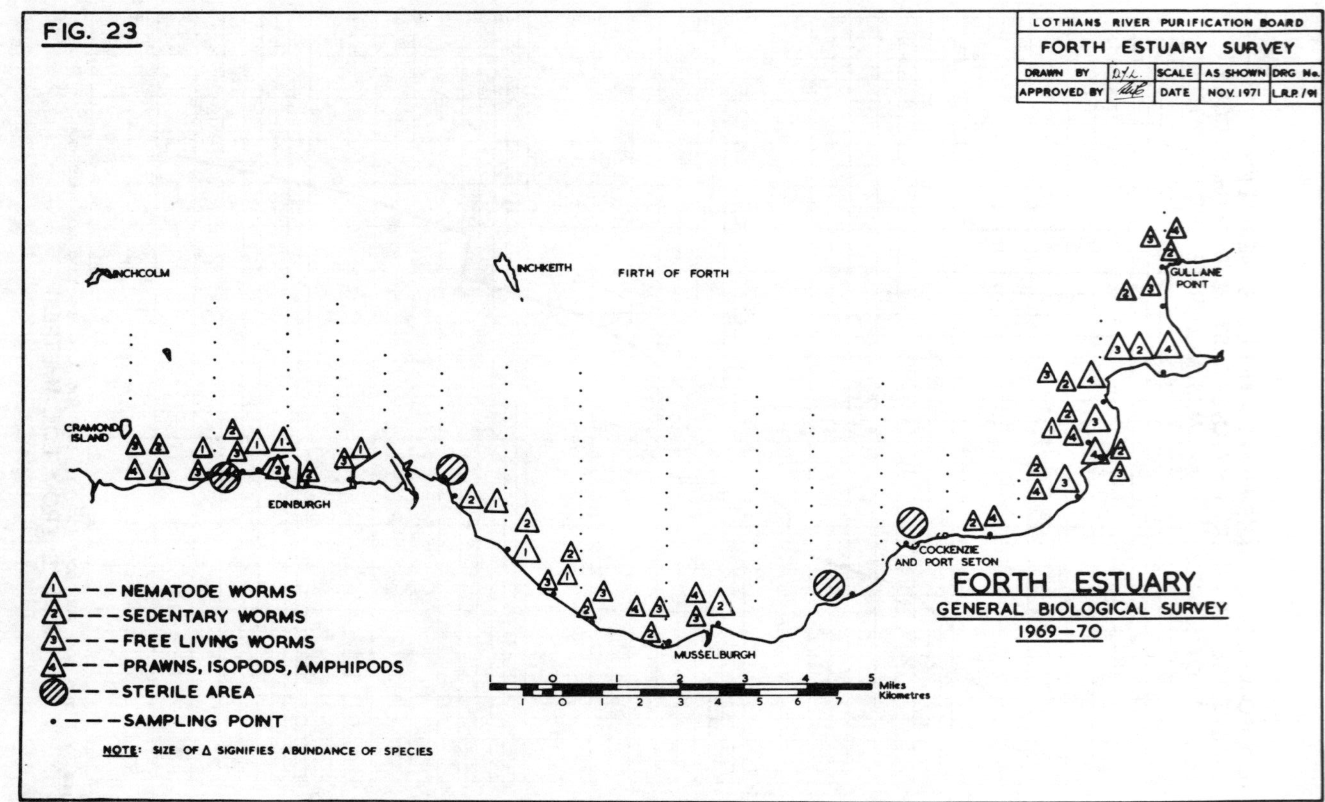
FIG. 23
LOTHIANS RIVER PURIFICATION BOARD
FORTH ESTUARY SURVEY
DRAWN BY
SCALE
AS SHOWN
DRG No.
APPROVED BY
DATE
NOV. 1971
L.R.P. /91
INCHCOLM
INCHKEITH
FIRTH OF FORTH
CRAMOND ISLAND
EDINBURGH
MUSSELBURGH
COCKENZIE AND PORT SETON
GULLANE POINT
FORTH ESTUARY
GENERAL BIOLOGICAL SURVEY
1969—70
Miles
Kilometres
1 ---NEMATODE WORMS
2 ---SEDENTARY WORMS
3 ---FREE LIVING WORMS
4 ---PRAWNS, ISOPODS, AMPHIPODS
---STERILE AREA
• ---SAMPLING POINT
NOTE: SIZE OF Δ SIGNIFIES ABUNDANCE OF SPECIES

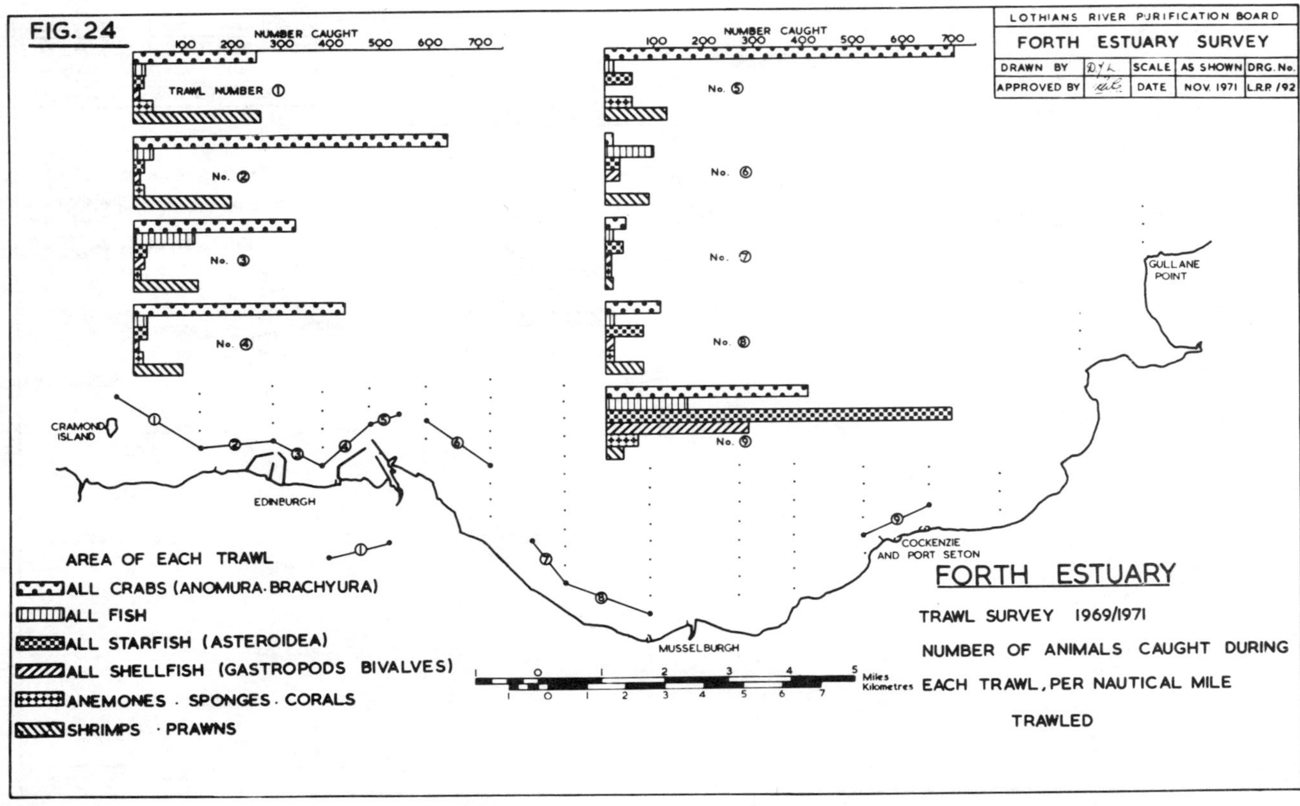

FIG. 24
LOTHIANS RIVER PURIFICATION BOARD
FORTH ESTUARY SURVEY
DRAWN BY
SCALE
AS SHOWN
DRG. No.
APPROVED BY
DATE
NOV. 1971
L.R.P. /92
NUMBER CAUGHT
100
200
300
400
500
600
700
TRAWL NUMBER ①
No. ②
No. ③
No. ④
No. ⑤
No. ⑥
No. ⑦
No. ⑧
No. ⑨
CRAMOND ISLAND
EDINBURGH
MUSSELBURGH
COCKENZIE AND PORT SETON
GULLANE POINT
Miles
Kilometres
AREA OF EACH TRAWL
ALL CRABS (ANOMURA. BRACHYURA)
ALL FISH
ALL STARFISH (ASTEROIDEA)
ALL SHELLFISH (GASTROPODS BIVALVES)
ANEMONES . SPONGES . CORALS
SHRIMPS . PRAWNS
FORTH ESTUARY
TRAWL SURVEY 1969/1971
NUMBER OF ANIMALS CAUGHT DURING EACH TRAWL, PER NAUTICAL MILE TRAWLED

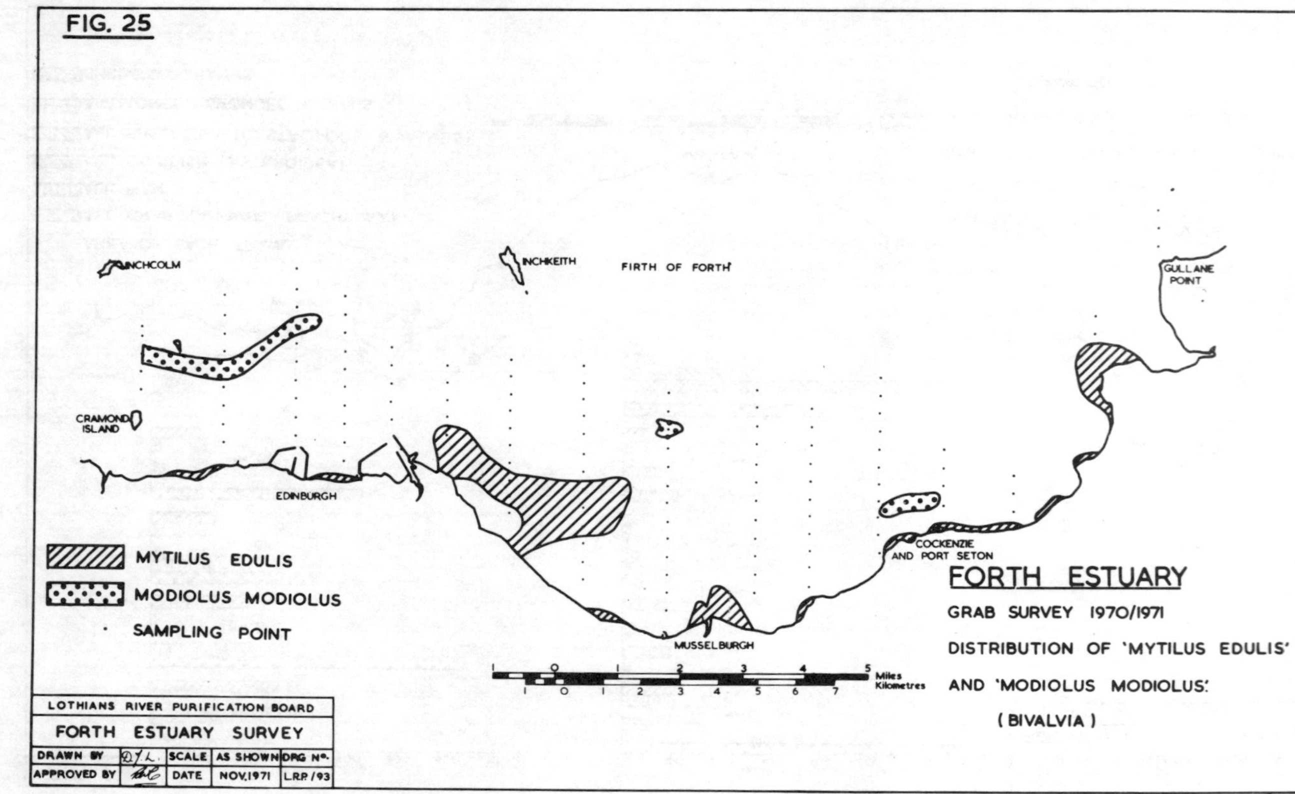

FIG. 25
INCHCOLM
INCHKEITH
FIRTH OF FORTH
GULLANE POINT
CRAMOND ISLAND
EDINBURGH
COCKENZIE AND PORT SETON
MUSSELBURGH
MYTILUS EDULIS
MODIOLUS MODIOLUS
SAMPLING POINT
FORTH ESTUARY
GRAB SURVEY 1970/1971
DISTRIBUTION OF 'MYTILUS EDULIS'
AND 'MODIOLUS MODIOLUS'
(BIVALVIA)
Miles
Kilometres
LOTHIANS RIVER PURIFICATION BOARD
FORTH ESTUARY SURVEY
DRAWN BY
SCALE AS SHOWN
DRG No.
APPROVED BY
DATE NOV.1971
L.R.P./93

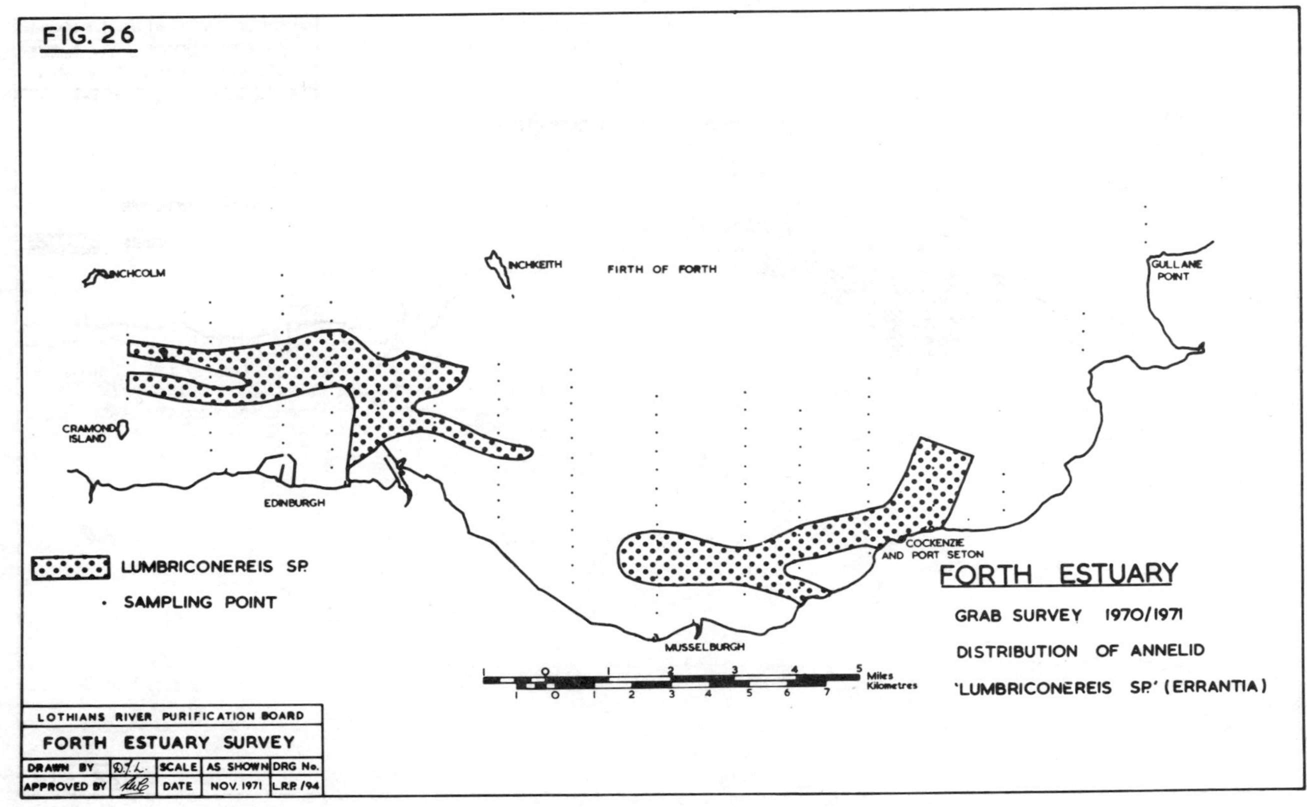
FIG. 26
INCHCOLM
INCHKEITH
FIRTH OF FORTH
GULLANE POINT
CRAMOND ISLAND
EDINBURGH
COCKENZIE AND PORT SETON
MUSSELBURGH
LUMBRICONEREIS SP.
· SAMPLING POINT
Miles
Kilometres
FORTH ESTUARY
GRAB SURVEY 1970/1971
DISTRIBUTION OF ANNELID
'LUMBRICONEREIS SP.' (ERRANTIA)
LOTHIANS RIVER PURIFICATION BOARD
FORTH ESTUARY SURVEY
DRAWN BY
SCALE AS SHOWN
DRG No.
APPROVED BY
DATE NOV. 1971
L.R.P./94

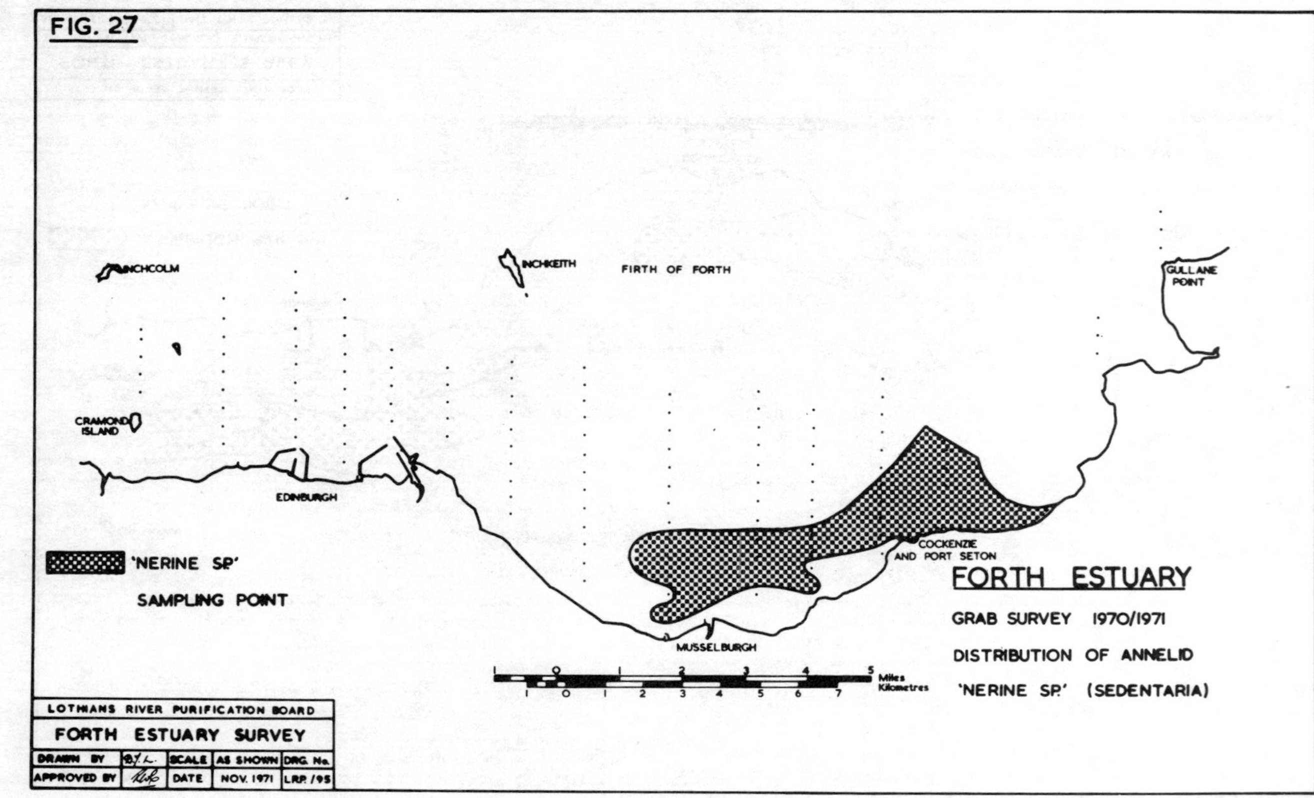
FIG. 27
INCHCOLM
INCHKEITH
FIRTH OF FORTH
GULLANE POINT
CRAMOND ISLAND
EDINBURGH
COCKENZIE AND PORT SETON
MUSSELBURGH
'NERINE SP.'
SAMPLING POINT
FORTH ESTUARY
GRAB SURVEY 1970/1971
DISTRIBUTION OF ANNELID
'NERINE SP.' (SEDENTARIA)
Miles
Kilometres
LOTHIANS RIVER PURIFICATION BOARD
FORTH ESTUARY SURVEY
DRAWN BY
SCALE
AS SHOWN
DRG. No.
APPROVED BY
DATE
NOV. 1971
L.R.P. /95

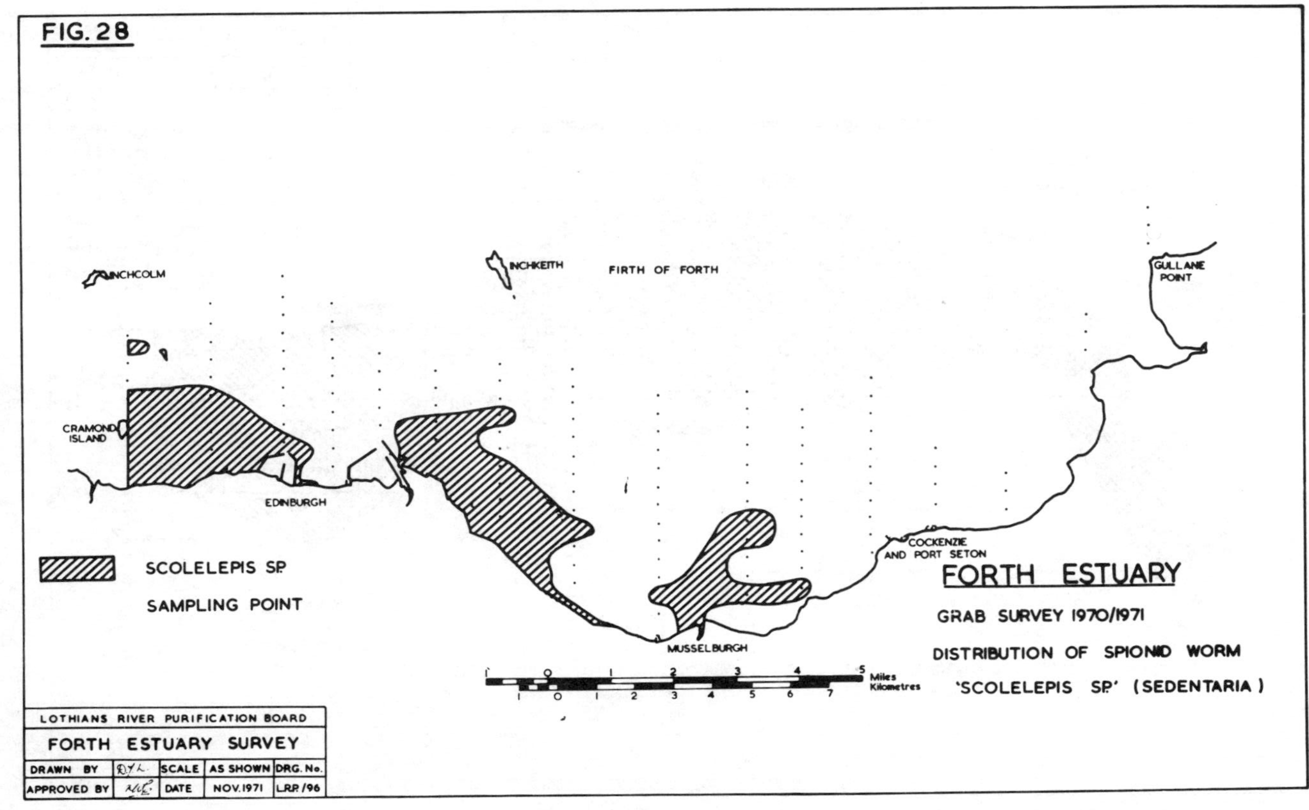
FIG. 28
INCHCOLM
INCHKEITH
FIRTH OF FORTH
GULLANE POINT
CRAMOND ISLAND
EDINBURGH
COCKENZIE AND PORT SETON
MUSSELBURGH
SCOLELEPIS SP.
SAMPLING POINT
FORTH ESTUARY
GRAB SURVEY 1970/1971
DISTRIBUTION OF SPIONID WORM
'SCOLELEPIS SP.' (SEDENTARIA)
Miles
Kilometres
LOTHIANS RIVER PURIFICATION BOARD
FORTH ESTUARY SURVEY
DRAWN BY
SCALE
AS SHOWN
DRG. No.
APPROVED BY
DATE
NOV. 1971
LRP/96

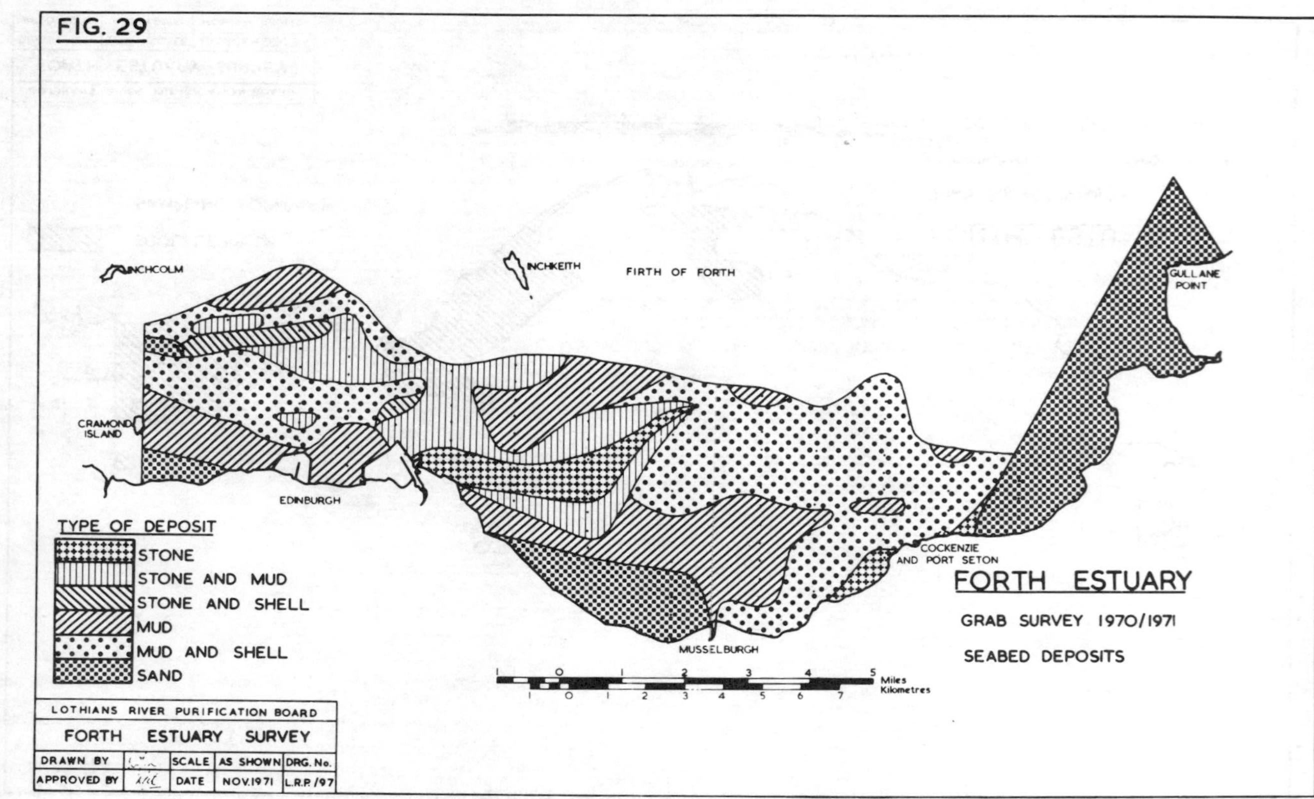
FIG. 29
INCHCOLM
INCHKEITH
FIRTH OF FORTH
GULLANE POINT
CRAMOND ISLAND
EDINBURGH
COCKENZIE AND PORT SETON
MUSSELBURGH
FORTH ESTUARY
GRAB SURVEY 1970/1971
SEABED DEPOSITS
TYPE OF DEPOSIT
STONE
STONE AND MUD
STONE AND SHELL
MUD
MUD AND SHELL
SAND
Miles
Kilometres
LOTHIANS RIVER PURIFICATION BOARD
FORTH ESTUARY SURVEY
DRAWN BY
SCALE
AS SHOWN
DRG. No.
APPROVED BY
DATE
NOV.1971
L.R.P./97

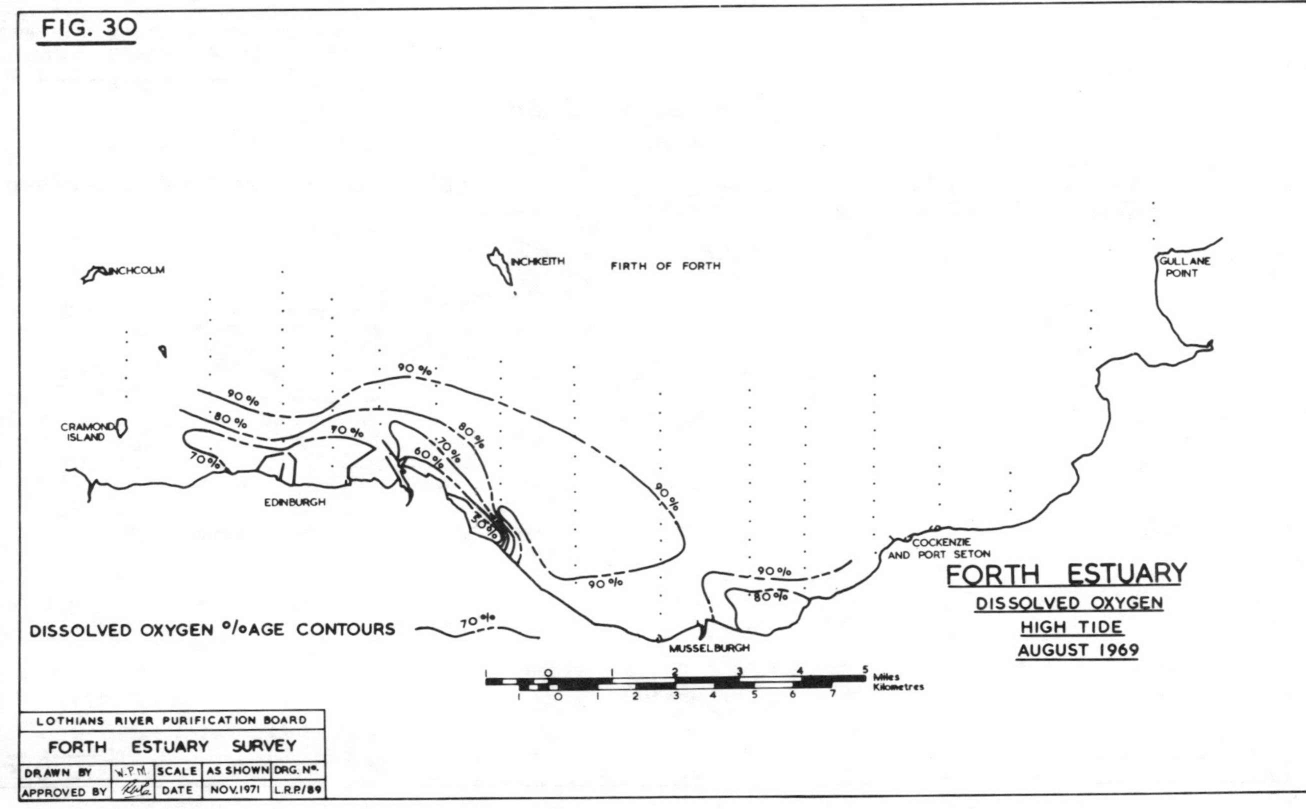
FIG. 30
INCHCOLM
INCHKEITH
FIRTH OF FORTH
GULLANE POINT
CRAMOND ISLAND
EDINBURGH
90 %
80 %
70 %
60 %
30 %
COCKENZIE AND PORT SETON
MUSSELBURGH
FORTH ESTUARY
DISSOLVED OXYGEN
HIGH TIDE
AUGUST 1969
DISSOLVED OXYGEN %AGE CONTOURS
70 %
Miles
Kilometres
LOTHIANS RIVER PURIFICATION BOARD
FORTH ESTUARY SURVEY
DRAWN BY W.P.M.
SCALE AS SHOWN
DRG. No.
APPROVED BY
DATE NOV.1971
L.R.P./89

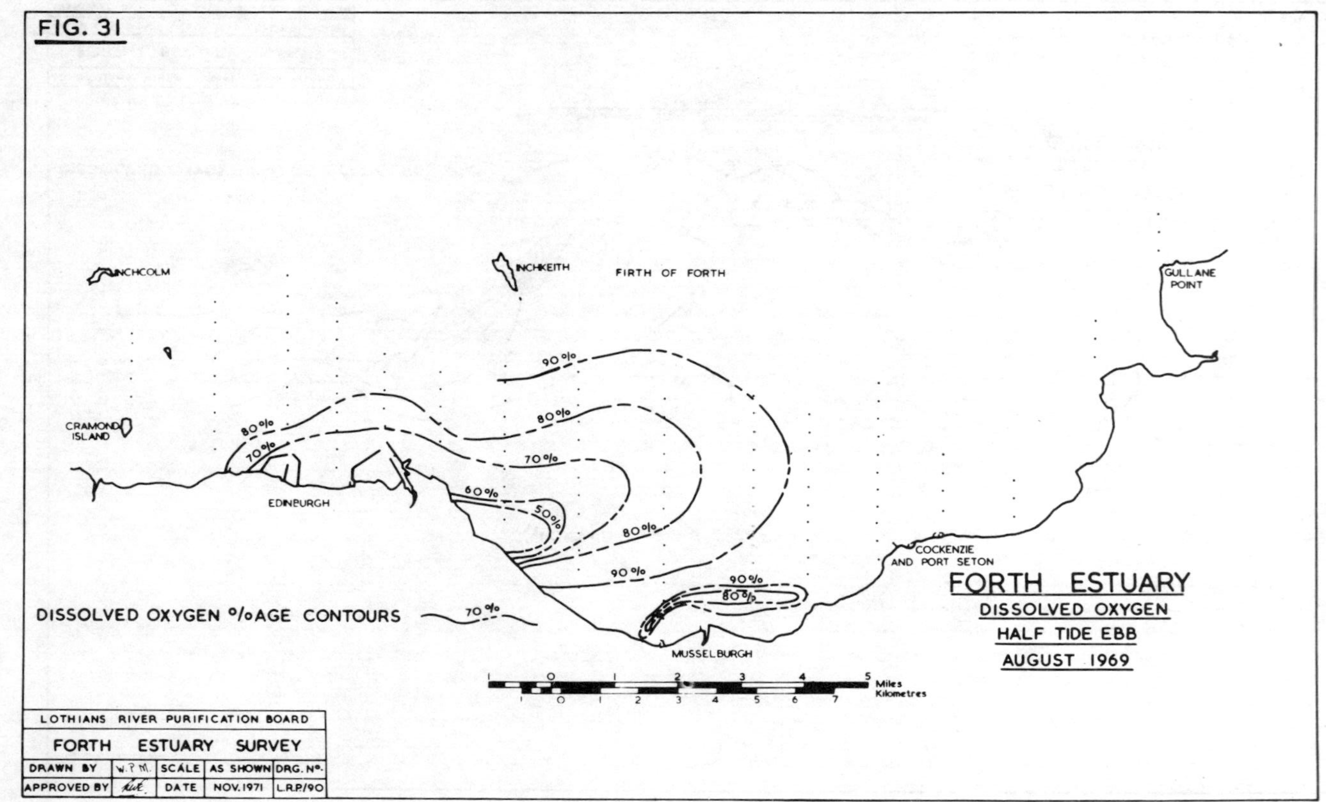
FIG. 31
INCHCOLM
INCHKEITH
FIRTH OF FORTH
GULLANE POINT
CRAMOND ISLAND
EDINBURGH
MUSSELBURGH
COCKENZIE AND PORT SETON
90%
80%
70%
60%
50%
80%
90%
90%
80%
DISSOLVED OXYGEN %AGE CONTOURS
70%
FORTH ESTUARY
DISSOLVED OXYGEN
HALF TIDE EBB
AUGUST 1969
Miles
Kilometres
LOTHIANS RIVER PURIFICATION BOARD
FORTH ESTUARY SURVEY
DRAWN BY W.P.M. SCALE AS SHOWN DRG. No.
APPROVED BY DATE NOV. 1971 L.R.P./90

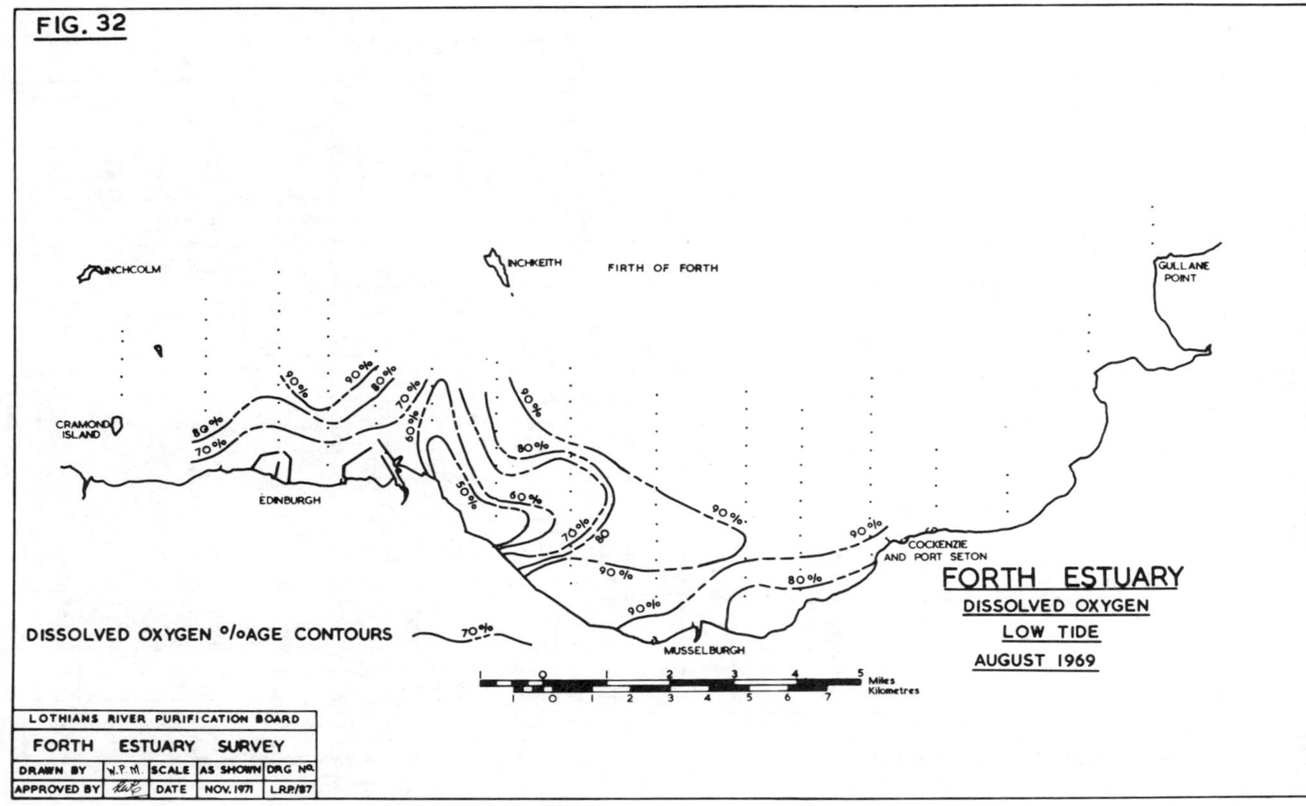
FIG. 32
INCHCOLM
CRAMOND ISLAND
INCHKEITH
FIRTH OF FORTH
GULLANE POINT
EDINBURGH
MUSSELBURGH
COCKENZIE AND PORT SETON
FORTH ESTUARY
DISSOLVED OXYGEN
LOW TIDE
AUGUST 1969
DISSOLVED OXYGEN %AGE CONTOURS
70%
90%
80%
60%
50%
Miles
Kilometres
LOTHIANS RIVER PURIFICATION BOARD
FORTH ESTUARY SURVEY
DRAWN BY W.P.M. SCALE AS SHOWN DRG No.
APPROVED BY DATE NOV. 1971 L.R.P./87

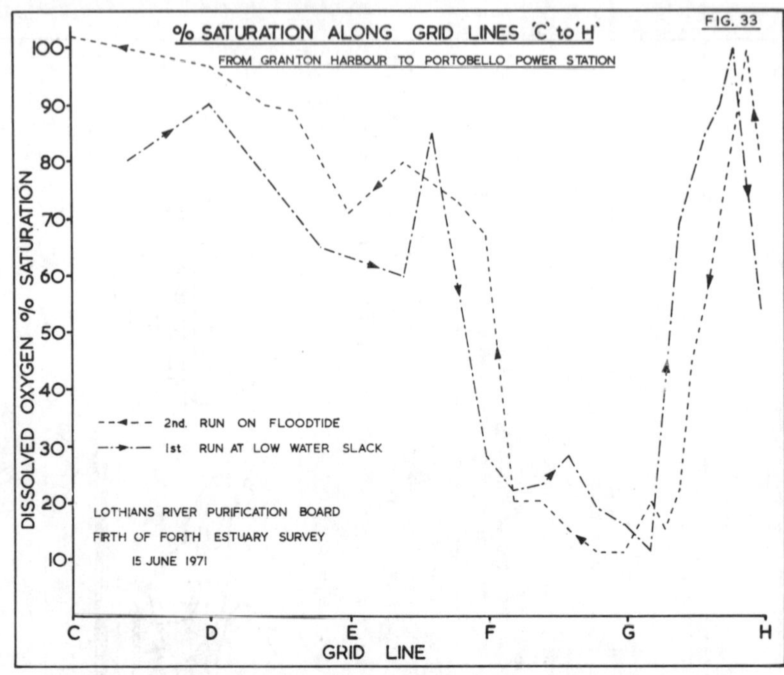
FIG. 33
% SATURATION ALONG GRID LINES 'C' to 'H'
FROM GRANTON HARBOUR TO PORTOBELLO POWER STATION
DISSOLVED OXYGEN % SATURATION
100
90
80
70
60
50
40
30
20
10
2nd. RUN ON FLOODTIDE
1st RUN AT LOW WATER SLACK
LOTHIANS RIVER PURIFICATION BOARD
FIRTH OF FORTH ESTUARY SURVEY
15 JUNE 1971
C
D
E
F
G
H
GRID LINE

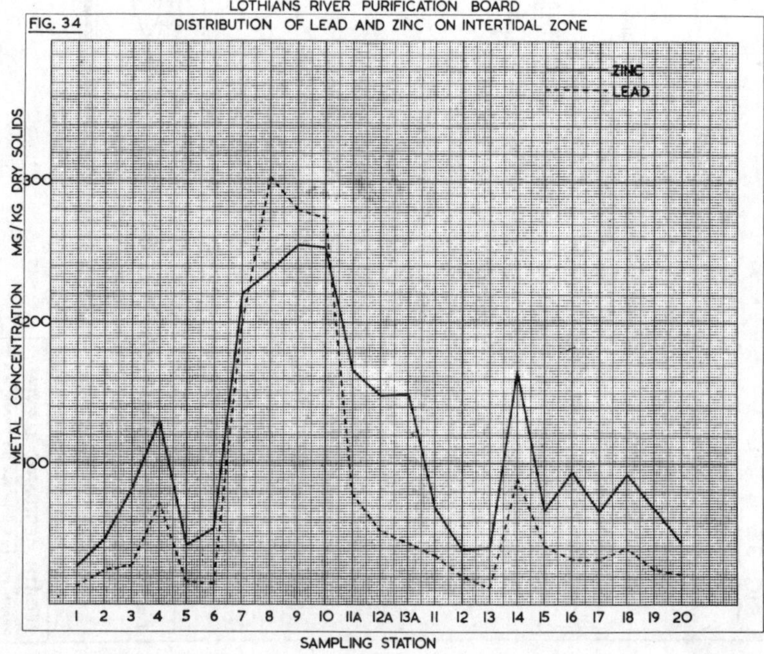
LOTHIANS RIVER PURIFICATION BOARD
FIG. 34
DISTRIBUTION OF LEAD AND ZINC ON INTERTIDAL ZONE
ZINC
LEAD
METAL CONCENTRATION MG/KG DRY SOLIDS
300
200
100
1 2 3 4 5 6 7 8 9 10 11A 12A 13A 11 12 13 14 15 16 17 18 19 20
SAMPLING STATION

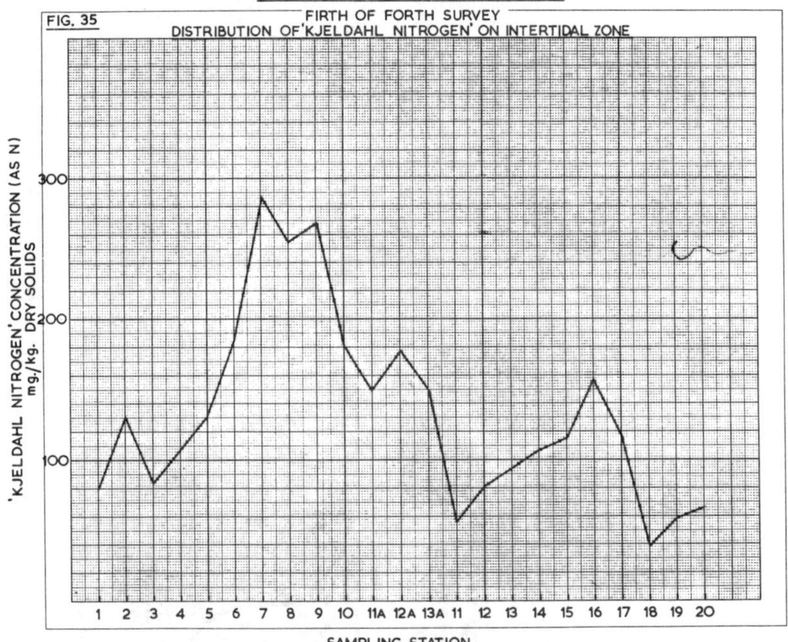
LOTHIANS RIVER PURIFICATION BOARD
FIG. 35
FIRTH OF FORTH SURVEY
DISTRIBUTION OF 'KJELDAHL NITROGEN' ON INTERTIDAL ZONE
'KJELDAHL NITROGEN' CONCENTRATION (AS N) mg./kg. DRY SOLIDS
300
200
100
1 2 3 4 5 6 7 8 9 10 11A 12A 13A 11 12 13 14 15 16 17 18 19 20
SAMPLING STATION

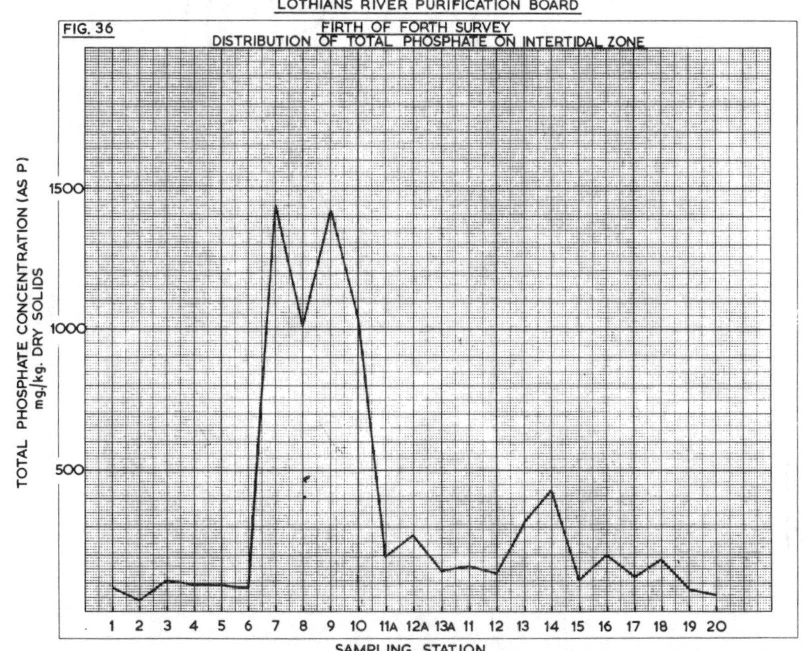
LOTHIANS RIVER PURIFICATION BOARD
FIG. 36
FIRTH OF FORTH SURVEY
DISTRIBUTION OF TOTAL PHOSPHATE ON INTERTIDAL ZONE
TOTAL PHOSPHATE CONCENTRATION (AS P) mg./kg. DRY SOLIDS
1500
1000
500
1 2 3 4 5 6 7 8 9 10 11A 12A 13A 11 12 13 14 15 16 17 18 19 20
SAMPLING STATION

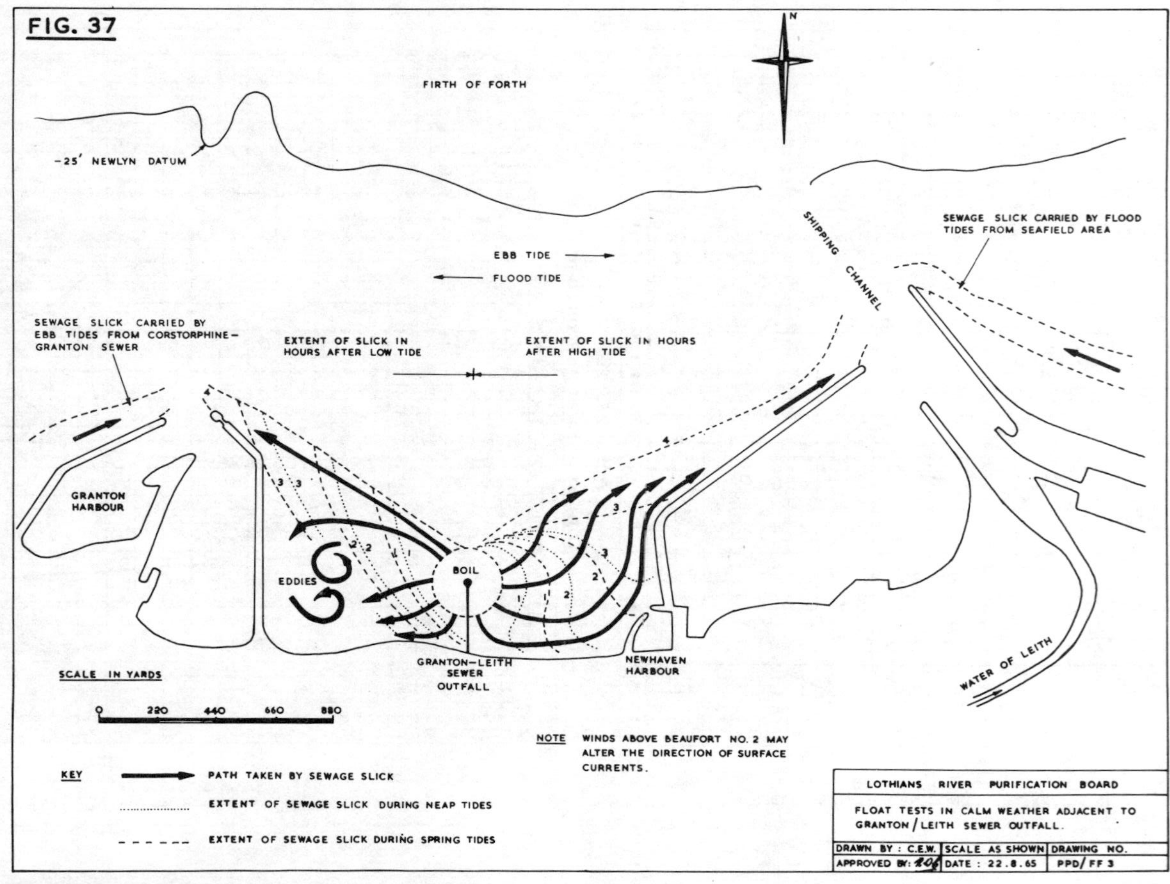
FIG. 37
N
FIRTH OF FORTH
-25' NEWLYN DATUM
SHIPPING CHANNEL
SEWAGE SLICK CARRIED BY FLOOD TIDES FROM SEAFIELD AREA
EBB TIDE
FLOOD TIDE
SEWAGE SLICK CARRIED BY EBB TIDES FROM CORSTORPHINE-GRANTON SEWER
EXTENT OF SLICK IN HOURS AFTER LOW TIDE
EXTENT OF SLICK IN HOURS AFTER HIGH TIDE
GRANTON HARBOUR
EDDIES
BOIL
GRANTON-LEITH SEWER OUTFALL
NEWHAVEN HARBOUR
WATER OF LEITH
SCALE IN YARDS
0 220 440 660 880
NOTE WINDS ABOVE BEAUFORT NO. 2 MAY ALTER THE DIRECTION OF SURFACE CURRENTS.
KEY
PATH TAKEN BY SEWAGE SLICK
EXTENT OF SEWAGE SLICK DURING NEAP TIDES
EXTENT OF SEWAGE SLICK DURING SPRING TIDES
LOTHIANS RIVER PURIFICATION BOARD
FLOAT TESTS IN CALM WEATHER ADJACENT TO GRANTON/LEITH SEWER OUTFALL.
DRAWN BY : C.E.W.
SCALE AS SHOWN
DRAWING NO.
APPROVED BY:
DATE : 22.8.65
PPD/FF 3

Grid Line Group	1 — 74 Group	1 — 20 Group
A	1, 2, 3	1
B	4	2
C	5, 6, 7, 8, 9, 10, 11	3, 4, 5
D	12, 13	6
E	14, 15, 16	—
F	17, 18, 19	7, 8, 9, 10
G	20, 21, 22, 23, 24, 25	11, 12, 13
H	26, 27, 28, 29, 30, 31, 32, 33, 34, 35, 36, 37, 38, 39, 40	14
I	41, 42, 43, 44, 45, 46, 47, 48	15
J	49, 50	16
K	51, 52, 53, 54	17
L	55, 56, 57	—
M	58, 59, 60, 61	18
N	62, 63, 64	19
O	65, 66, 67	—
P	68, 69, 70, 71, 72, 73, 74	20

Results and Discussion

12. (1) Bacteriological

(i) Grid Lines A-M, Stations 1-4, 1966-1972

(a) *Coliform Counts, Figure 7.*

As anticipated the graphs indicate peaks at the points of location of the major outfalls; however, the most significant aspect is the general coincidence of the graphs for the four points on the grid. This fact stresses the high degree of concentration of pollution in the inshore waters, a fact which in turn is influenced by the following circumstances;

(x) The general contour of the coast line within the area covered by the sampling points.

(y) The tidal and current patterns within the area producing a circulatory system.

(z) The lack of dispersion of the sewage and industrial effluents resulting from (x) and (y) above.

(b) *Plate Counts–Nutrient Agar 20°C and 37°C, Figures 8 and 9.*

The coincidence of the graphs for each station is significant for the counts at 20°C and 37°C, and the general pattern follows the location of the outfall sewers.

(c) *General.*

The emphasis on gross pollution in the inshore waters, is highlighted both by the coliform and plate counts.

(ii) Grid Lines A-M, Station 1–Mean Plate Counts, 1966-1972 Figure 10.

This graph shows the mean plate counts in respect of nutrient Agar at 20°C and 37°C, together with the MacConkey Agar count at 37°C. The Agar counts follow the anticipated pattern highlighting the major sewage and industrial effluent discharges, whilst the MacConkey Agar count major peaks coincide with the Leith and Seafield outfalls.

(iii) Grid Lines A-M, Station 2–Mean Plate Counts, 1966-1972 Figure 11.

The graphs for Nutrient Agar –20°C and 37°C, and MacConkey Agar–37°C follow, in general, a similar pattern to those for Station 1 above.

General

The plate counts at both Stations 1 and 2 on the Grid Lines A-M emphasise the constant concentration of pollution within the inshore waters.

(iv) Shore Stations 1-74, 1967-1969 Figures 12-14

(a) *Coliforms.*

This series of samples covered the shore line for the extent of the Tidal Waters Order 1960, implemented by the Board. The coliform curves clearly stress the effect on the estuary of the Edinburgh discharges and the peaks in the Prestonpans, Port Seton and Aberlady areas, emphasise the effect of sewage discharges in those locations. The maximum counts in the Seafield area give figures for coliforms per 100 ml exceeding 10^6 and even in the Port Seton area the corresponding figure exceeds 10^5 coliforms per 100 ml. Such figures demonstrate the inadvisability of bathing in inshore waters in those areas.

(b) *Plate Counts, Nutrient Agar and MacConkey Agar, Figure 15.*

The Nutrient Agar 20°C count shows agar peaks in the Seafield area and others corresponding to the sewer outfalls in the Edinburgh region. Further east minor peaks are evident in the Prestonpans–Cockenzie–Port Seton areas and also in Aberlady Bay. The Nutrient Agar 37°C count follows in general a similar pattern, whilst the MacConkey Agar curve shows some conformity at selected points and variations from the former curves at other locations. These differences could be accounted for by possible variations in the type of discharges, e.g., domestic sewage compared with sewage containing industrial effluents. In general all the counts emphasise the serious pollution existing in the inshore waters.

(v) Stations A-O on Shore Extensions of Grid Lines Period 1970-1971

(a) *Coliform Counts Figure 16.*

The graphs maximum, mean and minimum emphasise the pollution effect resulting from the major discharges between Cramond in the west and Kilspindie (Aberlady) in the east. In general the peaks correspond with the outfall locations. It is of interest to note the secondary peaks on the mean and minimum curves at sample station 'L' which corresponds with the Cockenzie Power Station cooling water discharge. No doubt the higher figures are related to the influence of the higher temperatures in that area.

(b) *Plate Counts—Nutrient Agar and MacConkey Agar, Figure 17.*

The effect of the cooling water discharge from the Cockenzie Power Station (point L) is reflected in the peaks for the Nutrient Agar 20°C and MacConkey Agar 37°C plate counts. However, the Nutrient Agar shows only a slight peak. The general pattern of the curves is in agreement with those of the coliform counts, with peaks corresponding to the major sewage outfall locations.

(c) *General.*

The graphs emphasise the desirability for maximum dispersion of both cooling waters and treated or partially treated sewage and/or industrial effluent, in order that bacterial counts may be at a minimum.

(vi) Grid Lines A-O Shore Extensions (Stations 1-20) 1972

(a) *Coliforms, Figure 18.*

The highest counts are indicated in the Seafield area. However, it is noted that the mean values between the Grid Lines A-L (Stations 1-17) are overall, higher than in 1970-71. This may be due to an increase in volume and strength of the discharge, or to better overall weather conditions. The coliform trend figures confirm an increase in volume and strength of the discharges. Again, it is interesting to note the peak in the area of Cockenzie, Port Seton, point L (Station 17) after which there is a marked decline compared with earlier years, and this is attributed to the increase in primary sewage treatment in the East Lothian Area. For example, the Aberlady and Gullane and Longniddry plants were commissioned late in 1969.

(b) *E.coli. Figure 19.*

The graphs conform in general with those of the coliforms; there is, however, one notable exception, and that is at Sampling Point L (Station 17) Cockenzie Power Station area where the E.coli peak equals that in the Seafield area.

(c) *Plate Counts—Nutrient Agar and MacConkey Agar. Figure 20*

It is interesting to note that the peaks in respect of the Nutrient Agar—20°C, and the Nutrient Agar – 37°C are in the Leith Docks area not the Seafield area. This is because the strongest sewages—in terms of Suspended Solids and B.O.D., discharge here. However, the MacConkey Agar peaks correspond with the Seafield and Cockenzie areas, similar to the graphs for *E.coli.*

(vii) Coliform Concentrations.

It is of interest, at this stage, to compare current American standards for the bacteriological quality of surface waters used for bathing, with the count figures obtained from the estuary survey.

Table 3 Current Standards for the Bacteriological Quality of Surface Waters Used for Bathing

State	Allowable Bacteriological Indices—No. of Coliforms
California	No more than 20% of samples at a Station >1 000 per 100 ml.
City of New York Class A.	Group I: Bathing allowed—average ≤ 1 000 per 100 ml—Epidemiological experience satisfactory. Sanitary surveys satisfactory.
	Group II. Bathing allowed—average > 1 000 but ≤ 2 400 per 100 ml—Epidemiological experience satisfactory. Sanitary survey shows exposure of increasing pollution.
City of New York Class B.	Not recommended for bathing—average > 2 400 per 100 ml with 50% of samples > 2 400 per 100 ml. Epidemiological experience satisfactory. Gross evidence of sewage pollution present.
Class C.	Beach closed. Epidemiology poor.
Ontario	No mandatory standards in force. Suggested standard for local health agencies M.P.N. <1 000 per 100 ml with 2 400 per 100 ml permissible, provided no evidence of increasing pollution, Sanitary survey satisfactory, and there is no significant Epidemiology
Florida	Median M.P.N. of ≤ 1 000 per 100 ml (Median of not less than 10 samples) with bathing prohibited if possible fresh pollution in area as determined by Sanitary survey.

From the Table it will be seen that coliform counts of less than or equal to 1 000 per 100 ml of sample, are considered desirable for satisfactory bathing conditions. Various graphs have therefore included a horizontal line at the concentration of 1 000 coliforms per 100 ml. The number of occasions on which sections of the graph exceed this figure undoubtedly emphasises the degree of gross pollution existing within the inshore waters of the Lothians River Purification Board.

(viii) Sewage Slick Surveys, Figures 21 and 22.

During 1968, sub-surface samples were taken along the lines of the Caroline Park and Seafield Sewer Outfall slicks at intervals varying between 100 and 500 metres. The purpose of the surveys was to determine, by means of the coliform concentrations at the different points, the fall off in numbers as the slick was traced towards mid estuary. Surveys were made under differing weather conditions and the accompanying graphs indicate clearly the fall off in the coliform counts. Extreme cases indicate an initial concentration of just under 10^7 coliforms per 100 ml at the 'Boil' (point of discharge from the source) falling to less than 10^3 just under 1 500 metres from the discharge point. It was noted that varying sea and weather conditions did not have a significant effect on the fall off in the coliform counts.

(ix) Sewage Samples—Examination for Salmonellae.

During the period 1970-1972, forty-one samples of crude sewage from the following outfalls, Cramond, Granton, Caroline Park, Trinity, Water of Leith, Seafield, Joppa and Eastfield were examined for salmonellae, and of these twenty-four were positive and seventeen negative. The following salmonellae were 'typed' by Dr. I. Campbell,(5) in samples from the Joppa, Trinity and Seafield sewage outfalls;

1. *S. paratyphi* B
2. Unidentified Group D. *Salmonella*
3. Group B. Salmonella—possibly *S. derby*.

Samples examined for the M.P.N. of Salmonella.

A number of samples were examined quantitatively in accordance with the procedure used by McCoy (1962). The examinations were carried out by enriching different volumes of the sample and observing the number of positive tubes at each dilution or volume. The M.P.N. was then obtained by reference to tables given in McCoy (1962). The procedure followed in the case of sewer swabs was as follows:-

(a) The fluid was squeezed from the swabs and collected; the swabs were then discarded.

(b) *Primary enrichment*

From the collected fluid, 1.0 ml was pipetted into each of ten tubes containing 10 ml Tetrathionate Broth (Oxoid), a further 0.1 ml of fluid was pipetted into each of ten further tubes containing 10 ml each of Tetrathionate Broth. The inoculated tubes were incubated at 37°C.

(c) *Secondary enrichment*

On the following day after incubation at 37°C 0.1 ml was pipetted from each previously incubated tube, into a fresh tube containing 10 ml of Tetrathionate Broth. It was suggested by McCoy that this secondary enrichment was essential to allow Salmonellae to outgrow Pseudomonas which, during the winter period, was the predominant organism in sewage. The secondary enrichments were incubated at 37°C and subcultured to Bismuth Sulphite Agar (Oxoid) after 24 and 48 hours incubation. The number of positive tubes out of ten at each dilution was recorded and the M.P.N. read from the following Table:

Table 4 Average number of organisms per 100 ml for dilution test with 10 tubes each containing 1.0 and 0.1 ml of sample

Tubes positive	1.0 ml Sample	0.1 ml Sample
0	<10.5	<105
1	10.5	105
2	22.3	223
3	35.7	357
4	51.1	511
5	69.3	693
6	91.6	916
7	120	1 200
8	160	1 600
9	230	2 300
10	>230	>2 300

Samples taken from Sewer Swabs from Seafield and Trinity sewers on 11th January 1971, gave the following results:

Seafield All tubes were positive with 1.0 and 0.1 ml of sample.
The M.P.N./100 ml was therefore >2 300.

Trinity All tubes were positive with 1.0 ml sample.
Nine tubes were positive with 0.1 ml sample.
The M.P.N./100 ml. was therefore 2 300.

(x) Sewage Samples—Examination for *Clostridium perfringens* and *Streptococcus faecalis*.

During the years 1970 and 1971, a number of crude sewage discharges were examined for *Cl. perfringens*. Tests were carried out on twelve occasions covering the major outfalls between Cramond and Eastfield. In all instances, *Cl. perfringens* was demonstrated by the production of acid, gas and 'stormy fermentation' in litmus milk medium and secondly by a method described in the Metropolitan Water Board, Director of Water Examination, Report for 1956. The medium used included nutrient agar with 2% agar, 100 ml; 20% solution Na_2SO_3, 10 ml; 20% solution of glucose, 5 ml; 8% solution of $FeSO_4.7H_2O$ (freshly prepared), 1 ml. The water sample was heated for 10 min at 75-80°C before filtering through a membrane. The membrane was placed face down on the surface of well dried poured plates and another layer of the medium poured on top. Incubation was from 1-5 days at 37°-45°C. *Cl. perfringens* produced colonies with large black halos. Samples taken from the major sewage outfalls during 1971, were all found to contain *Streptococcus faecalis*.

(xi) Beach Core Samples.

On occasions during 1971 and 1972, beach core samples were taken along the more polluted sections of the estuary. The samples were identified as given in Table 5.

Table 5

Date	No. of Samples	Location	Result
22.2.71	3	Grid Lines G, H and I Seafield to Fisherrow	Two samples demonstrated Salmonella, Clostridium and Streptococcus.
23.8.71	14	Stations 4-10 Granton to Seafield	Five samples-Salmonella present.
30.8.71	10	Stations 11-20 Seafield to Gullane	Five samples—do.
8.11.71	10	Stations 11-20	Three samples—do.
22.11.71	8	Stations 11-16 and 18 19, Seafield-West Pans, and Cockenzie and Port Seton Bay	Seven samples—do.
10.1.72	10	Station 7 Seafield	Eight samples—do.

(xii) Mussels *(Mytilus edulis)*

On two occasions in 1971, samples of mussels were taken from selected points along the shore between Cramond and Gullane Point. The mussels were examined for the presence of Salmonella and the results are as follow:-

Table 6

Date	No. of Samples	Location	Result
10.11.71	9	1. Gullane Point	Salmonella absent
		2. Aberlady Bay	do.
		3. Craigielaw Point	do.
		4. Bogle Hill	do.
		5. Seton Sands East	Salmonella present
		6. Seton Sands West	do.
		7. Cockenzie Harbour	Salmonella absent
		8. Prestonpans	Salmonella present
		9. Musselburgh	do.
23.11.71	4	1. Cramond	Salmonella absent
		2. Granton Harbour East	Salmonella present
		3. Trinity	do.
		4. Seafield	do.

13. (2) Biological

Surveys have been carried out in the inshore waters and intertidal zone on the shore line. Figure 23 illustrates the results of a general survey from the qualitative and quantitative aspects. It will be noted that there are a number of completely sterile zones, as well as areas indicative of polluted conditions as evidenced from the flora and fauna which are absent.

14. Figure 24 shows the information gained from a typical Trawl Survey. The trawl net was of the Granton Trawl type, scaled down to one-third of the normal size. The diagram indicates the distance covered during each trawl along the coastline and the numbers of animals found per unit of one nautical mile towed. The importance of scavengers (crabs etc.) is shown, and it is pointed out that there were no large edible fish caught between Granton Point and Fisherrow. The fall off in biomass off Seafield and Portobello is noted as well as the difference in fauna off Cockenzie, where the cleaner and warmer conditions (power-station cooling water discharge), are apparent.

15. It is also interesting to note the fall off in the number of shrimps and prawns from west to east and the increase in starfish towards the east.

16. Figure 25 shows the distribution of *Mytilus edulis* (Common Mussel) and *Modiolus modiolus* (Horse Mussel) as indicated by grab and shore surveys. It is interesting to note that the Modiolus sp. has taken over areas of the sea bed once inhabited by the oyster beds. In the west the specie lives on the sides of a gully extending west-east, where it probably feeds on the debris brought down by the River Almond. The gully may

be, in fact, the original bed of the river. The central bed of Modiolus sp. is again on the side of a slope, this time on a rock/stone face which falls steeply to a flat area of mud.

17. The largest bed of *Mytilus edulis* is on the stone/mud and stone area just off the present Seafield outfall—a very good reason for banning those mussels as being unfit for consumption. Similar pollution conditions exist in the areas of mussel beds at Musselburgh and Aberlady Bay.

18. Figures 26-28 inclusive show the distribution of the following species in the inshore waters:

Lumbriconereis sp. (Errantia)
Nerine sp. (Sedentaria)
Scolelepis sp. (Sedentaria)

Figure 29 shows the results of a Grab Survey, 1970-71, of seabed deposits. It is stressed that the diagram is only one interpretation which could be put on the results.

19. (3) Chemical

(i) Grid and Slick Samples

With regard to the chemical surveys, Figures 30 to 32 were typical of the dissolved oxygen percentage saturation conditions existing in the inshore waters at the various states of the tidal flow. It is to be noted, for example, that comparing the percentage saturation of dissolved oxygen in 1958 and 1969, there has been slightly more than a 9 per cent decrease. The lowest percentage saturation of dissolved oxygen recorded was 11 per cent. It will be observed also from the Figures that, in spite of the varying tidal conditions, there is little offshore movement of the pollution resulting from the discharge of sewage in the area between Granton and Portobello. Figure 33 indicates the Dissolved Oxygen profile taken on 15th June 1971, between Grid Lines C-H, i.e., from Granton Harbour in the West to Portobello Power Station in the East. The dissolved oxygen percentage saturation sag in the Seafield-Portobello area emphasises the effect of the sewage discharges.

20. Chemical analyses of 'Slick' samples have shown that the effect of changes in salinity, phosphate, dissolved oxygen and turbidity are only significant for a small area in proximity to the 'boil' (point of discharge). The results given in Table 7 of a survey on the Corstorphine-Granton sewer slick, illustrate this point.

21. The stations in the survey were approximately 200 metres apart and the wind direction was W S W on the day of sampling, (24th July, 1968). This factor stresses the importance of the work carried out by Edinburgh Corporation relative to the siting of the proposed new outfall sewer, which will discharge the total volume of partially treated sewage at one point just east of Leith Docks at Seafield.

Table 7 Station

	Boil	1	2	3	4	5	6	7	8	9
Salinity (p.p.th.)	19.0	30.4	30.4	30.8	31.5	31.4	31.8	30.8	32.0	32.2
D.O. (% saturation)	80.5	94.5	97.0	96.0	78.0	66.0	69.0	103.0	103.0	104.0
Phosphate (mg/l)	11.9	1.0	0.8	0.9	0.6	0.5	0.2	0.2	0.2	0.2
Turbidity	70.0	7.0	9.5	8.0	4.5	5.0	3.0	1.5	1.5	1.0

22. (ii) Core Samples Intertidal Zone

A programme of analyses of samples from 20 sampling stations in the intertidal zone between Cramond in the West and Aberlady Bay in the East, has been under way since August 1971. Figures 34-36 indicated the distribution of Lead, Zinc, 'Kjeldahl Nitrogen', and Total Phosphate in the intertidal zone. With regard to the Total Phosphate figures, it is pointed out that industrial discharges in the Leith area are the source of large concentrations of phosphate in the estuary, and in consequence the scope of tracing sewage by this parameter is limited.

23. Table 8 gives the average for 1972 of the core sample analyses.

24. The sampling stations 11A, 12A and 13A were discontinued as from 6th November 1972, due to reclamation work for the site of new Sewage Works at Seafield Bay.

25. Recently heavy metal analyses have been carried out on composite samples of the crude sewage discharges to the estuary, and details of these are given in Table 9.

The chemical data provided by the core and sewage analyses have proved helpful in determining the source of industrial effluents containing these elements.

26. (4) Hydrography of the Estuary

Figures 37-39 inclusive, summarise the 51 Float Surveys made in the areas of the existing outfalls at Caroline Park and Cramond, Wardie Bay and Seafield Bay sewage outfall areas. The surveys were carried out in wind conditions of less than or equal to Beaufort Scale 2, and indicate the general pattern of current flow in calm conditions, the mean velocity of the currents being of the order of 0.35 m/s. Each of the sewage outfalls is situated in a bay or indented coastline and the eddy currents produced tend to distort the general tidal flow pattern of the estuary. The sewage slicks tend to be contained in the bays and drift on to the foreshore. The freshwater flows via the rivers along the coastline also have an added effect in producing eddy currents within the bays and under

Table 8

	Location	Total Nitrogen (N)	Total Phosphorus (P)	Zinc	Copper	Lead	Chromium	Nickel
1.	Cramond	77.8	88.1	26.7	8.2	13.1	3.4	15.1
2.	Silverknowes	131.1	42.8	45.6	11.4	23.6	6.1	22.6
3.	Granton Gasworks	82.5	110.7	82.4	31.7	28.4	12.3	21.9
4.	Caroline Park	108.7	100.1	130.9	27.8	71.6	22.5	18.1
5.	East Breakwater Granton	130.7	100.4	41.6	6.9	16.0	6.5	19.1
6.	Trinity	184.7	88.7	53.5	8.2	14.6	7.5	19.5
7.	S.A.I. Works, Leith	289.3	1429.2	219.8	119.7	197.7	23.6	25.8
8.	Leith Docks	255.0	1010.6	237.0	132.6	303.7	24.8	30.1
9.	Leith Docks	267.9	1411.8	255.3	144.1	278.7	23.4	32.2
10.	Leith Docks	180.6	1047.1	255.2	150.4	275.2	22.8	28.3
11A.	Seafield West	147.5	195.8	165.8	29.7	77.9	12.1	19.1
12A.	Seafield Mid	177.6	271.2	148.4	25.9	53.4	10.2	18.0
13A.	Seafield East	150.7	149.0	148.8	42.7	43.7	14.0	18.9
11.	East of Seafield Outfall	54.5	160.7	70.0	18.5	35.0	10.0	12.5
12.	Portobello (King's Road)	82.9	140.8	38.5	5.7	20.0	7.3	9.7
13.	Portobello	95.3	318.4	41.3	6.3	11.7	7.7	12.4
14.	Fisherrow	106.1	430.7	166.1	54.7	88.0	13.8	19.7
15.	Prestongrange (Fly Ash Lagoons)	113.5	118.7	67.4	35.8	40.9	11.0	26.4
16.	Prestonlinks (Marina)	158.7	206.4	94.0	40.0	31.7	13.5	25.3
17.	Port Seton (Bathing Pool)	116.1	130.8	65.6	89.9	31.4	7.7	14.7
18.	Longniddry (Sewage Works)	39.1	186.1	91.7	16.0	40.0	49.0	42.7
19.	Gosford Sands	59.5	82.9	67.9	9.3	24.6	8.7	19.0
20.	Aberlady Bay	65.8	62.1	44.2	10.3	12.5	9.2	14.2

Table 9

Source of Sample	Date Sampled	Time of Sampling	Heavy Metal Content mg/l				
			Cr	Cu	Ni	Pb	Zn
Water of Leith 1864 Sewer	5.2.73	10.35-14.05 hrs.	3.75	1.75	Nil	0.15	0.42
Water of Leith 1889 Sewer	5.2.73	10.30-14.00 hrs.	0.22	0.31	0.07	0.20	0.50
Cramond Sewer	6.2.73	10.30-14.15 hrs.	Nil	0.23	Nil	0.25	0.96
Eastfield Comminutor Station	8.2.73	11.15-14.00 hrs.	Nil	44.65	0.02	0.07	0.28
Joppa Comminutor Station	8.2.73	11.20-13.50 hrs.	Nil	0.30	0.11	0.20	0.45
Granton Gas Works Outfall	12.2.73	10.50-14.30 hrs.	0.37	0.18	0.20	0.40	1.15
Granton West Pier Outfall	12.2.73	10.05-14.30 hrs.	0.03	1.60	0.04	0.10	2.20
Granton/Corstorphine Sewer	12.2.73	10.45-14.00 hrs.	0.03	0.18	0.04	0.13	0.30
Seafield Screening Station	12.2.73	10.00-14.30 hrs.	Nil	0.18	0.04	0.13	0.30
Trinity Screening Station	19.2.73	10.00-15.00 hrs.	0.02	0.16	0.10	0.25	0.68

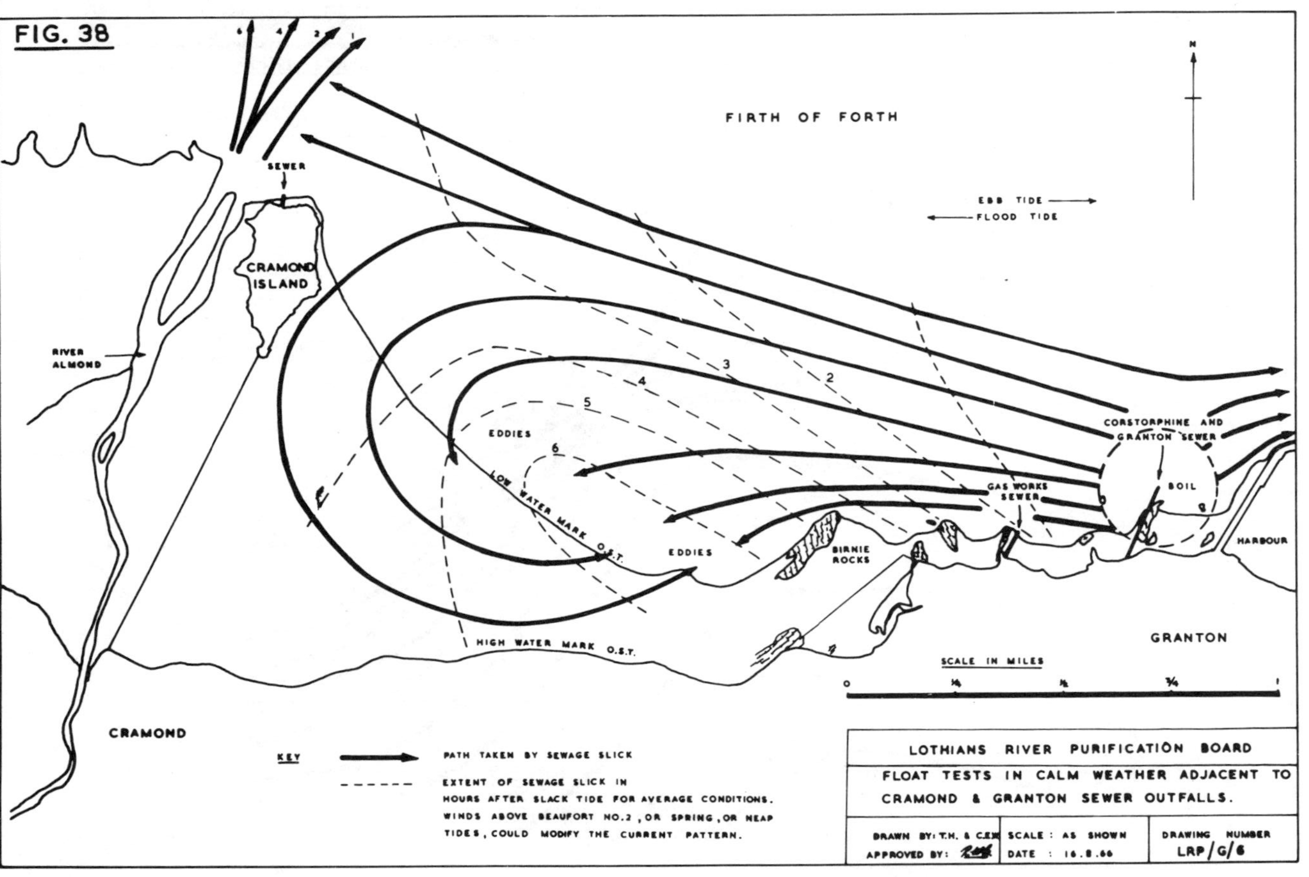

FIG. 38
FIRTH OF FORTH
N
EBB TIDE
FLOOD TIDE
SEWER
CRAMOND ISLAND
RIVER ALMOND
CORSTORPHINE AND GRANTON SEWER
BOIL
GAS WORKS SEWER
HARBOUR
EDDIES
EDDIES
2
3
4
5
6
LOW WATER MARK O.S.T.
HIGH WATER MARK O.S.T.
BIRNIE ROCKS
GRANTON
SCALE IN MILES
0 ¼ ½ ¾ 1
CRAMOND
KEY
PATH TAKEN BY SEWAGE SLICK
EXTENT OF SEWAGE SLICK IN HOURS AFTER SLACK TIDE FOR AVERAGE CONDITIONS.
WINDS ABOVE BEAUFORT NO.2, OR SPRING, OR NEAP TIDES, COULD MODIFY THE CURRENT PATTERN.
LOTHIANS RIVER PURIFICATION BOARD
FLOAT TESTS IN CALM WEATHER ADJACENT TO CRAMOND & GRANTON SEWER OUTFALLS.
DRAWN BY: T.H. & C.E.W
APPROVED BY:
SCALE: AS SHOWN
DATE: 16.8.66
DRAWING NUMBER LRP/G/6

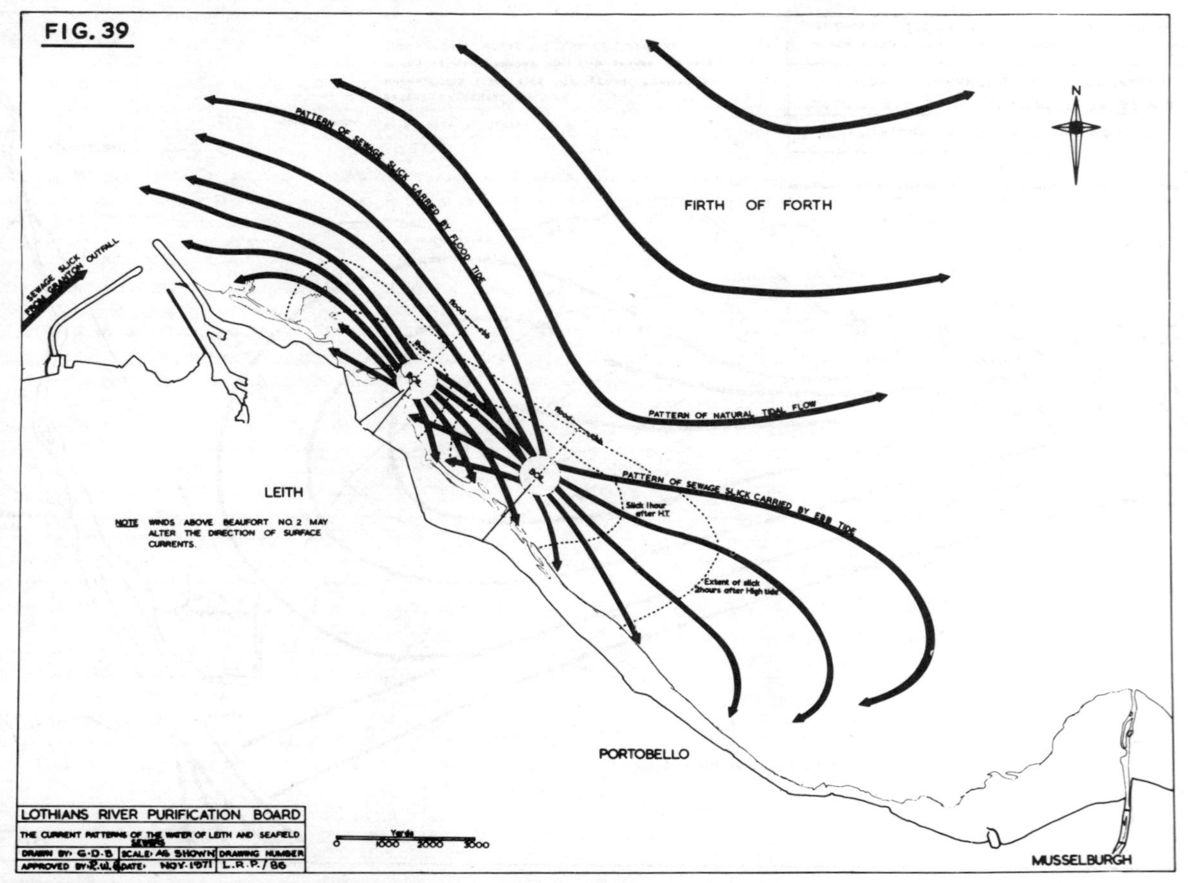
FIG. 39
N
FIRTH OF FORTH
SEWAGE SLICK FROM GRANTON OUTFALL
PATTERN OF SEWAGE SLICK CARRIED BY FLOOD TIDE
PATTERN OF NATURAL TIDAL FLOW
PATTERN OF SEWAGE SLICK CARRIED BY EBB TIDE
Slick 1hour after H.T.
Extent of slick 2hours after High tide
LEITH
NOTE WINDS ABOVE BEAUFORT NO.2 MAY ALTER THE DIRECTION OF SURFACE CURRENTS.
PORTOBELLO
MUSSELBURGH
Yards
0 1000 2000 3000
LOTHIANS RIVER PURIFICATION BOARD
THE CURRENT PATTERNS OF THE WATER OF LEITH AND SEAFIELD SEWERS
DRAWN BY: G.D.B
SCALE: AS SHOWN
DRAWING NUMBER
DATE: NOV. 1971
L.R.P./86

certain tidal conditions, act as a buffer to contain the sewage 'slick' within the bays. The hydrographical work has supported the fact that the sewage from existing outfalls is liable to drift on to the foreshore.

27. Recent surveys over the period 1971-1973, have been made with the object of establishing the current patterns from the proposed sewer outfall locations. This work has shown that the sewage discharge from the proposed westerly location will be diffused satisfactorily in the estuary water, and that there will be no tendency for the sewage to return to the immediate vicinity of the shore.

CONCLUSIONS

28. The gross pollution existing at the present time in the inshore waters of the estuary, within my Board's jurisdiction has been demonstrated.

29. Further detailed work will continue, which will enable my Board to appreciate the future conditions resulting from the discharge of partially treated sewage from the plants presently being constructed by Edinburgh and Midlothian County Council.

ACKNOWLEDGEMENTS

The Chairman and Members of the Lothians River Purification Board are thanked for their permission to present this Paper. The assistance of my Biological, Chemical, Hydrological and Administrative staff is fully appreciated and acknowledged.

REFERENCES

1. Lothians River Purification Board, 1958, River Forth Survey.
2. Lothians River Purification Board Annual Report 1967.
3. Lothians River Purification Board Annual Report 1971.
4. Covill, R.W., 1971, *Proc. R.S.E.* (B) **71**, 12, 1971-72.
5. Campbell, I., 1971, Private Communication.

DISCUSSION

COVILL It is of particular interest to note that over the last two years the Forth/Tay Research Group under the capable leadership of Dr. Hale of Edinburgh University has brought together for the first time many people of different disciplines who are interested in the problems of the Forth Estuary as well as the Tay. These disciplines cover civil engineering, chemical engineering, chemistry, biology, microbiology, and include representatives of four Universities including at least ten professors and their supporting staff, who now meet regularly. Proposals relative to effective survey work in the future have been and are being fully discussed. A new survey vessel is to be built which will cost approximately £100 000. This will be a seventy-footer and will enable fieldwork to be carried out much more effectively. Areas of investigation which suggest themselves include for example the energy flow in estuarine systems, in terms of quantities and importance of organic carbon entering via macro algal secretions and from terrestial communities; the role of bacteria in nutrient turnover; and a programme linked to studies on molluscan distribution in macro algal communities. These, together with basic hydrographic, geological, sedimentological, biological and microbiological data collection, will provide some of the answers to the many problems.

Primary sewage treatment works are being built or have recently been commissioned by a number of local authorities in the area. These will have a major effect in reducing pollution. When all these plants are completed as primary treatment units, the River Purification Board still reserves the right to ask for a greater degree of treatment should this be shown desirable by the extended survey work. It then becomes a question of cost. Are the public prepared to pay for this?

MARTIN (Mouchel & Partners). From the Author's slides it is obvious that there is clear visual evidence of gross pollution. The work undertaken involved detailed analytical and bacteriological study and served to confirm the results of the hydrographical survey as well as the visual evidence. This seems very elaborate.

The Author compares the conditions in the Forth Estuary with standards required in California, New York and other American states. These standards are extremely high and are not necessarily universally accepted as being necessary. The divergence between the conditions in the Estuary and those standards is so great that it seems very optimistic to consider one in the context of the other.

In the paper attention is drawn to the peaks in coliform counts at a point corresponding with the Cockenzie Power Station cooling water discharge and states that the higher figures are related to the influence of higher temperatures in that area. This effect is not acknowledged by Savage (Paper 4), who is possibly in agreement with the current C.E.G.B. view that warm water is not polluting. Is this really true? Why do *E.Coli* multiply in warm water? Surely they ought to be dying off, being affected by salinity, ultra-violet light and grazing by other organisms.

The Author expresses concern regarding the mussel beds at Seafield and at Musselburgh. Presumably, by the very name of the place, there must have been mussels at Musselburgh for a long time.

The Author discusses animal life with no real mention of plants or algae. Presumably there are very high nutrient levels in this part of the Forth Estuary. A large discharge of phosphate is mentioned, and the presence of crude sewage would ensure a high nitrogen content in the form of ammonia.

In discussing the Power Stations, no mention is made of any problems with cooling water intakes. Mouchel & Partners are involved in an investigation into the effect of the discharge of treated effluent into the Hayle Estuary, from which cooling water is taken. Investigations are being made by an inter-disciplinary team involving biologists, chemists and engineers. They are considering the growth of ulva and enteromorpha. My part as an engineer in the investigating team has been most rewarding, particularly in the contact with pure scientists.

Has the Author, or any other member of this symposium, information regarding the potential for weed growth at various nutrient levels and combinations, on the optimum conditions that can generate growth of this sort, or on the preference for nitrogen in the form of nitrates or ammonia?

COVILL More information is indeed required on the potential for weed growth, and should be available in the foreseeable future. With reference to the intakes of power stations, there have been no complaints from the South of Scotland Electricity Board relative to these. Power station intakes have to be cleaned regularly on the Forth Estuary. Johnson, of the Forth/Tay Research Group, is embarking on a study of the relationships between heat, effluent and increase in *E.Coli* counts at power stations.

When considering higher standards for sewage discharges, should sterilisation be specified? The costs are significant, and there will be objections because of this. But should we be content to ignore bacteriological standards? Surely bacteriological standards should be just as reasonable to impose and monitor as chemical standards, particularly bearing in mind the complexity of chemicals at the present day.

DOWNING (Water Pollution Research Laboratory). We are now making contact with the problem before the conference. It is in the area of definition of criteria where the real difficulties arise. The present thinking as set in train by the Medical Research Council Report (1959) is that one major criterion for coastal waters should be that visible pollution should be eliminated. The absence of correlation between bacterial counts and ill health when there was no visible evidence of pollution, means that one need not worry about health hazards provided that this other criterion is satisfied.

WAKEFIELD (Coastal Anti-Pollution League Ltd.). This is indeed the crux of the question. A number of States in the U.S.A. have adopted standards for bacterial quality, and these are not related to health. They are bacteriologically enumerated but they are really to establish an aesthetic standard. It is not possible to attach figures to the aesthetic side of this. You cannot say that because there are so many turds to the square yard that you should do something about it, but some sort of aesthetic standard can be achieved in terms of bacteriological standard. The Medical Research Council set their face against this and said that it was impossible to monitor.

There is then the question as to whether to apply these standards at the shore or whether there should be a standard for the effluent itself. There is yet no answer to this. One problem is that a very large number of samples are needed to establish median and 5 and 95 percentile values. A decision is then required as to which of these is the appropriate value to take. Sampling at the point of discharge must be much cheaper. One then has to rely on the mathematical calculations of dilution in the estuary to find the values at the shoreline.

COVILL It would be helpful to have a bacteriological standard on the effluent. Although normal sterilisation by chlorination will not of course eliminate viruses, it could take care of bacteria. This must be the solution. Whether or not authorities should be required to give primary treatment plus partial secondary treatment by high rate biological oxidation, followed by sterilisation, or whether they should have full conventional treatment followed by sterilisation is a matter to be decided upon.

Consideration must be given to the particular areas with which one is concerned. It would be thoroughly wrong to suggest that all discharges to coastal waters should necessarily receive such a high standard of treatment. One has to consider the use made of these waters. It has already been emphasized that recreational uses are becoming increasingly important. As far as Scotland is concerned it is recognised that Scotland has the largest area of remote country in the E.E.C. This will lead to tourist development on an increasing scale.

In the Forth Estuary area there have been from time to time carriers of *salmonella typhosis*. These organisms have been found in rivers discharging to the area. McCoy at Hull has emphasized for many years that it does not necessarily require *salmonella typhosis* to be present for a public health hazard to be present. He suggests as a member of the medical profession that any group of salmonella demonstrates a possible public health hazard, and therefore if such can be eliminated, all to the good.

Further information is needed on the relationship between eye, nose and throat complaints and bathing in polluted water.

GILSON (Committee for Environmental Conservation). We must not lose sight of the fact that there are only two ways of getting rid of an unwanted indestructible material. It can either be buried, enclosed where it will not escape as is done with radioactive materials deposited in the depths of the ocean, or it can be diluted. What we are engaged in trying to do with sewage effluents and trade effluents is to dilute them. In inland waters and upper reaches of estuaries the amount of water available for dilution purposes may be restricted, therefore for discharge to such an area it may well be necessary to treat the effluent before discharge. One of the most damaging properties of sewage effluent so far as rivers is concerned is its high biological oxygen demand because this results in the de-oxygenisation of the river and destruction of the fauna. Conventional sewage treatment is therefore aimed at reducing the oxygen demand of the effluent, and it does very little else beyond this. The dilution available in the sea sufficiently far off shore is enormous and this has the great merit that it dilutes all things in the effluent including the bacteria.

The concentration of bacteria is likely to fall off exponentially with distance from the point of discharge. It is therefore well worth trying to discharge effluents well off shore rather than on the beach. The high figures that are quoted for samples collected along the shores of the Forth are surely related to the fact that these effluents are all discharged on the shore. It does not matter very much whether the discharges are more or less crude sewage or sewage treated to Royal Commission standards. If the discharge is on shore there will still be a large number of bacteria. Discharge well off the shore gives much better chance of avoiding pollution of the beaches with bacteria or other pollutants. Furthermore, it will not then be necessary to spend large sums of money on treating the effluent to reduce its biological oxygen demand as it is not suggested that there is any shortage of oxygen in the Solent. This will incidentally eliminate the problems of sludge disposal which the ordinary sewage works creates.

COVILL In the Forth, the limited hydrographical studies have shown most emphatically that within reasonable distances of possible sites where outfalls may be located, discharges will return to the beaches. Consider an example from the English Channel where one would suggest that the tidal movement and the volume of water would be sufficient to ensure effective dispersion. A particular outfall had appeared to be satisfactory and the place in question had no trouble at all over a long period. Then matters began to change. It was not a question of a fracture of the outfall pipe but of the alteration of the current pattern within that section of the English Channel. The circumstances resulted in the necessity for complete treatment of the sewage.

Treatment of the sewage does not necessarily take care of all pollution in estuaries but it will be a 'must' for the future, in certain instances.

OFFORD (Rofe Kennard & Lapworth). It is no use having a standard for effluents unless it can be related to what happens at the shore line. The Author seems to be trying to do this in Figs. 21 and 22. There is some missing information however which would help in understanding these diagrams. This is the discharge quantities and the initial dilutions which seem to affect the initial counts of the *E.Coli.* 10^7/100 ml is quoted in the one case and 10^6/100 ml in the other. These may be also influenced by the proportion of storm water in these discharges and by the type of diffusers, if any.

COVILL It should be appreciated that Figs. 21 and 22 refer to present untreated sewage discharges and in particular two of the major discharges in the Edinburgh area.

The dry weather flow from these two outfalls is $50 \times 10^3 m^3$/day for the Caroline Trunk Sewer, draining Granton and Corstorphine, and $81 \times 10^3 m^3$/day for the Seafield Sewer.

AIRBORNE SENSORS FOR MONITORING POLLUTION

P. G. Mott, B.A., F.I.C.E., F.R.I.C.S.
Managing Director, Hunting Surveys Ltd.

1. In a report on Marine Pollution presented at the United Nations Conference on Human Environment it is emphasised that pollution has its maximum effect in estuaries where man's influence is greatest. "The open sea (90 per cent of the ocean and nearly three quarters of the earth's surface) is essentially a biological desert, but close to shore, in estuaries and where upwelling currents bring nutrients to the surface, productivity is 10-25 times higher. It is these areas where air, land and water meet, which produce the greatest amount of marine life. Small wonder therefore that man has established his habitats, expanded his population, and concentrated his industrial growth, within these areas. It follows also that this is where he has discarded his wastes which are thereby concentrated at the end of the food chain—in birds, fish and mammals".

2. Estuaries are usually large areas with complex environments resulting from tidal movements, ocean currents and climatic changes. By their nature they are difficult of access and therefore any detailed study on foot or by boat is likely to take a long time and may well be unable to keep pace with the rapidity of change taking place, and the urgency of the investigation. The alternative is to carry out such studies by remote sensing from aircraft and in the future perhaps from satellites. By such means the whole of an estuary area can be comprehensively surveyed within specific states of the tide and at critical periods of change due to seasonal and climatic effects. Of course ground studies will still be necessary to monitor the interpretation of the airborne sensors and to relate the information so obtained to physical and chemical analysis of corresponding ground-truth samples, which will then act as interpretation keys for the area as a whole. There will also be information that cannot be recorded by any remote sensor and which must continue to be obtained on the ground. There is, however, no doubt whatever that remote sensing in its present state of development can reduce by a very large factor the amount of ground-work undertaken, and will at the same time serve as a valuable pointer to those areas where detailed ground studies are needed, and of which the ground worker with his limited vision may be totally unaware.

3. Airborne remote sensors operate in many parts of the electromagnetic spectrum and are designed to detect and differentiate between surface features by differences in radiation characteristics. Most sensors are termed 'passive' because they measure the energy generated by ground objects or reflected solar radiation. An 'active' sensor, on the other hand, itself generates an energy signal and measures the radiation reflected back by the target. Air photography and infra-red thermal imagery are examples of passive sensors while radar is the most widely used of the active sensors.

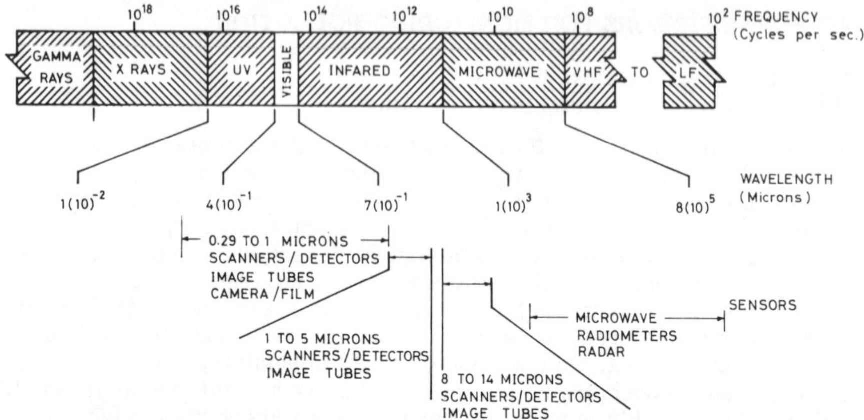

Fig. 1. Electromagnetic spectrum and sensors.

4. Although the atmosphere absorbs or disperses much radiation nevertheless there exist specific 'windows' where certain wave-bands pass through without serious attenuation and it is in these that the sensors are designed to operate. The spectrum of greatest interest is that from near ultra violet up to and inclusive of microwave as shown in Fig. 1. It includes the following best-known types of sensor:-

Conventional panchromatic air photography.
Colour air photography.
Infra-red (false colour air photography).
Multi-spectral air photography.
Infra-red thermal imagery.
Side-looking radar imagery.

5. Many other forms of sensing are in use or in various stages of development but the majority of these are non-imaging and therefore likely to be of restricted use in marine studies. This paper will therefore confine itself to a description of the sensors named above and to examining possible ways in which they can be of value in the context of marine pollution.

6. First it is necessary to say something of the operational and economic problems involved in the acquisition of airborne data. Apart from radar and passive microwave images none of the other sensors has the ability to penetrate clouds and haze. Therefore to achieve satisfactory aerial coverage of the whole of a large estuary by any other means than radar may well require several days or weeks of flying in order to reconcile the often conflicting requirements of clear weather, good lighting and the desired tidal periods. Clear skies are of course essential and to expect that such conditions will coincide with the particular days and periods of an hour either side of low-water springs (when the

maximum extent of the foreshore will be revealed) requires a large measure of optimism in this country. In any event it will mean that the aircraft has to stand by in immediate readiness to be airborne during the tidal periods and must therefore be charged for irrespective of whether any productive flying is done. Consequently tidal photography will always tend to be more expensive than that which is carried out at the operator's convenience. As an example the air photography of Morecambe Bay carried out in 1967 at a negative scale of 1/12 000 took six days over a period between 22nd July-14th September.

7. With some of the more sophisticated sensors such as Infra-red Imaging the equipment is expensive and because of its limited use is normally hired from the manufacturer rather than owned by the operator. Consequently arrangements have to be made in advance to secure the equipment and to fit it into the aircraft which is then restricted for other work.

8. Side-look Radar is an extremely costly installation and requires a large aircraft (such as a D.C.6 or D.C. 9) and for these reasons can only be made available in the case of a long-term and wide-ranging survey of pollution surveillance or for mapping large areas.

9. The operating problems are therefore considerable and will be referred to again later in the paper. A summary diagram of a pollution monitoring facility is shown in Fig. 2.

PANCHROMATIC (BLACK AND WHITE) AIR PHOTOGRAPHY

10. This is the conventional type of material used for the preparation of mosaics and maps and for specialised stereoscopic study. It is sensitive within the range 400-780 μm and within this range will record with a high degree of resolution. The modern air survey camera lens, which can be used either for panchromatic, colour, or infra-red emulsions, is virtually free of distortion and colour corrected.

11. Conventional air photography is taken with a 60 per cent forward overlap and 30 per cent sidelap so that all parts of the terrain can be examined three-dimensionally. Since the accuracy of ground heighting is largely dependent on the altitude of the aircraft above ground, in an estuary area it would require a low height (500 metres) to provide contours to the accuracy required. This would involve a large number of photographs and a close framework of ground heights which is neither practical nor economic in this type of survey. Therefore it is best to use photography as a means of preparing scaled mosaics which will serve

(1) As a basis for all subsequent information either obtained from remote sensors or on the ground.
(2) For recording changes in the channels, limits of mud and sand, marsh etc., from coverage flown in successive periods.
(3) To record main ecological boundaries, development areas and industrial workings where pollution is likely to be greatest.

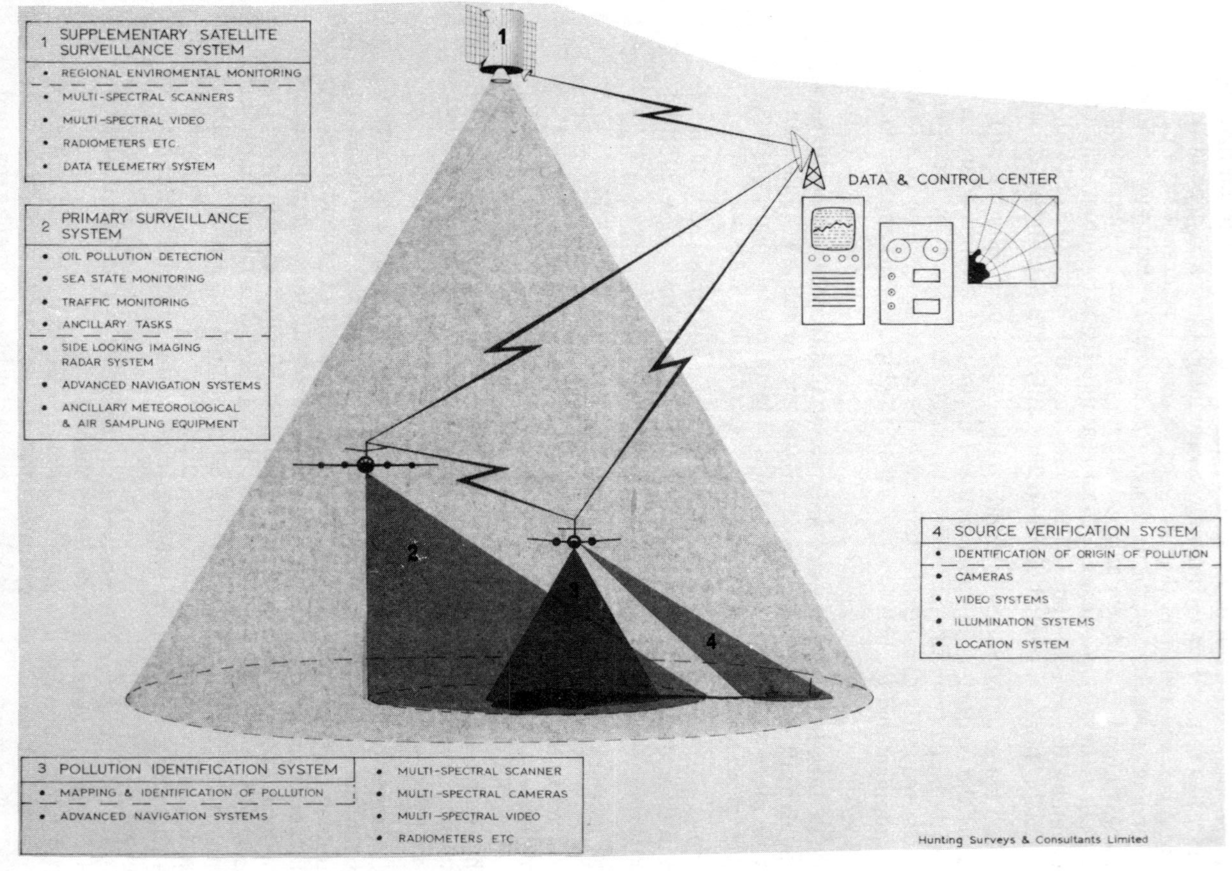

Fig. 2. Marine pollution monitoring facility.

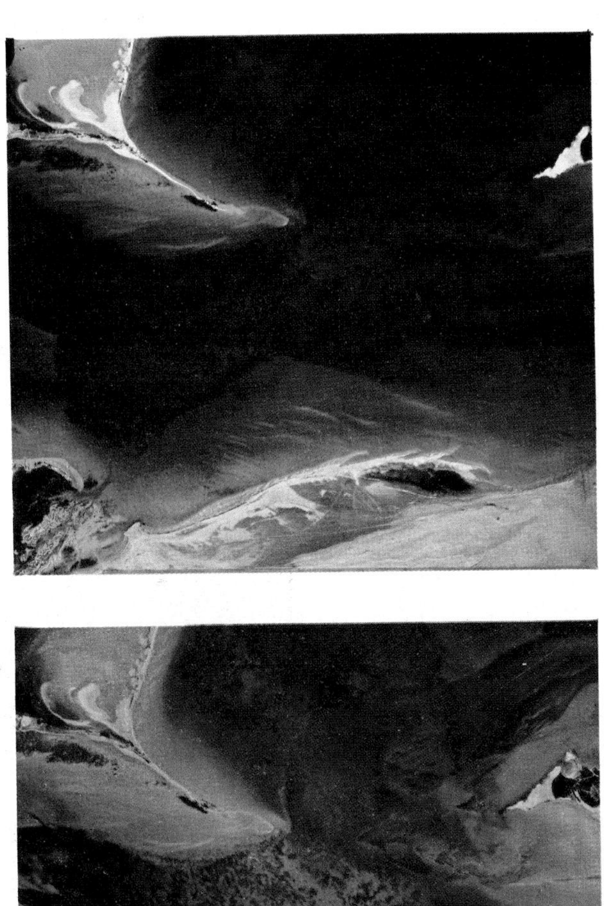

Fig. 3. The comparison of colour and false colour film in an estuary area.

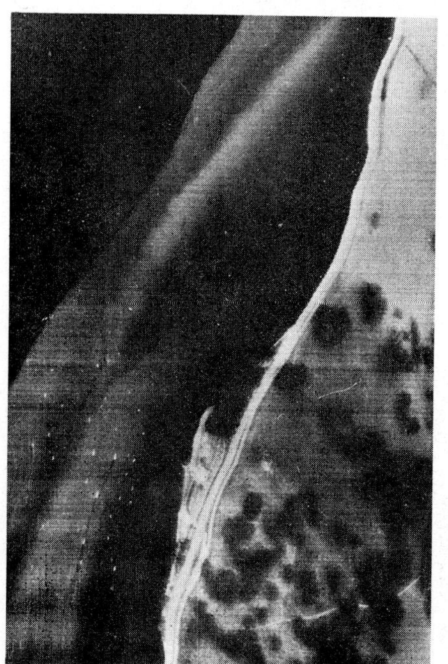

Fig. 4. 'By the process known as 'Density Slicing' it is possible to filter out objects appearing on a photograph within defined tonal limits, and to present these in enhanced colours. On the left, an infra red thermal line scan record with a stream of water emerging from the coastline and the corresponding image enhancement on the right.'

12. There is a tendency to underrate the value of panchromatic coverage in favour of more sophisticated and complex emulsions. Panchromatic has, however, many advantages over colour and Infra-red including superior resolution and speed, relatively low cost, and ease of handling and reproduction. The special emulsions should therefore in general be confined to limited selected areas, which have been defined on the panchromatic coverage.

COLOUR EMULSIONS (Fig. 3)

13. Standard colour film is composed of three dye-coupled halide layers each of which is sensitive to a different part of the spectrum comprising the primary colours. It has the best penetration in depth of clear water and is the optimum emulsion for detecting changes of colour due to pollution or sediment. For marine work colour is used for tracing the flow and distribution of effluent and natural current patterns by means of introducing dye tracing elements at the source of outflow or in cases where the effluent has an inherent colour discrimination from the surrounding water. Colour film has also been used for recording the relative movement of currents at varying depths by the introduction of coloured marker buoys attached to sinkers at different depths.

INFRA-RED PHOTOGRAPHY (Fig. 3)

14. For marine pollution infra-red false colour film has also a number of specialist applications. This departs from colour film by replacing the blue sensitive emulsion layer with one sensitised to infra-red radiation between 700-900 μm the other two layers (green and red) remaining unchanged. In addition the dyes produced in the different emulsion layers are not complimentary in colour to the effective sensitivities of the layers: hence the term 'false colour'. Table 1 illustrates the approximate visual appearance fo objects of different hues on an I.R. false colour film.

REPRODUCTION ON INFRARED COLOUR FILM. TABLE 1.

	REPRODUCTION ON FILM	
visual appearance of object	**infrared strongly absorbed**	**infrared strongly reflected**
red	green	yellow
green	blue	magenta
blue	grey to black	red
cyan	blue	magenta
magenta	green	yellow
yellow	cyan	grey to white
grey	cyan	grey

Objects having an intermediate level of infrared reflectance will be reproduced with a colour intermediate between the two extremes shown in the second and third columns. In any case, practical results may differ somewhat from the hues predicted in the table, because the reflected light from natural objects generally possesses some energy in other parts of the spectrum than that which gives rise to the visual appearance.

15. The importance of I.R. Ektachrome film for shorelines and estuary pollution studies lies in its ability to distinguish between live and dead vegetation and also for sharply defining the water-line as well as giving an indication of the relative depth of water in the channels. The chlorophyll content of healthy vegetation reflects infra-red radiation strongly. Thus healthy vegetation will in general show up as a strong red or magenta in comparison with dead or dying vegetation (i.e. that affected by pollution) which will appear green.

16. Water pollution studies will also be assisted by the detection of algae or seaweed floating on the water which is clearly picked out (if living) as patches of red. The distribution of seaweed on reefs or mud flats at low tide will also be clearly seen.

17. Water, because of its low reflectivity will show up as a strong blue, and on shallow beaches both the water-line and water-logged areas at low tide will be clearly defined.

MULTI-SPECTRAL PHOTOGRAPHY (Fig. 5)

18. Multi-spectral photography attempts to get the best of both panchromatic, colour and infra-red by covering the whole spectrum between 400 and 900 μm in one simultaneous exposure taken with a camera fitted with several lens/filter combinations each designed to cover a specific band of the visible spectrum and to filter out light of other wave-lengths. The I^2S camera has four lens combinations which expose on to one piece of Kodak I.R. 2424 aerial film. The dimensions of each multi-spectral frame are 94 × 94 mm so that a 2.4 times enlargement of any one exposure is needed to get back to the size of a normal aerial survey negative, and in that case from a given flying height the scale of the enlarged multi-spectral negative will be 1.5 times that of the survey negative.*

* Note: This is assuming the multi-spectral camera is fitted with a 100 mm focal length lens. With the alternative 150 mm focal length lens the scale would be approximately the same as that of the survey negative but the area of ground covered by the multi-spectral photography will only be 1/6th that of the survey photograph.

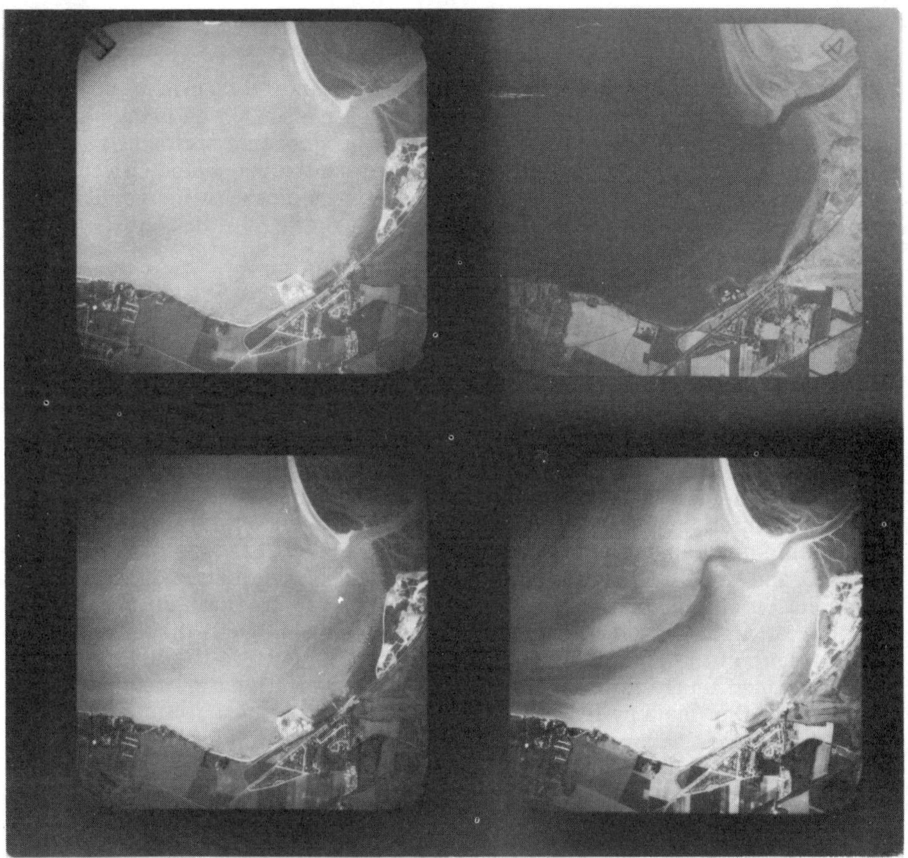

Fig. 5. One exposure of the multi-band camera provides 4 separate filtered images in the blue, green, red and infra-red bands of the visible spectrum on the same film.

19. Consequently multi-spectral photography is usually carried out separately from overall survey coverage and confined to areas of special interest.

20. Transparent positives made from the four multi-spectral negatives of a single exposure can be superimposed in a special viewer and each projected on a screen through a filter of different colour, the intensity of illumination being variable for the four images.

21. In this way the specialist interpreter looking for a particular type of feature, by varying the controls of the viewer, is enabled to bring

into play the band of the spectrum in which the feature appears in optimum contrast.

22. I.R. Sensors operate in the range 3-5.5 μm and 8-14 μm and are line-scanning instruments which collect radiation responses by means of a rotating mirror which scans the ground in progressive strips at right-angles to the direction of flight. The output can be presented on a cathode ray tube which in turn is photographed to give a pictorial image. Alternatively the output can be recorded on magnetic tape in digital form for later processing.

INFRA-RED THERMAL IMAGERY (Fig. 4, 6)

23. Since much of the pollution in rivers and estuaries arises from warm effluent such as that discharged from generating stations, oil refineries and manufacturing industries, the flow and distribution of these discharges can be traced from source of dispersal at sea by recording the temperature difference between the effluent and the surrounding water. Thermal anomalies are also produced by biochemical action at sewer outfalls which the I.R. scanner should be able to detect. Scanners now available are capable of recording temperature differences down to 0.2°C. and when operated from a flying height of 300 metres have a resolution of approximatcly 0.5 metre. They can be flown at any time of day or night but not of course through cloud which will disperse the radiation. In the collector system at present adopted in the U.K., the I.R. scanned image is not compensated for perspective distortion and therefore the imagery is not dimensionally correct except along the line of flight. More advanced systems however exist which employ electronic rectification of the imagery.

Fig. 6. Infra-red Thermal Linescan showing the discharge of warm water in an estuary area.

Fig. 7. Radar imagery covering an estuary in South America.

SIDE-LOOK RADAR (Fig. 7, 8)

24. Lastly there is side-looking radar which as already noted is an active system unlike those already described, which are all passive. Radar generates its own energy and measures the signals reflected back from the surface. SLR operates in the wave-bands 9 mm and 3 cms and is consequently able to penetrate all but very dense cloud. Its resolution as compared with photography is low and it is therefore more of value for large area coverage at small scales. As the aircraft moves, successive pulses are generated as a fan-beam at right angles to the direction of flight and cover a strip either side of the aircraft which is usable between 10°-70° slant from the vertical. Reflected signals are corrected in the processing for slant distances so that the cathode ray tube record, unlike that of I.R. Thermal, is reasonably true to scale.

25. Because of the obliquity of the signals radar imagery can differentiate clearly between surface roughness characteristics. Thus signals striking a smooth flat surface will be mostly refracted away from the aerial and there will be little or no return signal so that on the positive film record the area will appear dark. With rough water, the large number of steep faces at an angle to the incident radar beam will give high return signals and generate a light textural pattern on the film record. By this means it is possible to distinguish an oil slick from its surrounding rough water as illustrated in the photograph. (Fig 8).

26. The increasing magnitude of oil pollution along our coasts and the continuous danger of massive disasters like that of the Torrey Canyon, it is hoped, will eventually persuade the Government to create a permanent long-range surveillance system for continuous SLR coverage of

the main shipping lanes and for the accurate tracking of oil discharges by this means. This indeed should be an essential part of any plan to protect our estuaries and coastline.

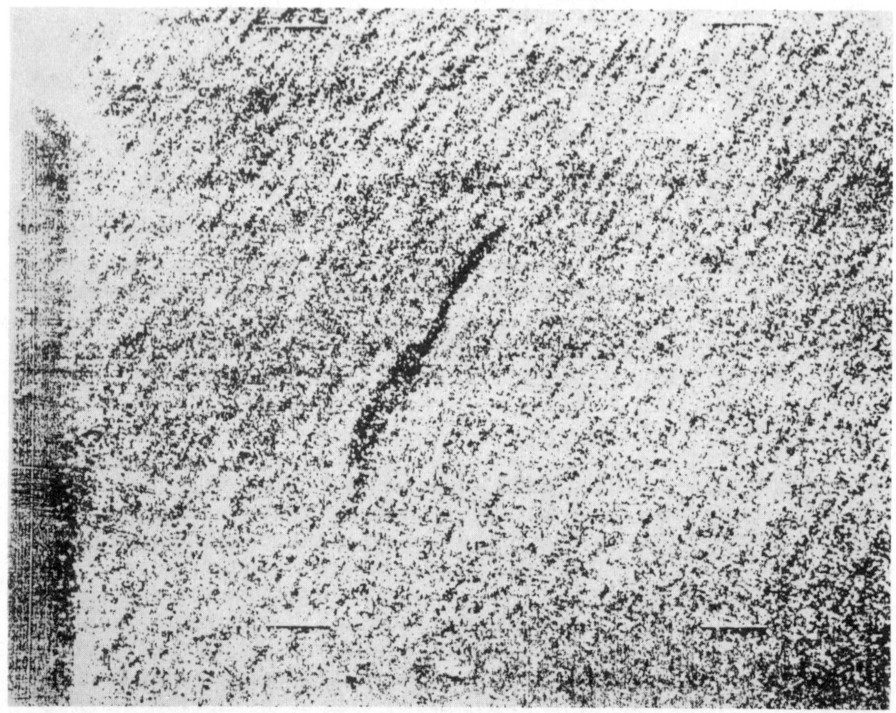

Fig. 8. Side-looking radar imagery, clearly showing the presence of an oil-slick at sea.

DISCUSSION

27. These then are the most common type of sensors applicable to the study and containment of marine pollution. The technology for using them and the development of new and improved forms of sensor is progressing rapidly as are also the means of handling and processing the data. Any data displayed as imagery or recorded on tape can be broken down into its component parts for analysis. For instance a conventional photograph is formed of a very large number of grey tones each of which may be associated with the reflectance of particular types of object (this is of course only partially true as aspect, sun angle and cloud shadow will affect the intensity of reflected light). Such tonal differences can be measured with a microdensitometer and digitised for computer processing. Alternatively by using a special type of film it is possible

to reproduce a series of masks each related to a particular range of grey tones. These masks can then be dyed in different colours and put together. Known as density slicing this process accentuates areas of small density differences so bringing out in sharp contrast small tonal variations that may not be discernible to the eye when examining a photograph or a photographic sensor image. Fig. 6 shows a density slice of an infra-red linescan image of warm water effluent from drains entering an estuary and being pushed upstream by the cooler waters of the incoming tide. More sophisticated computer analysis can identify targets by their signatures in terms of image pattern density and can 'clean-up' imagery by removing superfluous signal effects or electronic noise. Density Slicing can also be displayed through a television system up to 32 shades of grey and the areas of each density automatically recorded.

28. The high cost of the airborne equipment has already been mentioned and when to this is added the laboratory equipment for processing and analysis of the records it will be appreciated that to mount an effective remote sensing operation requires both careful and selective planning and a fairly substantial budget.

29. From the standpoint of an aerial survey, an estuary is a small area and the outflow of a power station even smaller. Economically therefore the best way to provide a service for the effective and continuous recording of marine pollution as well as for the specialist interpretation of the records would be to establish a series of regional centres throughout the country each equipped with a team of specialists and the necessary equipment for analysing the records both of the airborne sensing as well as for sampling on the ground. The airborne operation should be centralised on a national basis or alternatively flown by contract which would have to be of sufficient size and duration to enable the contractor to fit out and operate one or two special aircraft with the capability needed to provide any of the designated forms of sensor at extremely short notice. In this way the major cost of the unit would be distributed between the user centres throughout the country, and an individual centre would be in a position to call on the services of the unit at short notice and at a relatively moderate cost. A nationwide shared system appears to be the only viable method of approach and with proper interchange between the regional centres, the pollution problem could be tackled as a whole with all the advantages of an integrated approach and exchange of information on a national scale.

30. Finally looking a little further into the future it seems likely that orbiting space platforms will play an increasing role in providing sensor information. The ERTS 1 Satellite has now been orbiting the earth for nearly a year and has already produced some remarkable results with its multi-spectral scanners operating on the same principle as I.R. Scanners. The negative scale of the ERTS photography is however 1/2.5 million so that it is hardly surprising that one cannot expect to use such photography for the detailed detection of small effluents. Nevertheless

the potentialities of satellite photography are far greater than may at first be apparent. 1973 will see the launch of the first manned space laboratory Skylab and this in turn will be followed by a series of space shuttles in which man and equipment will be ferried back and forth between the earth and an orbiting space station. We shall soon be viewing space photographs taken with long-focal length cameras and very high resolution film brought back to earth for processing. By this means and by the further improvement of sensors and processing techniques it is certain that a whole new field of remote sensing will be opened up with less reliance on conventional aircraft, fewer records to handle, and thereby the cost of pollution control greatly reduced.

DISCUSSION

JAMES (Department of the Environment). The Author has set out very clearly both the potentialities and also, with great frankness, the limitations in the use of aerial surveys. The high cost of airborne scientific equipment necessitates careful and selective planning and a fairly substantial budget. Objectives must be clearly defined before embarking on any survey work and particularly so with aerial survey work.

When pollution survey is required it is usually fairly obvious that pollution in some degree exists. In this country, the major points of discharge are known together with at least some knowledge of the general hydrographic patterns in the area; for example, we do not need an aerial survey to pin-point the Gosport boil. However, a more useful application may be in tracing the effects of discharges rather than merely pin-pointing where they are. Such problems as the rate of dilution of any discharge in the sea, the area affected, the production of algal growths on the surface or on the beaches, and the deposition of organic residues might be investigated. This will generally mean regular monitoring at similar tidal and seasonal conditions, although the weather may not always be suitable for flying at such times.

To get the most out of aerial surveys much data should be obtained before flying. For example, many sewage and industrial outfalls are regulated to discharge only for a few hours after high water and often sludge is discharged over a much shorter period around high water. So it is necessary to know at the time at which the aircraft is flying exactly what is being discharged so that the effect of the maximum, minimum, normal or abnormal conditions can be seen. It is important to know the meteorological conditions, but such data is normally obtained as routine in these surveys.

From studies in Liverpool Bay, it was concluded that by far the greatest pollution load came not from individual points of discharges, but from the Mersey. In this case of course the pollution is not a point load but is diffused and enters the Bay in the transitional region between fresh and salt water. Could the Author comment on whether either at present or in the future we could expect remote sensing to assist in comparing the effects of a diffused load coming in say from a polluted river with a load coming in from a point source?

In para. 11 the Author mentioned the use of panchromatic air photography to define the limits of mud and sand. One of the current studies being undertaken by the Author's firm for the Working Party on the Disposal of Sludge in Liverpool Bay is an attempt to map the mud in the intertidal area, initially off the Wirral coast. If this initial trial is successful, it is hoped to establish a regular monitoring exercise at, say, two-year intervals to detect any changes. Would the Author like to comment on the sort of scale or the degree of accuracy which might be achieved? The air survey is of course supported by ground survey, but ground survey in intertidal areas is difficult and limited by tidal conditions. There is also the question of how much time one can justify on a ground study to back up an aerial survey.

To what extent does individual judgement come into photo-interpretation? Would, for example, any two equally experienced interpretors arrive at exactly the same interpretation.

This is a most useful and timely paper, because it does seem that developments in this field are proceeding rapidly. Improvements in remote sensing and development of new techniques should ensure a great future for this type of work.

MOTT (Hunting Surveys Limited). If one is measuring temperature or colour it should be possible to compare the effect of a diffuse load and a point source. If it is a point source, it will be seen almost like a freshwater spring which can be measured or located that way. A diffuse load would show rather like a cloud of varying density. As it spreads out, smaller and smaller temperature differences occur as the warm effluent disperses in the colder water. This can be measured by air survey techniques.

From the preliminary work in the Mersey, it seems that there are differences in internal texture between mud and sand and doubtless also in the texture of the surface of the mud and sand. One tends to be rather smoother and the other tends to develop runnels. The effect may differ from area to area and interpretation will depend on ground samples as well as photographs. The accuracy of such an exercise is difficult to assess, but there is no reason to suppose that the method is not sound and will not work. Soils mapping is based on the same type of approach.

There is no such person as a photo interpreter. What one does is to take a specialist in a particular field, whether he be an ecologist, geologist or a forester, or a mud expert perhaps, and teach him how to use air photographs. He remains however primarily a specialist with interpretation skill. Given two sound specialists, generally speaking both, once they have learned to interpret, will interpret the same things on the photographs.

CHARLTON (University of Dundee, Tay Estuary Research Centre). On the point of differentiating between mud and sand, in the Tay Estuary this has been accomplished satisfactorily by comparing field observations and actual aerial photography.

What degree of penetration have these methods of air photography? For instance, an indication of silt load in a deep channel may be due to silt in the surface waters only, possibly as little as the top one or two centimetres. This may impose limitations on the use of the method.

MOTT In clear water, with colour film, a penetration of 7 to 10 metres can be achieved but the water surface must be smooth and the sun's angle such that there is little reflection. With silty water there would be little or no penetration.

BOTTOMLEY (Trent River Authority). Since the Author mentioned the Trent specifically perhaps some amplification should be given. We co-operated with Hunting Surveys on a thermal imagery survey being carried out on the River Trent. We were trying to check the results with temperatures actually measured in the river but it was only at the third attempt that the flight time of the aircraft was made to coincide with the ground survey and then there were discrepancies. The most important conclusion which emerged was that because the thermal imagery survey only measures surface temperatures, the practical application of technique in river management is limited where stratification occurs, as for instance in the immediate vicinity of cooling water discharges from power stations. The second main disadvantage is that the aircraft track cannot follow the course of a naturally meandering river. The two only intersect fairly infrequently and not always at the desired points.

TUCK (C.E.G.B.). There is no need to monitor warm water effluent from power stations all round the coast every day. The infra red imagery technique can be applied for planning purposes where an existing station discharges cooling water, and there is possibly additional capacity on the site for further power generation. The overall picture of the plume of warm water emerging from the existing station could be obtained and the effect extrapolated to represent the total site capacity. By this means the cooling water system designer can ensure that recirculation could be avoided or minimised.

The Author had said that the heated water was the greatest polluter of estuaries. This could be because the infra red techniques pictured warm water discharges more vividly than other effluents. The general order of temperature differences induced in the long term, however, were quite small; 0.6°C had been quoted in Mr. Savage's paper on Southampton Water, and it is debatable whether this should be considered as a pollutant detrimental to the health of the estuary, or serving to enrich it.

Infra red imagery has been found to have fairly severe shortcomings, although the C.E.G.B. has not employed Hunting Surveys for this work. The commercial system used by C.E.G.B. has not been capable of reliable resolution to 0. 2°C. Out of three trials only one was a success. At £5000 a time, this is a poor risk to take. The ground back-up required is considerable. Can the Author say whether a higher success rate can be expected?

MOTT It was not intended to imply that power stations were rated as major pollutors!

The point was stressed in the paper that remote sensing requires a very considerable exercise in co-ordination. This becomes a particular problem when the equipment is on hire, which is the case with infra red thermal sensing. In this country there are only two equipments made. One of them can only produce an output on film and cannot produce a fully calibrated magnetic tape record. Neither of them is image-corrected, the reason being that it has so far not been commercially worthwhile to develop equipment to the degree that has been done in the United States. Consequently, to obtain the most sophisticated equipment, it is necessary to hire equipment from the U.S. The U.K. results so far obtained have been quite good. Parts of the South coast of England have been recently covered with I.R. thermal sensing to locate offshore springs. This survey also recorded a number of effluent sources.

DODD (Berridge International Consultants). What we 'see' is often dependent on the viewing technique used.

In 1970 and 1971 oil spillage trials were carried out in the Eastern Atlantic in which some 100 tonnes of Kuwait crude oil were released and the physical, chemical and biological parameters of the oil slick studied. In a typical example, the slick had attained an area of about $10km^2$ in 26 hours as measured by convention photography from a helicopter, and also by flying the helicopter around the periphery of the slick and plotting its circuit on the accompanying ship's radar. The results of the trial were subsequently reported to a NATO Committee in Brussels and were followed by comment from a US representative who produced satellite photographs of the sea area concerned. The latter, obtained by microwave scan, showed the slick to cover an area of some $25km^2$, i.e. $2\frac{1}{2}$ times that recorded by conventional photography and by pilot observation.

The explanation given was that surrounding the optically visible slick there was a considerable area covered by a film below interference colour thickness—probably down to a monolayer—which nevertheless suppressed the capillary ripples on the water surface, and it was this latter effect which was recorded by the microwave technique. Other evidence has indicated that this very thin layer is produced by the rapid migration of surface-active agents present in the oil—an observation which has also been made in this conference in relation to sewage slicks.

TECHNIQUES FOR POLLUTION CONTROL IN ESTUARIAL WATERS

D. W. Mackay, M.I.Biol., P.A.I.W.E., A.M.B.I.M.

Estuary Survey Officer, Clyde River Purification Board

1. Major studies on levels of pollution in the estuaries of the British Isles are still relatively few in number and, for obvious reasons, have been concerned with those cases which have been recognised as highly polluted for many years.

2. Techniques for pollution measurement have tended to match the scale of the problems to be dealt with, that is, they have been concerned with such factors as gross faecal, and therefore presumably bacterial and viral, contamination, substantial de-oxygenation of waters, large scale discharges of acutely toxic industrial effluents, and substantial oil spills. This work, often involving the physical or mathematical modelling(1,2,3) of the estuarine circulation system and its capacity to absorb polluting loads, has been of first priority, and is essential for the rational planning of remedial measures.

3. Such an approach was adopted for dealing with the highly polluted estuary of the River Clyde on the West Coast of Scotland, when work commenced in 1965. The estuary receives the drainage from a highly industrialised area, with a population of 2.2 million, and is subject to a wide variety of pollutants, resulting in extremely bad conditions, by any standards, in the upper reaches. The measurement of pollution in such an area is relatively simple, but further seawards, although we know that there are large quantities of pollutants present, they are so well dispersed that detection is not easy, and an assessment of the possible harmful effects is extremely difficult.

4. As the organisers have stated, the purpose of this conference is to discuss pollution in those areas of high commercial, amenity, and recreational value which the general public would like to feel are being maintained in a state of near pristine purity. Such areas include commercial shell-fisheries, bathing beaches, wild life sanctuaries, and those areas of great natural beauty on which the tourist industry depends. Pollution in such circumstances is rarely drastic, but it can be insidious, and the remainder of this paper records some of our experiences and unresolved problems in measuring and controlling pollutants in the cleaner parts of the Clyde Estuary.

5. The term 'estuary' has been used rather loosely in this paper to embrace the Firth of Clyde, as shown in Fig. 1., including the fiordic sea lochs and Ayshire coastline.

6. The problems to be encountered within this area include pollution by untreated domestic sewage, sewage sludge dumped from vessels, toxic elements, organochlorine compounds, and oils.

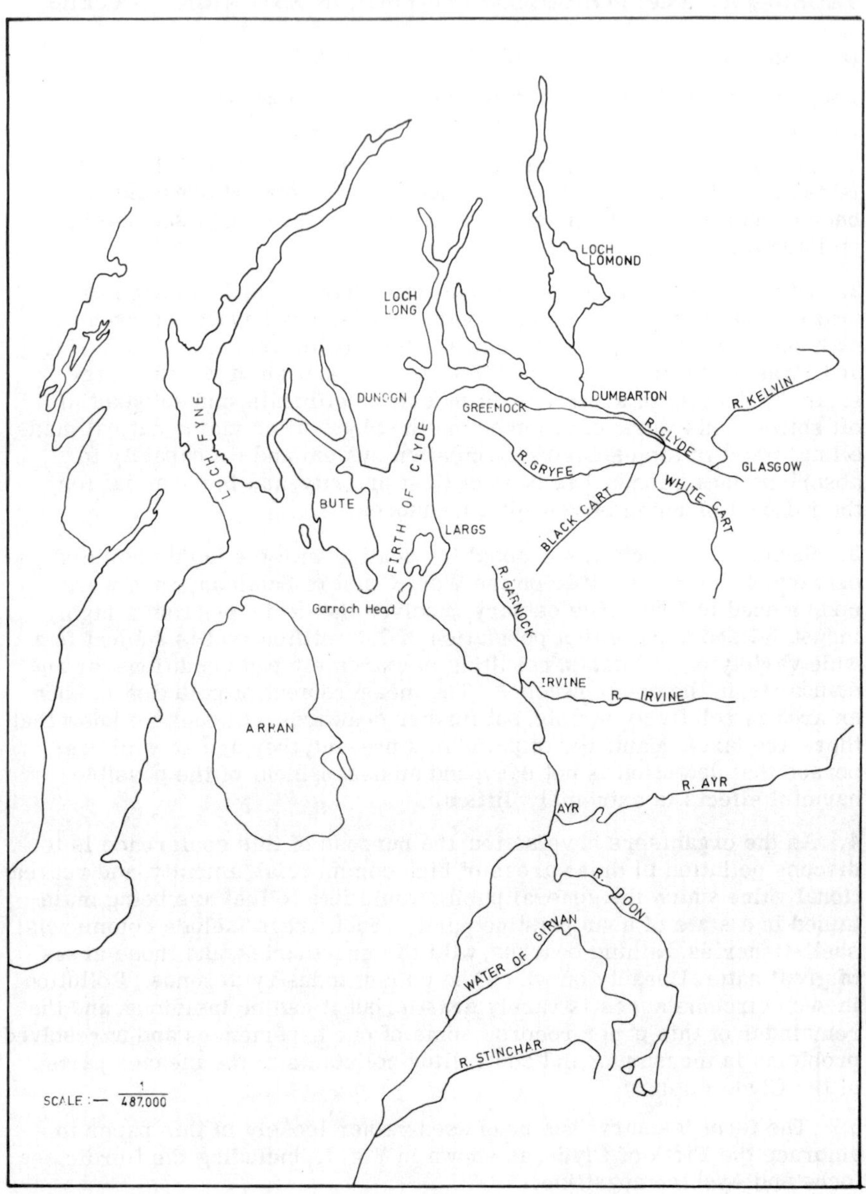

Fig. 1 The Firth of Clyde.

7. A typical monitoring exercise is that carried out monthly between Arran and Glasgow when, at 14 stations spaced over a distance of 90 km, the water mass is sampled at surface, mid-depth and near-bottom.

8. Seventeen parameters are currently determined for each water sample, and these are shown in appropriate grouping in Table 1.

Table 1 Parameters determined on Clyde Major Traverse

Physical and Biochemical	Nutrients (Dissolved Inorganic)	Trace Metals
Salinity	Nitrate	Zinc
Temperature	Nitrite	Copper
Dissolved Oxygen	Ammonia	Nickel
Biochemical Oxygen Demand	Phosphate	Lead
Suspended Solids (Organic)	Silicate	Cadmium
(Inorganic)		
Chlorophyll		

9. The survey is time-consuming, often arduous, and requires skilled personnel for the analytical procedures. A typical set of results is shown in Fig. 2, but the point which it is desired to illustrate is that, while the influence of major discharges in the upper estuary is easily discernible, there is no obvious indication, by any of the parameters considered, that more than a million tons of sewage sludge per year are dumped in the vicinity of station 2.[4,5,6]

10. This perhaps serves to illustrate the degree of sophistication which is required for measuring other than gross pollution. Some examples are given below.

DOMESTIC SEWAGE

11. Traditionally measured by relation to levels of bacteria in the receiving water, the difficulties of providing a clear picture by these methods are only too well known. Viruses are now an important subject of investigation in sea water, but routine monitoring, or indeed interpretation of results, has a long way to go. The dilutions which can be achieved between the discharge point of sewage and the adjacent shoreline are a matter of considerable interest to anyone concerned with the question of how long a marine outfall should be. On the West Coast of Scotland, Rhodamine B dye has been used successfully in several cases[7], but for general purposes, and especially for measuring the suitability of existing installations, there is a real requirement for identifying some component of sewage, preferably some component of faeces, which is reasonably conservative and can be measured at great dilutions.

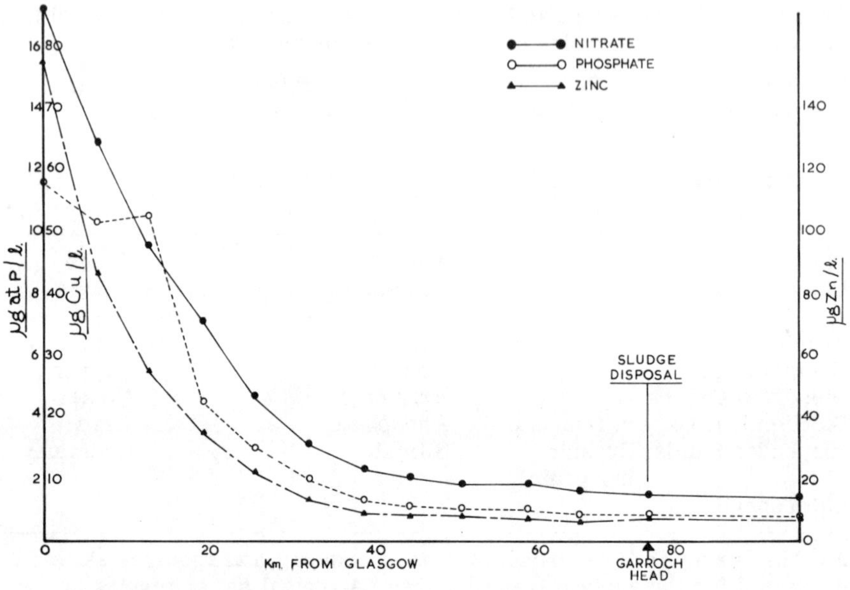

Fig. 2 The Clyde Major Traverse.

12. Once, and if, such a component can be identified, the analytical techniques must be streamlined so that many samples can be dealt with speedily and economically, before the method will have substantial practical usefulness.

13. The tracing of coastal water mixing patterns on a larger scale has, in the past, been attempted by artificial injection of dyes, radioisotopes or micro-organisms, followed by observation of the equilibration of the input within the marine system.

14. A more elegant technique towards the same end, at present being applied by M. S. Baxter and A. McKenzie of Glasgow University, in collaboration with the Clyde River Purification Board, involves the study of the natural isotope pair $^{228}Ra/^{226}Ra$. Here the tracer source is represented by estuarine sediments from which both ^{228}Ra ($t_{1/2}$ 6.7 yrs., β-emitter) and ^{226}Ra ($t_{1/2}$ 1620 yrs., α-emitter) are leached from their insoluble parent species ^{232}Th and ^{230}Th respectively.(8) Removal of the relatively soluble radium isotopes is accelerated in estuarine systems by disturbance of the sediment regime by turbulence and biological mixing. In addition, because of the relatively short half-life of ^{228}Ra compared with ^{226}Ra, loss to the water of the former isotope is much more rapidly regenerated. Estuarine waters are therefore both

rich in total radium and high in ratio $^{228}Ra/^{226}Ra$. In effect, the radium content and $^{228}Ra/^{226}Ra$ ratio are a direct function of the residence time of the water mass. In contrast, surface ocean waters which incurse into estuarine systems are low in $^{228}Ra/^{226}Ra$ since the mean time of vertical transport of ^{228}Ra from its source in pelagic sediments is very long in comparison to the decay half-life. Similarly, pure river water of low salinity can be expected to be of low radium content, due to its very limited exposure to sediment material. Thus, in an estuarine or coastal marine system, there is a marked radium concentration gradient between the component water bodies, so that detailed evaluation of isotope variations can yield data of value in establishing the rate and direction of water movement.

15. Experimentally, the measurement of the two radium isotopes requires careful chemistry and nuclear counting procedures. The overall process has, however, been revolutionised recently by the development of Mn-impregnated acrylic fibres which quantitatively remove the minute levels of radium from sea water. ($\sim 10^{-17}$ g/l for ^{228}Ra). Thus pumping of 500 l of water through 250 g of acrylic fibres permits extraction of sufficiently high isotope levels for subsequent radio assay. ^{228}Ra is extracted from the fibre with HCl, precipitated with $BaSO_4$ carrier, which is converted into $BaCl_2$ via $BaCO_3$, and finally the ^{228}Ra is measured by low-level β-counting of its ^{228}Ac daughter product in a TTA extract. ^{226}Ra is measured by α-counting of its gaseous ^{222}Ra decay product emanated from solution.

16. In essence, the technique provides a new and natural parameter for tracing water transport. Combined with the conventional measurements of salinity, temperature, dissolved O_2, etc., radium isotope assay can provide information not only on current directions, but also, most significantly, on their rates.

ORGANOCHLORINES

17. Pesticides and chlorinated bi-phenyls are typical examples of materials which, initially present in very low concentrations, can nevertheless produce disastrous effects in the long term. The Clyde estuary has the unhappy distinction of having had in 1971 higher levels of pesticides in Mussels (*Mytilus edulis* L.) than most European estuaries, with values for Dieldrin as high as 0.13 mg/kg in a relatively unpolluted area. Additionally, Holden[9] calculated that 1 tonne per year of P. C. B's were being discharged in sewage sludge to the outer firth, and Bogan and Mackay (unpublished data) have shown enhanced levels of P. C. B's in the fauna and sediments underlying the deposit area.

18. At first sight, the problems of pesticides and organochlorines might appear to be among those most difficult to deal with in terms of monitoring in estuaries, but our practical experience has led us to believe otherwise. The reason is that, having accepted that these materials are

potentially very dangerous, effective action at the only possible sites, that is, the sources, can be undertaken. Once a substance such as dieldrin was identified as a pollutant in the estuary, even at sub-lethal levels, a survey of industrial and agricultural practices in the area was carried out, several probable sources identified, and action taken to reduce the discharges, either by better industrial practice or the substitution of a less harmful material capable of doing the job specified. Regarding P. C. B's, a similar programme was organised, and in the Clyde Board's area a British Rail depot was identified as the major source[10] and an alternative material substituted. Coupled with similar action elsewhere, and voluntary restrictions by the manufacturers, the calculated output of P. C. B. to the firth of Clyde has been reduced by 60%. Nevertheless, the monitoring of the dumping area continues.

LOW DENSITY MATERIALS

19. One particular aspect of estuarine pollution which has not received much study, and may be of significance, is the surface film between water and air. Oil and discharged sewage have at least two things in common— they float, at least temporarily, and they are unpopular with bathers. Sewage, carried in fresh water and discharged to the denser waters of an estuary or marine area, rises, in most cases, to the surface, and three phases can be readily identified. Large lumps of material protruding above the surface may sail along, mainly under the influence of wind. Suspended solids may be transported in the water mass or, subject to such phenomena as accretion and flocculation, settle out and become part of the bottom sediments. A third phase, which appears to have been little studied, are those components of the sewage which become entrapped in the surface film. Almost all seawater surfaces contain measurable quantities of fatty acids and higher alcohols, and anybody who has bathed in even lightly polluted inshore waters will have realised how efficient an instrument the human body, with its covering of fine hairs, is for collecting and concentrating scum. It has been demonstrated[11] that a fine layer of surface active material can influence rates of evaporation and rates of gas transfer through the water surface. Very few, if any, pollution control authorities regularly sample the surface film, but, in an era when sewage is screened and finely minced before being discharged to marine waters, the surface film may be expected to assume some degree of importance. Sampling is an obvious problem, and Liss[12] has reviewed several methods, including the surface skimmer described by Harvey 1966, a variety of fine meshed metal screens, and the use of rising air bubbles. Probably of more practical use is the method devised by Levy and Walton[13] for sampling dispersed and particulate petroleum residues. The sampler was a modified neuston sampler-basically an open-ended aluminium box, to which a nylon plankton net (mesh opening 243 μm) is attached. It is constructed to be towed at the side of ship, clear of the wake. Levy and Walton towed the device

for distance of 1 nautical mile at speeds of 5-7 knots. Approximately 740 m^2 of sea surface were thereby sampled to a depth of 15-20 cms.

20. The Clyde River Board are developing this technique for the investigation of both sewage slicks and oil pollution.

OIL POLLUTION

21. In the Clyde estuary, oil pollution takes two forms. There is the identifiable 'incident'—when a spill is large enough to result in the mobilisation of the highly organised 'Clydespill' unit, a consortium of local authorities, the Navy, the police and other relevant bodies, whose activities are coordinated by the Director of the River Purification Board. Appropriate action can be taken in such cases, and several fairly major spills, including the sinking of a small tanker, have been contained within recent years by the rapid and efficient use of booms, pumps and, where advisable, dispersants of the least harmful type. However, every year there are many unreported incidents, spillages of no more than a few gallons, the dumping of waste oil from car sumps, the small quantities carried from road surfaces and garage forecourts, individually trivial, but additively perhaps of more significance than the major spills.

22. To quantify this material, both in, and on, the receiving waters, is a major problem, and, hopefully, the skimming device described in the previous section can be utilised for this purpose.

BEACH POLLUTION

23. How does one objectively describe the degree of pollution on a stretch of beach? A great deal of thought has been given to this problem, but no general quantitative method has been developed. My own Board is experimenting with the subjective scale of Garber(14), as illustrated below:-

WATER APPEARANCE ANALYSIS
(Beach or Ocean Sampling)

Code	Identification	Application
CONTAMINANTS		
C	Clear Water	Clear, no off-colour water or particulate matter.
D	Ocean Debris	Driftwood, marine organic trash not otherwise classified.
K	Sewage Debris	Match sticks, hair, sludge floc, some garbage.

Code	Identification	Application
H	Human Faecal Matter	Intact faeces must be differentiated from animal waste. Must determine whether from pleasure boat.
CC	Rubber Goods	Condoms and rings, old or new.
R	Refuse, including garbage from Beach and Land Use	Domestic trash such as cartons, cans, boxes, bottles and garbage from use of beach recreation areas.
TR	Floating Trash and Garbage from Boats and Ships	Similar to R above, but judged to originate from boats or ships.
S	Seaweed	Any kind of seaweed.
B	Dead Bird	Any dead marine bird.
ML	Dead Marine Life	Usually dead sharks, sea lions, or seals. Sometimes fish or other marine animal.
P	Plankton Blooms or Rafts	Plankton bloom discolouration of water. Confirmed by plankton counts and analysis.
SP	Spores	Usually kelp spores that appear as a surface scum or film. Confirmed by analysis
O	Oil	Mineral oil from ships or other sources. Ship bilge pumping, fuel spills, etc.
OS	Mineral Oil Scum	Mineral oil slicks associated with natural oil seeps.
G	Particulate Grease, Sewage Origin	Grease particles or balls near waste outlet containing 40% saponifiables.
GS	Grease Scum, Sewage Origin	Slick appearing to originate at a sewage discharge point.
T	Tar	Floating mineral oil tar. Collected and examined physically and chemically to ascertain possible origin.
N	Noxious Odours, Fumes or Gases—Non-sewage	Mercaptans, sulphides, smog odours from industrial activities.
NS	Noxious Odours, Fumes or Gases—Sewage	Sewage or treated sewage odours present in water or along beach.
M	Murky-Dirty	Water dirtied by causes other than plankton blooms, M_1 approx. 5 ft. Secchi, M_2 approx. 2.5-4 ft. Secchi, M_3 approx. 2.5 ft. Secchi.
F	Outflow of Water to Ocean from Land	Usually storm drain outflow which can affect ocean water condition.

QUANTITY

1	Small Amount	Traces of the coded materials.

2	Moderate Amount	Some of the coded materials at intervals. Usually not objectionable.
3	Large Amount	Enough of coded materials to be objectionable.

WATER COLOUR

B	Blue	Predominant colour as noted by the observer. Usually coded with reason for unusual colour, if any.
BG	Blue-Green	
G	Green	
OD	Olive Drab	
BR	Brown	
R	Red	

TOXIC ELEMENTS

24. Metals as pollutants have received widespread attention in the last few years, and the quantities which can be discharged to an industrialised estuary are considerable. (Table 2)

Table 2 Estimated Discharges of Five Metals into the Firth of Clyde*
Tonnes per year

Source	Zinc	Lead	Copper	Chromium	Cadmium
Rivers	160	80	25	10	1.0
Sewage Effluent	30	25	10	45	0.6
Sewage Sludge	60	25	25	40	0.5
Total	250	130	60	95	2.1

25. However, large as these quantities are, there are other sources; for example, natural lodes in the area, the atmosphere, and incursion of open-sea water, which can complicate the situation to a degree where results are extremely difficult to interpret.

26. Metals can be analysed in water or sediments, with water samples providing the record for the immediate present, and sediment samples providing an integrated representation of long-term conditions. Trace metal analysis of sediments is relatively easy, either by atomic absorption spectrophotometry or by emission spectroscopy. Either direct-reading or photographic methods provide a rapid means of determining a large number of elements simultaneously in a solid sample. Although the precision is not currently greater than 15%, new techniques such as replacing the arc by a high temperature plasma are being developed. The principal advantage of optical emission spectroscopy is that it is possible to rapidly screen large numbers of samples for a wide range of elements.

* Based partly on data received from the Dalmarnock Laboratory of Glasgow Corporation.

27. Once a good deal is known about the natural mineralisation of an area, then selected elements can be used as tracers of components of discharges. For example, silver has been used as a tracer,[6] and detrital minerals can be used as indicators of deposition patterns in coastal waters where chemical analysis provides a more rapid, if less precise, picture of sedimentary sources than time-consuming granulometry and mineralogy.

28. The use of the mass spectrograph for isotopic discrimination can be of considerable help in elucidating sources of polluting elements. In the Clyde, for example, the lead present probably has at least three sources, e.g. industrial-dissolved, atmospheric aerosols, and natural ores, and isotopic ratios may be used to determine the proportions present from each source.

ORGANISATION OF SURVEYS

29. At the present time, the assessment and monitoring of pollution in estuaries is attracting a good deal of attention, and this section deals with how one might organise the investigation.

30. Several facts are worthy of mention, as starting-off points:-

(1) An estuary, more than most other bodies of water, is naturally variable, and the rapid fluctuations which can occur, temporally, geographically, physically and chemically, mean that sampling has to be comprehensive, both in space and time. In the same way, computations of steady-state, or average conditions are of little practical value.

(2) In the context of estuaries in Britain, pollution has no relevant baseline characteristics. The major industrial estuaries cannot realistically be expected to be restored to the condition of those few estuaries which remain virtually un-polluted. However, they can be vastly improved, and the standards for each estuary will depend on the historical pattern of use, and the value of certain characteristics. For example, it would be illogical to set a bathing water-standard, if such existed, on an estuary which, for other reasons, e.g. heavy urbanisation and industrialisation, is unlikely to be used for that purpose.

(3) The object of monitoring surveys should be to assist in the reduction or avoidance of pollution. It is useless if the surveys only serve to record grave damage after it has been done.

31. Based on the above points, the surveying of an estuary might include the following:-

(a) The calculation of available dispersion, and residence time of pollutants. An understanding of circulation in the estuary is of first importance, and probably basic to the establishment of efficient sampling procedures for chemical and biological data.

(b) To measure every possible pollutant at a sufficient number of stations to provide effective overall screening would require an enormous team of workers. Therefore, parameters and sampling stations both must be selected on the basis of predictable hazard, and although some pollutants might be common to all estuaries, the correct course would be to review the types of industrial, agricultural and domestic practices in the hinterland, together with sources of entry to the estuary, and design the sampling programme accordingly.

(c) Unfortunately, while the measurement of identified harmful chemicals in the environment, and in animals, is very useful for practical pollution control, it leaves the problem of unrecognised pollutants very much to the fore.

(d) Biological monitoring—that is, measuring pollution in terms of its total effects on the biota, is perhaps the only long term safeguard for estuaries. The accent must be on *long term* because natural fluctuations, such as, for example, were caused by the severe winter of 1962-63, can be very large, and estuarine communities are notoriously unstable. In spite of this, biological monitoring provides the only valid basis for measuring the sum and product of effects produced by the many pollutants added to most estuaries.

32. Measurement of the efficiency of biological processes, e.g. by C^{14} fixation, or the feeding rates of copepods, may, in the future, provide the most reliable short term evaluation of environmental quality. Specific ion electrodes are now being developed to the point where continuous automatic monitoring of at least a limited range of chemical parameters is now practicable, and major developments in this field are to be expected.

33. Finally, it can not be stated too often that pollution investigations in estuaries have no practical virtue if they are not linked with an executive authority over the hinterland. The pollution of estuaries results mainly from activities on land, including industrial processes, methods of sewage treatment, and specific planning policy, and it can only be reduced and controlled at these levels. This simple and obvious fact seems to be frequently overlooked by those who assume that if we identify and measure pollution, we have virtually cured it.

ACKNOWLEDGEMENT

This paper is published with the permission of J. I. Waddington, Director, The Clyde River Purification Board, whose guidance and advice is gratefully acknowledged.

REFERENCES

1. Mackay, D. W., and Fleming G. Correlation of dissolved oxygen, fresh water flows, and temperatures, in a polluted estuary. *Water Research,* 1969, **3**, 121-128.

2. Mackay D. W. and Waddington J. I. Quality predictions in a polluted estuary. Advances in Water Pollution Research. Vol. II Paper III 7. Pergamon Press, 1970.
3. Mackay, D. W. and Gilligan, J. The relative importance of fresh-water input, temperature, and tidal range in a polluted estuary. *Water Research*, 1972, **6**, 183-190.
4. Mackay, D. W. and Topping, G. Preliminary report on the effects of sludge disposal at sea. *Effluent and Water Treatment Journal*, 1970, **10**, 11, 641-649.
5. Waddington J. I. and Mackay D. W. Pollution control in estuaries and inshore waters (1972) Proceedings of the Institute of Water Pollution Control, Symposium on River Pollution Prevention. Edinburgh, 1972, pp 17-34.
6. Mackay, D. W., Halcrow, W. and Thornton, I. Sludge dumping in the Firth of Clyde. *Marine Pollution Bulletin*, 1972, **3**, 1, 7-10.
7. Mackay, D. W. (1968). Sea outfall studies; the use of temporary outfalls and fluorimetry. *Surveyor*, 1968, **82**, 3982, 14-18.
8. Moore, W. S. Measurement of ^{228}Ra and ^{228}Th in Sea Water, *Journal of Geophysical Research*, 1969, **74**, 694-704.
9. Holden, A. V. Source of polychlorinated biphenyl contamination in the marine environment. *Nature*, 1970, **228**, 1220-1221.
10. Waddington, Best, Dawson and Lithgow. P. C. B's in the Firth of Clyde. *Marine Pollution Bulletin*, 1973, **4**, 2, 26-28.
11. Jarvis, N. L., Timmons, C. O. and Zisman, W. A. The effect of monomolecular films on the surface temperature of water. Retardation of evaporation by monolayers. Transport processes, V. K. La Mer, (editor) Academic Press. 1962, pp 41-58.
12. Liss, P. S. A survey of the organic material present at the surface of the oceans. Paper presented to Marine Chemistry Discussion Group. 'Organic Chemistry in the Marine Environment' Imperial College, London, 1972.
13. Levy, E. M. and Walton, A. Dispersed and particulate petroleum residues in the Gulf of St. Lawrence. Presented at the Aarhus Symposium on the Physical Processes Responsible for the Dispersal of Pollutants in the Sea. Denmark 4-7 July, 1972.
14. Garber, W. F., Receiving Water Analyses. In Proceedings of the first International Conference on Waste Disposal in the Marine Environment. (E. A. Pearson, Ed.), 1960, pp 372-403.

DISCUSSION

MACKAY There is nothing absolute about pollution. You cannot with any justification say that if man changes something, it is necessarily pollution. For example, if the dissolved oxygen level in the tidal River Clyde could be raised to 60% saturation so that migratory fish could pass, and smell nuisance and problems with ships boilers were eliminated, this would be an adequate improvement. Colour, high bacteria counts and high levels of chronically toxic substances would not matter, as this area will never be a recreational resource. Over the last decade of so the mud flats of the Clyde Estuary have become enriched by organic material which has been shown to be derived, at least in part, from the material discharged in sewage effluents. These mud flats of decomposing organic material have become established as one of the most important wintering grounds in Europe for waders and ducks. The productivity in the central belt of the estuary is naturally high, but is increased by the sewage discharges. Are we dealing here with pollution or enrichment? It is not possible to draw absolute lines for universal application.

Having established that dissolved oxygen levels (D.O.) are a useful indicator of pollution in the upper estuary, we can immediately demonstrate other situations where this is not the case. In one of the sea lochs, which is completely unpolluted, the D.O. levels from time to time during the year drop down to just less than 1mg/1. It is then practically devoid of oxygen, but still supports a good range of fish. The low D.O. levels in parts of the loch are a natural occurrence due to the shape of the loch basin. If the shores had been built on, and then investigations of D.O. had been carried out, low D.O. levels might well have been attributed to pollution rather than to natural circumstances. All standards are subjective.

It is quite unrealistic to plan a monitoring survey until the objectives are defined. Why do we want clean water? There are obviously several possible answers. One is to protect public health. In this case clean water is characterised by having low levels of harmful bacteria. But how few? People frequently quote the figure of a most probable number of 1000 coliforms/100ml. Various agencies have adopted this standard as a useful indication of quality. If one reads the literature carefully, it is usually made clear that it is considered to be a useful indication of the presence of sewage. It is too easy to get the idea that this is some level below which there will be no disease and above which there will be incidence of disease.

There are two weaknesses in human nature that bring this about. One is the desire for simplicity, the love of a nice number. The second is our instinctive detestation and fear of our own excreta. The scientific evidence regarding the health risk from bathing in sewage-contaminated water is not proved. In this respect, Dr. McGregor (Paper 3) rendered his colleagues a disservice in the way he referred to the investigation of the Medical Research Council (1959) and the later work which has taken place, as a white-washing exercise. This work has been re-examined critically from time to time. Why should these workers have any reason to disguise or twist the facts to make them other than they found them? It *is* possible that their conclusions are wrong because the evidence is not always as much as one would like to have. Nevertheless, the conclusions must be accepted as honestly reached.

Risk of typhoid and cholera have been referred to in earlier discussions at this conference. (McGregor and Covill). These are diseases whose names produce an emotional response. However, we must examine the situation logically. Reduction of bacterial counts to a significant degree in sewage polluted estuaries would require chlorination or some similar process. The cost would be many hundreds of million pounds. How many cases of typhoid or cholera would then not occur? Are these two diseases present at significant levels anyway. Would it not be far better in terms of cost-effectiveness to spend the money on improvement of hygiene in shops, restaurants and operating theatres? (It is reported that there is a 30% risk of post-operative infection in hospitals). On the evidence available it is not possible to justify cleaning up sewage pollution on health grounds. If, despite this, we accept that it is necessary, then standards must be laid down.

Is a level of 1000 coliforms/100ml a reasonable target for bathing waters? Seagulls frequently produce levels in excess of this.

Are fisheries being ruined by pollution? There is very little evidence that commercial catches round the coastline are being seriously diminished by pollution. The oyster fishermen in Southampton Water might welcome the pollution levels presently found there. These seem to favour oyster growth. If the reverse had obtained, if the oyster population had been decreasing rapidly, which it might well have done for natural reasons, pollution would have been regarded as the main culprit immediately, without worrying about high correlation between the two. In any case contaminated shellfish can be cleansed before marketing. Even if all contaminated beds have to be rigorously guarded the financial cost would be trivial compared to the alternatives of sewage treatment that have been suggested.

The conclusion is that pollution by domestic sewage is a social problem. It is a matter of what people desire rather than what they really need. Some of us like living in the country surrounded by trees and fields rather than in a concrete city. Industrial pollutants are more of a potential hazard than an actual destroyer of our commercial fisheries and other natural resources in coastal waters, except in a few estuaries. This is not to say that pollution does not exist. It exists from the point of view that the majority of people in this country do not want to be by, or play in, water containing sewage whether it is a health hazard or not. It exists also from the point of view that industrial effluents increase every year in volume and variety and that catastrophies have happened elsewhere and may happen here.

Our objectives then are two-fold. Bathing beaches and recreational waters must look and feel right to those using them. If the quality of the water is such that people use the beaches without complaining, it is fair to say that, 'If it looks all right, it is all right', as regards bathing.

Secondly, the effects of industrial effluents must be checked continuously to ensure that no rapid or sudden changes in the environment are taking place.

How is this to be done? Each case must be taken on its merits and the best available survey techniques applied. Two case studies will be described. The first is mentioned in para. 7 of the paper. The sampling programme takes about 9 hours to complete and of course several extra man days are involved thereafter in analysis. Results are presented as in Fig. 2 (p.11.4). It is possible to see the influence of the pollutants entering at Glasgow, and the major sewage works on the estuary, but there is very little to show further down the estuary. Glasgow Corporation dump about one million tonnes of sewage sludge a year just outside Garroch Head, approximately twenty-thousand tonnes a week containing quite a substantial quantity of industrial effluent. This does not show up in the results although one of the sampling stations is on the dumping area. In fact, it *is* possible to detect it, but only by looking very hard. For example, there is quite a buildup in total organic carbon in the actual dumping area. Core samples from the mud in the dumping area also show increased metals content.

The question now arises as to whether there is economic damage to the system. The nearby beaches are not affected, so the tourist industry is not affected. It is a very productive fishery and the fishermen fish over the dumping area. They do occasionally have problems in that their nets become enfouled with gross solids from the sludge, but the fish population is apparently unaffected. The levels of metals in the fish have been thoroughly checked and found to present no health hazard.

The second survey to be discussed is that of the condition of the beaches along the coast from Ardrossan to Ayr. This area is somewhat like the Solent in that it is quite clean. It could not be described as filthy. Neither is it perfect. Most sewage is discharged at low water mark. The area has recreational and fishery functions and is developing rapidly in terms of both industry and population. The study was planned less than a year ago in the light of previous experience as a practical and useful monitoring programme within the normal constraints of finance, staff and resources. It resulted from the concern of a group of local authorities along 30km of shore about the effects of this environment of rapid industrial development and reorganisation of main drainage. Long sea outfalls have already been planned and the task now is to advise on what degree of treatment is necessary before discharge.

The local authorities have allocated the finance, which is quite considerable, to carry out the survey for three years in the first instance. A longer time than that will be needed to provide really useful results. Steady laborious checking over many years is going to be be needed to show the trends. On any short-term basis the separation of the effects of pollution from natural variations is going to be very difficult.

The survey team comprises the Author; four project leaders, who also have additional responsibilities in other fields, and who are a marine chemist, two biologists, and a hydrographic surveyor; plus supporting staff of whom four are of graduate calibre; together with the normal back-up services available to a pollution control authority.

It is a big step from river pollution to marine pollution. This has caused some of the confusion that is obviously present. Marine work is not just a slightly enlarged version of river work. It takes a different set of skills in terms of scientific staff and perhaps a different approach in terms of administrators. The team described is not large enough for the task so the support of the I.C.I. Brixham Laboratory has been enrolled in partnership.

How was the problem tackled? Firstly objectives were established. These are as follows. The first priority will be the correlation, mainly from existing data, of the nature and quantity of effluents entering the area from whatever source. It is necessary to select for study, from the infinite possibilities, the high risk components. If discharges of cyanide or D.D.T. are expected, for example, then these will be included. The composition of land drainage must also be known. As Irvine Bay is only one small part of the Clyde Estuary account must be taken of pollutants entering the area from the sea as well as from land sources. It is therefore essential to establish the exchange rate between the study area and the main Firth of Clyde before the capacity of Irvine Bay to mitigate the effects of pollution can be assessed. The main objective will be to establish levels of pollution on the beaches. The approach to this is described in the paper (para.23). It is not as analytical as one would like but it does give an assessment of the levels of pollution on the beaches. This is what matters to the users of the beaches.

The effects of existing discharges, re-located discharges, and new or increased discharges will be assessed when the capacity of the area to absorb polluting material has been established. An overall policy for the most economical and satisfactory method of dealing with waste will be evolved. It is essential to know how an effluent discharged to the system will be dissipated and dispersed in space. The study of water circulation including local transport from selected sites will be required.

A specific programme of action is then set out, including what is to be measured, how often, and what it will cost. It is realised that it will almost certainly be necessary to modify the programme in the light of experience gained during the first year.

Beach surveys show great variation along the length of beach considered but the pattern tends to repeat in various areas. Some areas are definitely good collectors of debris, but it may be cleared by a change of wind conditions. It will tend to come back to the same position again and clear again, and come back. Other areas remain fairly clean all the time.

Nutrients are important in this area because there are very large discharges from various factories. A very large concern will be coming into the area to be engaged in the production of anti-biotics. The interaction of these components must be studied also.

SMITH (University College, Galway). On the question of bacteriological pollution, it is useful to look at the aim of the Medical Research Council Survey of 1959. This was to establish whether or not poliomyolitis was being transmitted on beaches. It was then enlarged to deal also with the major waterborne diseases, typhoid and cholera. The results of this survey, with respect to water in the British Isles, were that in the absence of gross visible pollution there seems to be very little risk of catching the major killer diseases of typhoid and cholera. In general, microbiologists agree with this finding. The survey did not show that there is not a chance of catching minor infections particularly of the ear, eye, nose and throat and any open cut, from water which is sewage-contaminated. Are these minor ailments caught, and if so is there a level of contamination above and below which these are not caught? Work on this problem is extremely difficult to do because it requires a very large survey of people with illnesses that are frequently not taken to the Doctor. They are not notifiable diseases. The only large scale survey of any statistical significance whatsoever was one carried out by Stephenson(15) in America which has subsequently been re-analysed by Geldreich(16) who suggests that at coliform levels of 2400/per 100 ml there is a statistically significant increase in minor illnesses. This is the survey on which the Americans have based a safe level of 1000 coliforms/100 ml. They consider this to be a compromise between what is totally desirable and what is economically feasible.

Subsequent to that work the Americans have found problems with waterborne virus diseases. The incidence of these may or may not have any correlation with the coliform count. It is clear that sewage-contaminated water is a health risk due to micro organisms which happen to be viruses. Counting viruses is a much more difficult matter than counting bacteria and standards are therefore more difficult to set.(17)

The problem of anti-biotic resistance is important. 25% of all the *E. Coli* isolated from sea water by the speaker are resistant to one or more of six of the major anti-biotics and are capable of donating that resistance to any drug-sensitive pathogenic organism with which they come into contact. The control of transferable drug resistance is a major problem for the medical microbiologist. It seems to be irresponsible for any major survey to be carried out within this area without establishing *E. Coli* levels and without establishing the frequency of transferable drug-resistance factors. This information is needed so that the level and nature of anti-biotic resistance in the flora of the population living and bathing in these areas can be established. From this it will be possible to establish whether sewage-contaminated waters are significant vectors of anti-biotic resistance factors.

It seems therefore that with respect to faecal coliform counts, a figure based to a certain extent on experimental work in America of 1000 coliforms/100 ml is a reasonable standard to aim at for recreational water. In dealing with waters which are extremely polluted like the heavily industrialised areas of the Clyde, it is quite obvious that a standard of 1000 coliforms/100 ml is unobtainable. In other areas it is a feasible target. It is not a fixed limit above which everybody will die and below which nobody will die. It is one where a significant rise in minor infection occurs. The community has a right to decide whether it wants to take the risk of a significantly higher rise in minor infections.

A faecal coliform (*E. Coli*) count is preferred to a coliform count. This is a slightly technical point but faecal coliforms are definitely of faecal origin; coliforms may come from other sources. This demands a very minor difference in the testing. If a conversion factor of 2.5 to 1 is accepted, the standard would be 400 faecal coliforms (*E.Coli*) as opposed to 1000 coliforms/100 ml.

MACKAY The emphasis given by Mr. Smith is somewhat different from that of McGregor and Covill. We are now told that if we bathe we are not liable to contract killer diseases but that there is a risk of suffering minor infections. There is also a risk of gaining minor infections in cinemas, dance halls, etc. The public must have a clear picture of the degree of risk encountered in sea bathing as opposed to all other risks to which they are subjected.

The feasibility of large-scale monitoring of viruses at the present time is doubtful. In the immediate future there will not be enough skilled people for this sort of task. The money needed would be very great indeed.

WAKEFIELD (Coastal Anti-Pollution League Ltd.). The public wants a clean beach. Yet all the time the administrators seem to be sheltering behind the scientists and the doctors. The doctors have been arguing about the health risk for the last 15 years. The scientists are now doing the same. In the meantime the public are paying quite a lot of money and very little is being done. The great need is to get the sewage off the beach. Fix some sort of standard and then look and see what is happening. See if there is any improvement. But let us have action instead of argument.

MACKAY The public, by throwing rubbish, cans, bottles and plastic containers are among the main pollutors of beaches. If the public want oil pollution stopped, they must be prepared to pay a few more pence a gallon to have the oil brought into this country in safer ships. The practical clean-up problem from the public's point of view is simple. It just requires the expenditure of money, and this is very much in the hands of local rate payers everywhere.

BAKER (Oil Pollution Research Unit). The Author states that biological monitoring is perhaps the only longterm safeguard for estuaries. What should be monitored in a biological survey? Is there any validity in the idea of using indicator species or must the whole range of species be looked at to get a good idea of the system?

MACKAY One way of assessing the effect of a multiplicity of pollution sources is to look at the community of flora and fauna and see whether this has been seriously disturbed by the combined effects. There is one major problem, in that the community may change drastically without any interference at all from man. Our approach to this is to set up stations and look at the whole communities for long enough, over a wide enough area, to pick up unusual events which can then be investigated in more detail. The other type of problem is the gradual change in the system when it is modified by gradually increasing pollution. The only hope here is to pin-point the sub-areas where the changes are different from those that are occurring over a wider area. Everything will be changing, but it may be possible to pick out changes within localised areas which do not follow the normal pattern of change.

The idea of using indicator species is very attractive, but difficult in practice. An animal which is sensitive to one pollutant is a good indicator for that pollutant, but it may be highly resistant to another. The search for animals or plants with broad-band sensitivity is worthwhile, but is still at an early stage.

SUTTON (Esso Petroleum Co. Ltd.). There is a possible parallel between coliform bacteria as water quality indicators and sulphur dioxide as an air pollution indicator. Urban air pollution has been measured primarily in terms of sulphur dioxide, as SO_2 is easy to measure. Correlations have been made between morbidity, ill health and sulphur dioxide, and clear statistical correlations have emerged. Standards for sulphur dioxide pollution have been set. They have now been found to be largely irrelevant and erroneous because urban air pollution results from a mixture of substances of which sulphur dioxide is only one, and the total effect is not a consequence of sulphur dioxide alone.

We are saying here that if there is a coliform bacteria count of more than 1000/100 ml, there are likely to be cases of various virus and other diseases, and that therefore if the coliform count is kept below this value then all will be well. This is not a valid argument, and if accepted could lead to wasteful expenditure of very large sums of money. Unless presence of other harmful organisms is related to presence of coliforms regardless of treatment given, coliforms are not a good indicator. Is this true? Do we know? If coliform bacteria are not shown positively to be a good indicator, then they should be dropped from the standards.

MACKAY There cannot be universal standards set that will meet every situation. The people in the Solent area must be able to decide what they want from the Solent water, what they are going to use it for, and what they are prepared to pay for it, and then set the standards accordingly. The two probably will not match at all. Proponents of coliform standards maintain that they are a good indicator of the presence of recently discharged sewage. This is fair and reasonable, and it is unfortunate that this is translated by some into a public health risk without due qualification.

SMITH (University College, Galway). It is freely admitted that there is a problem in the use of coliform counts as a measure of health risk. On the other hand, it must be recognised that there is only one factor which has ever really affected the length of life in this country and that has been the improvement in the hygienic quality of water and food. This has been much more significant than any other advance in medicine. As far as water quality goes, this has been maintained on the basis of coliform counts. The justification for the reliance on this test is its success. It cannot be rejected simply because coliforms cannot be shown to cause disease. We are dealing with intestinal pathogens, which are cellular organisms. There is probably much more similarity between two intestinal organisms than between, say, two chemical gases.

15. Stevenson, A. H., Studies of bathing water quality and health, American *Journal of Public Health,* 1953, **43**, 529.
16. Geldreich, E. E., 89th Annual Conference, American Water Works Association, San Diego, Calif., 1969.
17. Shaval, H. I. and Katzenelson, E. The Detection of Enteric Viruses in the Water Environment, pp 347-361 in Water Pollution Microbiology, Ed. R. Mitchell. Wiley Interscience, New York, 1972.
18. Geldreich, E. E., 'Sanitary Significance of Faecal Coliforms in the Environment'. Water Pollution Control Series, Publ. WP-20-3. FWPCA, USDI Cincinnati, Ohio, 1966.

FACTORS AFFECTING SLICK FORMATION AT MARINE SEWAGE OUTFALLS

J. R. Newton, B.A.

Water Pollution Research Laboratory of the Department of the Environment

1. In any scheme for discharge of sewage to the sea an important aim must be to prevent deterioration in amenity; in many cases this may be the main consideration. Any judgement of what is an acceptable effect on amenity will inevitably be subjective. Nevertheless it seems clear that the greater the extent to which objectives of design can be specified in quantitative terms, the greater will be the precision with which these objectives can be attained, and thus the greater the chance of keeping costs to a minimum.

2. This paper discusses briefly several adverse effects of sewage discharges on amenity and the difficulties of describing these quantitatively. Such difficulties seem most likely to be overcome in the case of slick formation. A long series of observations of the slick at an outfall at Bridport indicates some of the parameters which determine whether a slick forms or not. Discharges were observed from each of two experimental outlets on the outfall and from the design diffuser, situated 430, 680 and 1370 m offshore respectively. Experiments to measure the areas of slicks formed by known quantities of sewage released into the sea are also described. These observations are then discussed in the light of theoretical considerations to indicate the manner in which the conditions causing slick formation may eventually be defined.

ADVERSE AESTHETIC EFFECTS OF SEWAGE DISCHARGES

Sewage Solids

3. Floating or stranded solids perhaps constitute the most objectionable result of marine discharge of sewage, and the only acceptable criterion for their incidence is that none should be visible. The best way to ensure this is by pretreatment to remove or disintegrate the solids; numerical measures of their occurrence are therefore inappropriate.

Discoloration and Turbidity

4. Reports by other observers[(1)] suggest that turbidity due to sewage diluted by clean sea water is barely detectable to the eye at concentrations of less than 1 part in 100. Measurements by this Laboratory[(2)] (some unpublished) of the optical densities of diluted sewage and of sea water support this conclusion and show further that at some sites the natural turbidity of the sea water can be equivalent to a concentration of more than 5 per cent settled sewage in filtered sea water. In some 350

cliff-top observations of the sewage boil of the 430-m outfall at Bridport, discoloration was detectable on only 13 occasions, although on 80 occasions the calculated initial dilution (I.D.) was less than 40. It thus appears that a discharge of domestic sewage is unlikely to cause visible discoloration if the I.D. is greater than about 50; a lower I.D. may well be satisfactory in this respect at some sites, especially in naturally turbid waters.

Odour

5. The maximum dilution at which sewage can be smelt is likely to vary with different individuals and with the composition of the sewage. Stander *et al*[1] consider that smell is noticeable at a dilution of 500 to 1000. Rawn *et al*[3] comment on the improvement achieved by fitting diffusers to two Los Angeles outfalls. Without the diffusers the initial dilution was about 40. In the boil areas there was a faint smell of sewage and industrial wastes, sometimes wind-borne for as far as 1½ km. With diffusers giving an I.D. of 170 at one outfall and 300 at the other, no smell was detectable except during calms, and with the slightest breeze it was dissipated within 300 m.

6. An attempt to examine the relation between odour and dilution was made when the old Bridport outfall was operating. This outfall, discharging crude sewage about 20 m below the average low-water mark, is now disused. For a period of a fortnight the presence or absence of a sewage smell was noted when routine samples were taken at the water's edge. The association was then examined between these results and the minimum sewage dilution found within 200 m of the sampling point, as estimated from the coliform counts in the sewage and the sea. Grouping the results according to the dilution values the relation shown in Fig. 1 was obtained. However, it is possible that overriding factors might include the dilution of sewage at the outfall itself. Casual records of smell noticed in the boil area of the new outfall during sampling to determine I.D. are compatible to the extent that a very faint smell was detectable when the I.D. was about 1000 and a much more distinct smell could be detected when it was 50-200.

Sewage Slick

7. A slick is a patch of water which appears smooth in comparison with the surrounding water. Slicks can occur naturally but one is frequently seen at an outfall, extending as a long streak up to several kilometres in a direction determined by wind and tidal currents. It does not necessarily mark the course of the most heavily polluted water because its movement will be mainly influenced by wind, while that of the underlying water will be more strongly affected by tidal currents[2]. The cause of slicks is an accumulation at the sea surface of a layer (probably a monolayer) of insoluble, surface-active material which damps out capillary ripples and thereby alters the reflection of light from that region. It

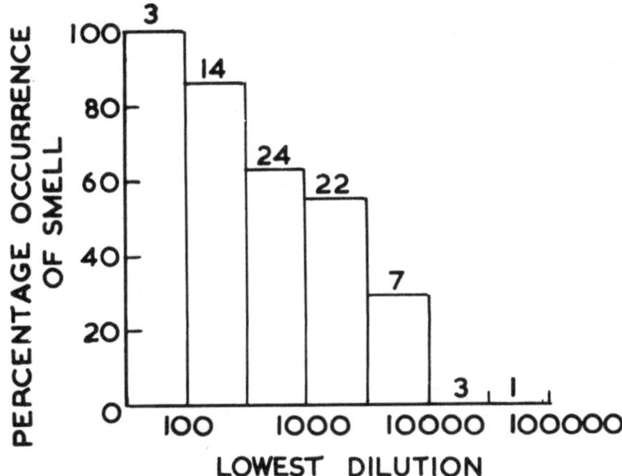

Fig. 1 Frequency of occurrence of smell on beach in relation to lowest dilution of sample within 200 m of old outfall at Bridport. (Numbers show total observations in each range)

should be pointed out that a slick which results from a spill of mineral oil is a different phenomenon from a sewage slick; the former involves much greater quantities of contaminating material usually present as an emulsion or thick layer rather than as a thin film. The remainder of this paper reports investigations into the causes of slicks at sewage outfalls and the conditions under which they form.

INVESTIGATIONS INTO FACTORS AFFECTING SLICK FORMATION

8. The observations and experiments described here were carried out in the course of field-work at Bridport between August 1970 and May 1972. The outfall, commissioned in 1970, serves a population of about 10 000 and carries mainly domestic sewage. Further details of the terminal diffuser and the experimental outlets are given in Table 1. The observations covered two periods of about a fortnight with each of the 1370-m and 680-m outfalls in use, and three similar periods with the 430-m outfall operating.

Routine Slick Observations

Method

9. The sewage discharge point in the sea was observed from a point on the cliff about 30 m above sea level and immediately inshore of the outfall. Observation times were at roughly hourly intervals, but varied more or less randomly according to pressure of other more urgent

Table 1 Outfall characteristics and percentage frequencies of slicks for each outfall length

Length from LWMST (m)	Depth below MSL (m)	Outlet type	No. of observations	Frequency of slick (per cent)
430	6.7	1 port facing seawards	358	58
680	8.4	1 port facing seawards	181	61
1370	16.1	5 port, Y-branch diffuser	198	12

tasks. Thus the number of observations per day was typically 7 but varied from 3 to 10 on occasion. The observer was almost always the author. The sewage slick was recorded as 'clear' (a marked slick clearly originating from a particular region) or 'faint' (a smoothing effect detectable but not originating from an obvious source) and its position and shape were sketched; if no slick was detectable the fact was reported. At or near the time of each observation, the sea state, sewage flow, the tidal level were noted.

10. The number of observations for each outfall length is given in Table 1 together with the frequency of occurrence of a slick with that outfall. It is clear that a slick was much less common with the 1370-m discharge point than with the shorter outfall. Little of this reduction in frequency is thought to be attributable to the increased viewing distance. To analyse the results further, the times of occurrence of a slick must be related to variations in relevant factors. The considerable uncertainties which arise in measuring or estimating these factors are now described.

Uncertainties in treatment of results

11. Sea state was visually estimated on a 4-point scale of wave heights: 1, 0-0.5 ft; 2, 0.5-1 ft; 3, 1-2 ft; 4, ⩾2 ft. Estimates by different observers usually agreed and seldom differed by more than one point. Outfall depth was calculated from the reading of a tideboard in the harbour entrance.

12. The sewage flow, read from the flow recorder of the pumping station, is that entering the pumping station. At low tidal levels this equals the gravity flow through the outfall. At high tidal levels a tide flap closes the gravity outlet and the sewage backs up. Siphons are actuated at a certain sewage level and the sewage flows into a pump well of volume about 60 m^3 whence it is pumped to the outfall, usually at 0.076 m^3/s (1.44 mil gal/d, 1 pump). Electrodes in the well control the pumps so that at higher flows two or more pumps may operate. Thus

typically over the high-water period the sewage is discharged via the pumps in about 6 periods of around 10 min, separated by longer periods when there is no flow through the outfall. Observations made during these periods of storage/siphoning and pumping were rejected (except those made with the 1370-m outfall in operation—see below) whenever there was any uncertainty as to whether the flow was zero, 0.076 m^3/s, or 0.152 m^3/s.

13. Tidal currents were estimated from the mean current curve for 700 m offshore allowing for tidal range, and are the greatest source of uncertainty. Substantial variations from the mean current are known to occur (e.g. predicted and observed slack water may differ by up to one hour) and each individual assumed current speed is thus suspect; however, as the quantity of data considered increases the effects of the errors will be reduced.

14. The results discussed below are those remaining after rejecting observations which involve obvious discrepancies or uncertainties but some undetected errors almost certainly remain. It should be noted that for some of the observations with the 1370-m and 680-m outfalls it was not possible to reject observations for which the sewage flow was probably zero, because no flow data were available. Furthermore, with the 1370-m outfall, 9 of a total of 14 slicks for which flow records are available occurred when the sewage was being pumped. Since this fact in itself is significant these observations were not rejected, as was done for the shorter outfalls, but were assumed to be associated with a flow of 0.076 m^3/s. On some of these occasions two pumps may have been operating but this would not alter the qualitative trends shown by the results.

Results

15. In Fig. 2 are shown the distributions of observed slicks from the 1370-m and 430-m outfalls with respect to three parameters thought likely to influence slick formation. Results for the 680-m outfall were very similar to those for the 430-m outfall, but exhibited the same trends rather more clearly. The following comments can be made:

(1) Sea state—Fig. 2(a). There is a tendency for fewer slicks to occur at sea states 3 and above. Observations suggested that at sea state 4 the roughness of the sea was the dominant mechanism preventing slick formation, and the histograms in Fig. 2(b) and (c) therefore refer only to observations at sea state 3 and below. Replotting these histograms for observations at sea state 2 and below showed no closer relation between the variables. This seems to indicate that up to sea state 3 the effect of sea state on slick frequency is small compared with errors and the effects of other parameters.

(2) Tidal currents—Fig. 2(b): The histograms show a tendency for fewer slicks to occur at higher currents.

(3) Sewage flow—Fig. 2(c): A tendency to more frequent slicks at

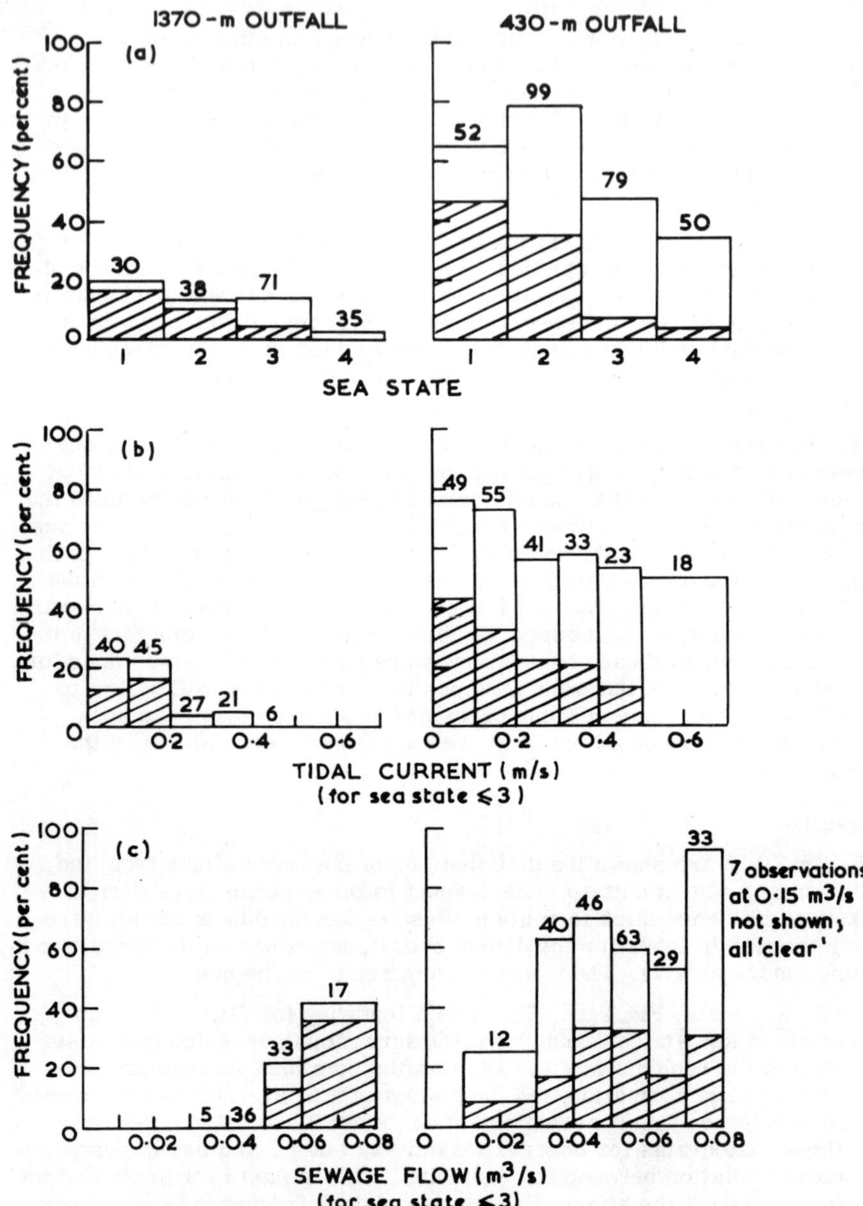

Fig. 2 Relation between percentage frequency of observation of slick and (a) sea state, (b) calculated tidal current, (c) sewage flow, for 1370-m and 430-m outfalls. Numbers show total observations in each range shaded areas indicate slicks recorded as 'clear', open areas as 'faint'

higher flows is apparent. Only dry-weather flows are considered, any observations relating to wet-weather flows having been rejected for all histograms.

16. The only other variables (apart from initial dilution which is discussed later) that seem likely to be important are wind speed, outfall depth, and the concentration of slick-forming material ('grease') in the sewage. Wind speed has not been considered because its effect seems likely to act directly through its influence on sea state. No plot against outfall depth has been included because depth is correlated to some extent with current (smallest and greatest depths occur at spring tides when currents are largest) and flow (pumped sewage discharges occur at times of large depths). No information is available on variations in the grease content of the sewage. Further discussion of these routine results will be made later in conjunction with other observations on the areas occupied by slicks.

Observations on Areas of Slicks Formed by Discharge of a Known Volume of Sewage

17. On 10 occasions the area of slick formed by a known volume of sewage was measured. The volume of sewage discharged was noted from times of pumping or from readings of the flow recorder. To estimate the area of the slick the Decca Navigator radio-fixing system was used. A boat using this system at Bridport can fix its position by means of two coordinates each given to an accuracy of about 10 m. An observer on the cliff sketched the shape of the slick and directed the boat by radio to four or more points on the edge of the slick. These points were marked roughly on the sketch and their Decca coordinates noted. Subsequently the points were plotted on a Decca chart, a line was drawn through them as near as possible to the shape of the slick sketched from the cliff, and the area of the slick was found.

18. This method is clearly not very accurate but gave reasonable agreement between the shape defined by the plotted points and that drawn by the cliff observer. The error in area is probably not more than about 40 per cent for slicks of area greater than about 10 000 m^2; all the slicks except one exceeded this area. The area of slick formed by one cubic metre of sewage ranged from 70 to 800 m^2 with a median value of 200 m^2.

Discussion of Results

Concentration of slick-forming material in sewage and slick

19. There is general agreement in the literature on surface chemistry[4,5], that surface films cause ripple damping, and in oceanographic literature[6,7,8] that such damping is the cause of a visible slick. While the details of the factors affecting the damping coefficient seem to be a matter of some dispute among surface chemists, the possibly over-simplified description of the behaviour of a surface film given below,

from Ewing[8], is adequate for the present discussion.

> 'Such a film, since it reduces the surface tension of water, spreads until it is one molecule thick (about 6×10^{-6} mm). If spread still further by some extraneous agency, the film breaks up, probably into discontinuous patches or islands of coherent material some millimetres in diameter; at the same time it loses its damping effect on ripples and its ability to depress the surface tension (Fig. 3, point C). If the area occupied by the over-expanded film is reduced by steady contraction, the film first begins to resist the contractions in an elastic manner (Fig. 3, point B); next it regains its ripple-damping ability and begins to reduce the surface tension; and then, as the islands fuse to form a continuous film (Fig. 3, point A) it abruptly becomes nearly as incompressible as the same material in bulk, since its molecules are then standing close packed like sticks of chalk in a box.'

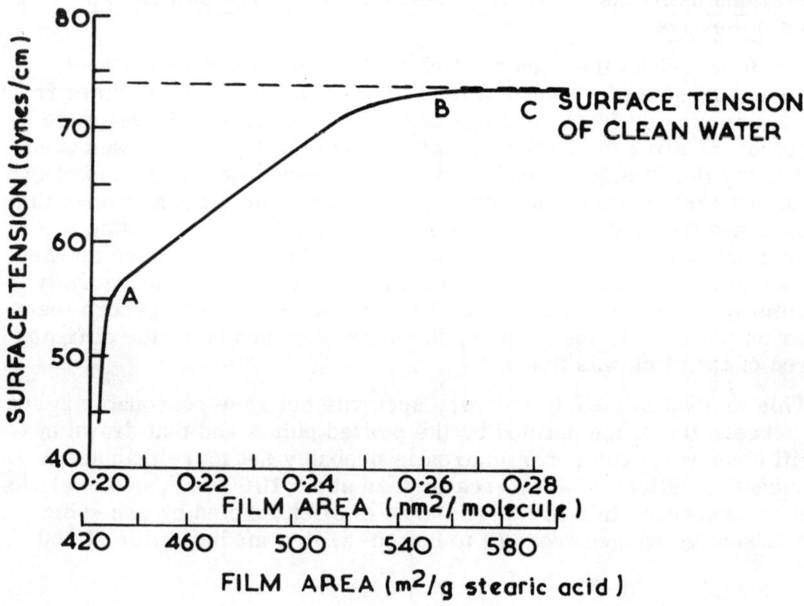

Fig. 3 Variation of surface tension with degree of extension for a typical film of fatty acid on water (derived from Ewing (8))

20. For a substance to show this sort of behaviour it must be both surface-active and insoluble in water. In domestic sewage most of the insoluble, surface-active material is in the form of fatty acids, their salts (soaps), and their esters (fats). Thus these are the substances

most likely to be the cause of a slick. Stearic acid ($C_{17}H_{35}COOH$) is a typical example. The curve in Fig. 3 is largely independent of the nature of the fatty acid for carbon chain lengths of 14-22 atoms(9).

21. A film area of 0.26 nm^2 per molecule can be shown to be equivalent to about 2 mg/m^2 (1 g-molecule stearic acid = 284 g = 6×10^{23} molecules). On these arguments, therefore, it is to be expected that a slick will be visible if the surface concentration of fatty material is greater than about 2 mg/m^2. It is possible that the slight alkalinity (pH 7.8-8.2) of sea water could cause ionisation of fatty acids in a monolayer and thereby change its properties(10) but it seems unlikely that any such change will be large enough to affect the general nature of the behaviour described.

22. The concentration of material in a sewage slick may be estimated from the measurements of slick area if a value can be obtained for the concentration of grease in the sewage. For Stevenage domestic sewage analysed by Painter and Viney(11) the major insoluble surface-active constituents were higher fatty acids (96 mg/l expressed as stearic acid) and their esters (42 mg/l expressed as tristearin). Thus 140 mg/l grease in the sewage may be a reasonable rough estimate in the absence of any information relating specifically to the sewage at Bridport. Applying this figure to the range of slick areas per cubic metre of sewage given above gives grease concentrations in the slicks of 2 to 0.18 g/m^2 with a median of 0.7 g/m^2. These values are high compared to the calculated minimum of 0.002 g/m^2, suggesting either that the sewage slick is disintegrated by wind and waves long before it becomes a monolayer or that much of the grease in the sewage does not contribute to the slick.

Factors affecting slick formation

23. The parameter which most clearly influences slick formation is sewage flow. It can be said with some certainty that the two shorter outfalls always form a slick if the sea state is 3 or less and sewage is being pumped. This fact does not emerge from Fig. 2 because most such observations were rejected because of uncertainty as to exactly when the pumps switched on and off. However in 46 out of a total of 61 such rejected observations a slick, clearly attributable to the last period of pumping, was observed either at the outfall position or drifting away from it.

24. Any general expression which predicts the tendency of an outfall to cause a slick must take account of differences in outfall configuration, water depth, and tidal currents (and also concentration of grease in the sewage if different sewages are considered). The most obvious parameter is I.D. In Fig. 4 is shown the relationship between slick frequency and the I.D. values calculated for each observation from Abraham's(12) curves for slack water and Agg and Wakeford's(13) correction for the extra dilution caused by a tidal current. The histogram shows a clear

decrease in slick frequency at higher I.D. but there is no clear upper limit above which slick formation does not occur. A lower limit of about 90, below which slick formation will usually occur is suggested by the observation above that the shorter outfalls usually form a slick when sewage is being pumped.

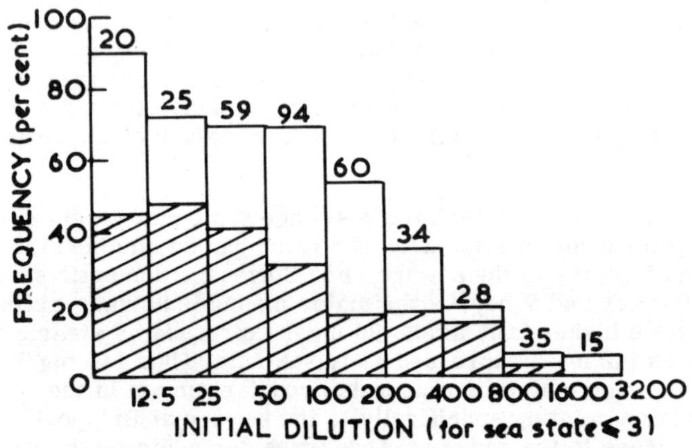

Fig. 4 Relation between percentage frequency of observation of slick and calculated initial dilution for combined data for all outfalls. Numbers show total observations in each range. Shaded areas indicate slicks recorded as 'clear', open as 'faint'.

Factors affecting area of slick

25. A slick formed at the boil will spread as it drifts away until a monolayer is formed and will then remain as a coherent film until broken up by wind or waves. Thus the area of any slick from a continuous discharge will be determined by how long the monolayer can resist disintegration. It is to be expected, therefore, that at higher sea states disintegration will be more rapid, resulting in a smaller slick, or none at all. To test this suggestion an examination was made of the sketches of slicks for the five periods with the shorter outfalls in operation. By measurement of the sketch the slick was classified as 'large', 'medium', or 'small' according to the length of its largest dimension (0-50 m small, 50-250 m medium, >250 m large). The results shown in Fig. 5 support the hypothesis that slick size is reduced as sea state increases.

SUMMARY AND CONCLUSIONS

26. Aesthetic criteria which might be used to indicate deterioration of amenity by pollution from sewage outfalls include discoloration and

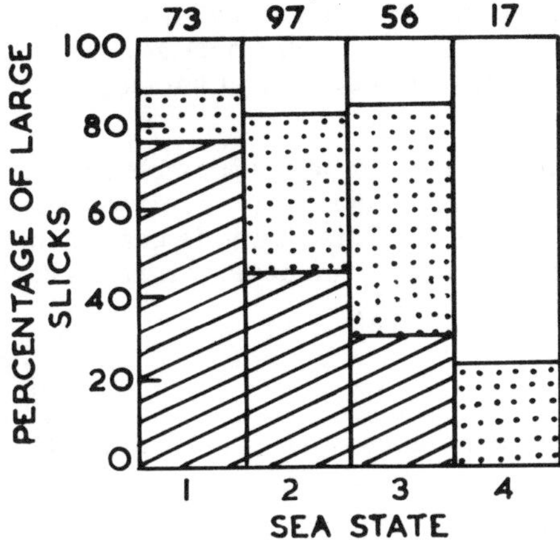

Fig. 5 Relation between percentage of large (hatched areas), medium (dotted areas), and small (clear areas) slicks and sea state for all observations with 680-m and 430-m outfalls for which a slick was seen. Numbers show total numbers of slicks seen at each sea state

turbidity, the presence of floating or stranded sewage solids, odour, and the formation of slicks. Attempts are made to describe some of these effects quantitatively and relate them to the characteristics of the outfall producing them.

27. At one outfall site (Bridport) studied, a detectable, though not necessarily unpleasant, sewage odour seemed to occur at dilutions of up to 1000 or more.

28. In the relatively clear water at the same site, the sewage plume was never more than just detectably discoloured (as seen from the cliff-top) on a few occasions although the initial dilution (I.D.) was frequently less than 50. In naturally turbid waters, discoloration by domestic sewage would probably not be detectable at a I.D. of 20.

29. The formation of a 'smooth-water patch' or slick is the adverse aesthetic effect that seems most amenable to quantitative treatment. A slick is caused by the ripple-damping effect of a surface film of insoluble, surface-active material.

30. The effect of increasing wave height is to decrease the frequency of occurrence of a slick at wave heights of greater than 0.3 m and to decrease the area of the slick formed by a continuous discharge of sewage.

31. For a given outfall configuration the frequency of slick increases as sewage flow increases and current decreases.

32. Frequency of occurrence of slick increases as I.D. decreases. For sewage with a grease content similar to that of Bridport sewage a slick will usually form if the I.D. is less than 100 but will form only infrequently if the I.D. is greater than 400.

33. The substances in sewage most likely to be responsible for slick formation are higher fatty acids and their esters ('grease') whose total concentration is of the order of 140 mg/l. This figure is consistent with the relevant surface-tension theory and with the observations that 1 m^3 sewage forms 70-800 m^2 slick in the sea.

ACKNOWLEDGEMENT

Crown copyright. Reproduced by permission of the Controller, H. M. Stationery Office.

REFERENCES

1. Stander, G. J., Oliff, W. D., and Livingstone, D. J. Problems in using the sea as part of a waste-disposal system. *Wat. Pollut. Control,* 1968, **67,** 143. See also the printed discussion of this paper.
2. Wheatland, A. B., Agg, A. R., and Bruce, A. M. Some observations on the dispersion of sewage from sea outfalls. *J. Proc. Inst. Sew. Purif.* 1965, 291.
3. Rawn, A. M., Bowerman, F. R., and Brooks, N. H. Diffusers for disposal of sewage in sea water. *J. sanit. Engng Div. Am. Soc. civ. Engrs,* 1960, **86,** SA2, 65.
4. Vines, R. G. The damping of water waves by surface films. *Australian J. Phys.,* 1960, **13,** (1), 43.
5. Goodrich, F. C. On the damping of water waves by monomolecular films. *J. Phys. Chem.,* 1962, **66,** 101.
6. Lafond, E. C., and Lafond, K. G. Sea-surface slicks. U.S. Naval Undersea Research and Development Centre, San Diego, California, Technical Paper TP 215, 1971, pp. 75-103.
7. Garrett, W. D. Damping of capillary waves at the air-sea interface by oceanic surface-active material. *J. mar. Res.,* 1967, **25,** 279.
8. Ewing, G. Slicks, surface films and internal waves. *J. mar. Res.,* 1950, **9,** 161.
9. Adam, N. K. The physics and chemistry of surfaces. 3rd edition. Oxford University Press, London 1941, p. 47.
10. Gaines, G. L. Insoluble monolayers at liquid-gas interfaces. John Wiley and Sons Inc., 1966, p. 228.
11. Painter, H. A., and Viney, M. Composition of a domestic sewage. *J. biochem. microbiol. Technol. Engng,* 1959, **1,** 143.
12. Abraham, G. Jet diffusion in stagnant ambient fluid. Delft Hydraul. Lab. Publ. No. 29, 1963.
13. Agg, A. R., and Wakeford, A. C. Field studies of jet dilution of sewage at sea outfalls. *Instn publ. Hlth Engrs J.,* 1972, **71,** 126.

DISCUSSION

BAKER (Oil Pollution Research Centre). Reports of slicks are received from aeroplanes quite frequently. They are usually reported as oil slicks but they can be anything; sometimes sewage, sometimes plankton. If visual criteria are used to measure pollution and remote sensing devices are employed to detect the pollution, it is doubtful whether the techniques described by Mott (paper 10) can distinguish between slicks of different origins.

NEWTON It is unlikely that a natural slick and a sewage slick can be distinguished because they are not very different. A natural slick is caused by the presence of insoluble surface-active material which is always there on the surface of the sea. If wind or currents act to sweep all this material together and compress it into a condensed film then a slick will be visible. A sewage slick and an oil slick might be distinguished, however, since the former consists of a monolayer of fatty acids while the latter is mainly a thick layer of oil/water emulsion.

CHARLTON (University of Dundee). In para. 7, the Author says that a slick does not necessarily mark the course of the most heavily polluted water. This is something which must be borne in mind when interpreting photographic images of slicks. The slick is essentially a surface phenomenon created by a small component of the total sewage content and the rest of the sewage may be somewhere completely different. The surface pattern is essentially governed by wind conditions rather than by water movement.

ROBERTS (Whitstable Urban District Council). This paper is very useful, and the results agree with some observations at the site of a new outfall at Whitstable. We too are conscious of the fact that the slick does not necessarily reflect what is under the surface. In conjunction with the Fisheries Laboratory at Burnham-on-Crouch, trials have been made using the bacterium *Serratia indica.* After the initial problems of tracing were overcome it proved quite useful in plotting and monitoring the levels of pollution present.

NEWTON The Water Pollution Research Laboratory has used the bacterium *Serratia indica* in experiments at Bridport with some success although its use is not without problems. It is a sensitive tracer in that a single bacterium in a 100-ml portion of sample can be detected thus allowing detection of substantial dilutions of tracer. However, when used in a study of average dispersion conditions in the sea over a period of a fortnight difficulty was experienced in maintaining the bacterial culture at an adequately high concentration for the whole of that time. A second difficulty which arises when interpreting results of such experiments is that the bacterial concentrations found are the result of both dilution and die-off processes. The mortality rate is uncertain and varies with sea and weather conditions, and it is thus difficult to separate the effects of the two processes.

BAKER (Oil Pollution Research Unit). Can visual estimates replace coliform bacteria as a criterion. Or are both needed?

NEWTON There is no necessary connection between the visual conditions and the coliform count, except insofar as where there are faecal solids lying on the beach, there will be a high coliform count. As to what criteria one chooses in defining acceptable conditions, caution is necessary. The criteria should be agreed between those who pay for achieving these conditions and those who suffer when they are not achieved. As a scientist I can only say 'If you take this course, this will be the result'. Somebody else must take the decision as to whether that result is worth paying for. This is a political-social-economic decision which must be made by the normal processes of public administration and politics.

DOWNING (Water Pollution Research Laboratory). The available evidence from the study of a number of sites in the South of England suggests that an outfall terminating about 400m below the low water mark taking sewage from a population of 10 000 would comfortably meet a standard of 1000 coliforms/100ml, as a median value or possibly even as a 90-percentile value. At Sidmouth, for example, the median count from the population of about 10 000, which is typical for the average small coastal community, was about 70/100ml. The 90-percentile would be about ten times greater. So even in these terms a standard of 1000/100ml would be satisfied.

I have no brief for a standard of 1000/100ml mentioned by Smith in discussion on Mackay's paper, because the arguments from which this standard were derived do not appear to be inviolable. However, the point is that should such a standard be desired, it can be achieved quite readily using the best modern design practice. The outfall at Sidmouth is about 400m long. A study of the statistics of outfall design and the lengths of outfalls under construction shows that nowadays 400m is about the minimum used. The one at Bridport is 1400m. At Bridport it is not possible to detect the effect of the outfall itself against the background of the input from seagulls and the discharge of polluted rivers in the vicinity of beaches.

It appears also from the Author's photographs that the sewage slick formed at Bridport is innocuous, although this is a purely subjective judgement and is open to argument. Many people might object to an outfall, the boil from which can be readily seen from the shore, no matter what assurances were given that it is harmless. However, at about 400 metres range it is doubtful if the boil could be seen. Neither at this distance is smell likely to be a problem, and if the sewage is properly screened or macerated there should be no visual evidence on shore if the outfall has been properly sited.

400m is not necessarily the right minimum outfall length for every circumstance but it might be right in many cases for populations of the order of 10 000.

WAKEFIELD (Coastal Anti-Pollution League Ltd.). There are plenty of seagulls off the end of the Gosport outfall and the sewage there is supposed to have been macerated. Do seagulls still abound around the end of an outfall with a diffuser on?

Could it be that seagulls are carrying the coliforms from the end of the outfall back in shore in cases where high coliform counts are attributed to the presence of gulls?

NEWTON There were always a lot of seagulls around the Gosport outfall during two days which I spent sitting in a boat at the site. The boil at Bridport is a very much smaller discharge. It does attract a few seagulls although not as many as the Gosport one. Whether this is the result of the diffuser or the much smaller discharge is hard to say. The Bridport outfall discharges about 3.5×10^3 m^3/d. The Gosport one is five or six times larger.

We have no information from our field studies to indicate whether or not gulls can carry bacteria from the outfall to the shore.

RODDA (Department of the Environment). The paper describes a study of a discharge to sea, not an estuary, and I wonder whether the Author has any evidence that sewage slicks in estuaries move differently or have different characteristics to those at sea?

There are many estuaries in Britain that receive a freshwater flow already mixed with sewage effluent discharged into the non-tidal reaches of rivers. There is therefore already likely to be a level of bacterial contamination present in estuaries as a result of sewage discharges to these upper reaches.

DOWNING (Water Pollution Research Laboratory). The Laboratory has not yet had an opportunity to examine slicks at sites other than Bridport by the methods adopted there. There seems no reason to suppose that a slick in an estuary will behave very differently to a slick on the sea, except for the fact that the rate at which the greasy material is dispersed may be slower.

Bacterial contamination from inland rivers was one of the problems at Bridport when trying to ascertain the effect due to the outfall alone. In particular the River Brit which emerges at the harbour not far from the line of the outfall sewer was a significant source of contamination of the nearby beaches. One method that has been used to distinguish contamination from the river from that from the outfall is to add *Serratian* as a tracer. Even if the sewage were removed completely there would be some background contamination from the river.

BAKER (Oil Pollution Research Unit). How much is known about the ultimate fate of constituents of sewage in estuary waters? How rapidly do bacteria die off? What is the fate of heavy metals and pesticides? What happens to grease before it is degraded?

NEWTON Bacterial mortality has been mentioned. Coliform bacteria in the sea die exponentially with time so we generally measure their die-off as T_{90}, the time for 90% of them to be removed. We have carried out a number of experiments in which a discharge of bacteria to the sea was followed and sampled for some hours. The shortest T_{90} found was half an hour; more typically it was about three hours and was seldom over twelve hours. The rate seems to be influenced by light, so it is slower at night.

I have no detailed knowledge of the ultimate fate of heavy metals and pesticides. Grease presumably disperses on the water surface and to some extent through its depth until it is degraded.

RAYMER (Southampton Corporation). One worrying thing about direct outfalls is the tendency for heavy industry to congregate around these areas because trade effluent control will be rather less efficient than it would be if the sewage went to a treatment works. The reason for trade effluent control when the flow goes to treatment works is to protect the treatment processes and to a certain extent to protect farm land if sludge is disposed of to land. At present there is unlikely to be such close attention to industrial discharges which are effectively directly discharged into either estuaries or into the sea. There will have to be a broadening of the control of industrial discharges generally to sewers for the reasons that have become apparent at the conference.

MARSTRAND (University of Sussex). The discharge of sewage to the sea seems to be an appalling waste of water. Most of it must have been fresh water before it passed into the sewer and in the normal course of events would be going back through the water cycle in one way or another. If all coastal towns followed the long outfall policy, the demands on the water being re-cycled by inland communities might become excessive. We have been told that the cost of building a new sewage works on an inland area or a coastal land area near here was roughly the same as the cost of a long outfall. It would be worth looking at this from the point of view that if land treatment is installed with a short outfall for a period, then the option remains open for re-cycling at a later date.

PLANNING THE POLLUTION BUDGET OF AN ESTUARY

R. E. Lewis, B.Sc., Ph.D., and R. R. Stephenson, B.Sc., Ph.D.

I.C.I. Brixham Laboratory

INTRODUCTION

1. One of the effects of legislation protecting our rivers and inland waters has been increased industrialisation of estuaries and coastal waters. Far less stringent control is necessary for the safe discharge of sewage and industrial wastes in these regions because of the greater dilution available and the neutralising effect of seawater. The objectives in disposing of wastes to tidal waters have been clearly stated in the Third Report of the Royal Commission on Environmental Pollution.[(1)] They suggest that 'pollution budgets' should be drawn up for each major estuary and the general aim of these should be to exploit the estuaries for waste disposal up to a level which does not endanger aquatic life or transgress the standards of amenity which the public need and are prepared to pay for.

2. Clearly then there are two aspects to be considered when determining what are acceptable levels of discharge—the effects on biological processes and the effects on amenity interests. The following account describes some of the procedures which must be followed by controlling authorities and industry when faced with the problem of deciding what levels of discharge are acceptable.

3. In assessing the biological consequences of a discharge there are usually two main considerations. One is the oxygen demand exerted as organic constituents of the discharges are broken down by micro-organisms or by chemical reactions, thus reducing the oxygen content of the receiving waters. The second is the toxicity of the constituents of the discharge to aquatic life. Lowering of dissolved oxygen will, of course, enhance the toxicity of any poisonous waste constituent in the water. Both the oxygen demand and the toxicity are dependent upon the concentration of the various materials present. Mathematical models of estuarine processes enable predictions of these concentrations to be made prior to the discharge of an effluent. What is more, in situations where pollution already exists, these models allow predictions to be made of the value of any improvement in effluent quality and the effects of additional discharges. The mathematical model does in fact permit the 'pollution budget' for each estuary to be calculated.

PREDICTION OF TOXIC CONCENTRATIONS

4. We shall now consider the approach most generally adopted for establishing the 'pollution budget' for discharges of soluble chemicals to an estuary. The methods for developing a predictive model are given,

without recourse to the mathematical details, and the application of this to waste disposal problems is discussed with reference to the techniques involved in allowing for biological effects.

5. To take a specific example:
Consider the problem of disposing of several thousand gallons of waste liquid to an estuary on a continuous daily basis. The waste liquid will probably be mostly freshwater but it contains a proportion of a soluble toxic chemical which does not degrade to any marked extent during its residence time in the estuary. Also let it be assumed that the estuary is relatively free from other pollutants.

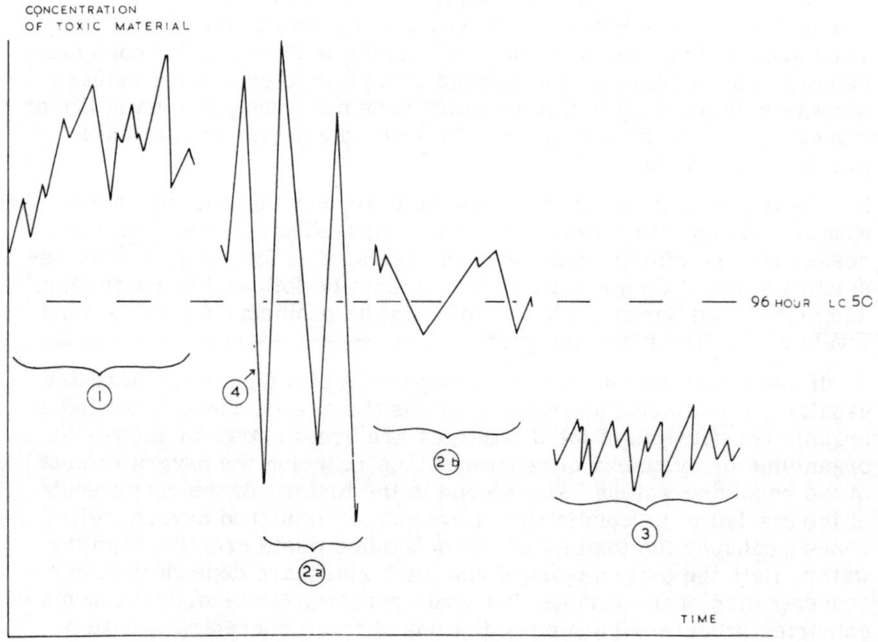

Fig. 1 The types of fluctuation in concentration of toxic material which may occur at a point in an estuary over a period of time.

(1) Acutely toxic—consistently above the 96 h LC50
(2) Acutely toxic—(a) wide variation about the 96 hour LC50 over short periods
(b) small variations about the 96 hour LC50 over long periods
(3) Chronically toxic or producing bio-accumulation.
(4) Short term non-lethal stresses due to rapid changes.

6. Oscillations in concentration will be observed at any fixed point in the estuary. Figure 1 shows a number of types of fluctuation which might be observed and relates these to an acute toxicity level.

7. It would be desirable to be able to predict the concentration of a chemical in the estuary at any position and at any time. Because of the complexity of water movements this is not possible and certain simplifying assumptions have to be made in order to make the problem capable of solution. The usual approach is to assume that the conditions are uniform over an interval of several minutes and also across the width or through the depth of the estuary.

8. In order to be able to assess the 'type' of estuary under consideration, it is necessary to study the distribution of a tracer, such as the salinity. A uniform salinity distribution across the estuary would be taken to imply that a pollutant would also become very evenly distributed after a number of tidal cycles. A marked difference in salinity between the surface and the bottom could mean that a pollutant would be subject to different flow conditions at different depths and, therefore, the depth variation would have to be taken into account explicitly in a representation of the estuarine system.

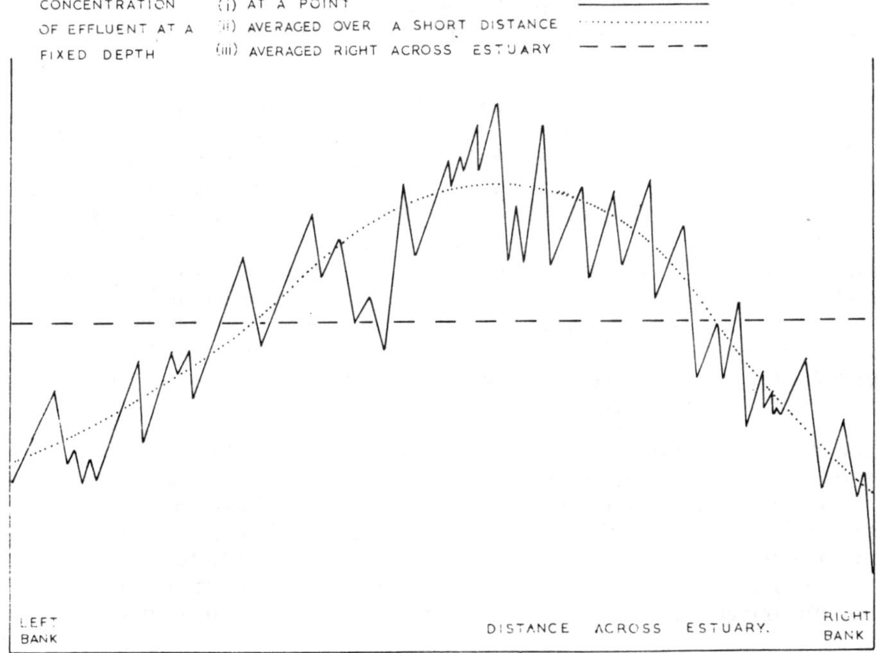

Fig. 2 The variation in concentration across the estuary which may be observed at some fixed time.

9. Figure 2 shows how the concentration of effluent material may vary across the estuary at some particular time after the chemical has intermixed with the bulk of the estuarine waters. It can be seen that there are small scale variations in concentration produced by the natural turbulence and larger scale variations produced by the velocity gradient across the estuary.

10. Once the 'type' of estuary has been decided, a suitable form of model can be chosen. Originally hydraulic models were used to describe estuarine systems but these suffer from some great draw-backs, particularly due to the difficulties in scaling down the physical processes of estuaries.(2) The development of high speed digital computers has meant that a mathematical model of an estuary can be constructed and predictions made simply by altering the input information to suit a particular situation.

11. The various types of mathematical models have been described by previous authors.(3,4) In their most basic form, these models assume a balance between the removal of pollutant material by the current and diffusion, and the replacement with new material from the discharges. Allowance is normally made for the effect of tides, freshwater input from tributaries and waste discharges with large volume flows. The particular cross-sectional shapes of the estuary at positions along its length are inserted into the model, together with details of the water depth at high and low tide.

12. Next, an assumed salinity distribution and assumed diffusion coefficients are introduced and the model run for a number of tidal cycles, until equilibrium has been reached. Equilibrium is said to have been reached when there is very little change in the salinity distribution between one tidal cycle and the next. The output predicted salinity distribution at the end of the run is compared with the actual distribution found by a detailed hydrographic survey. The magnitude of certain parameters in the model, such as the change in tidal range along the estuary and the diffusion coefficients, are known only approximately and these are adjusted until there is good agreement between the model and the prototype. The model is now capable of making satisfactory predictions of the natural salt distribution in the estuary.

13. In order to predict the distribution of each pollutant, diffusion coefficients have to be assumed for the pollutant. It is normal to use the diffusion coefficients determined for the salt unless there are strong grounds for believing that the pollutant and the salt respond to the mixing processes in completely different ways. Once the coefficients have been chosen the model is run to equilibrium from the clean river state, with the appropriate quantity of pollutant inserted at the point of discharge.

USING THE PREDICTIVE MODEL

14. The predictive model can be used to estimate the concentration

caused by a number of effluent discharges, making allowance for many of the time varying factors, such as freshwater flow and tidal height changes. Indeed, it is possible by reference to toxicity data, to make a preliminary assessment of the effects which a stable chemical may have on the aquatic life. However, most chemicals and all sewage wastes undergo degradation by physical, chemical and biological action. An answer to this problem is usually sought by introducing coefficients which can take account of rates at which materials are broken down. Because degradation of pollutants occurs, e.g. ammonia break down to nitrate, the basic mathematical model has to be combined with biological and chemical models. Predictions of the concentration distribution of a degradable material are then made. These breakdowns often involve a consumption of oxygen from the estuarial waters and have obvious biological implications.

15. Unfortunately, there is still an inadequate knowledge of the magnitudes of the reaction rates for many of the biological and chemical processes. Many more experimental investigations are required, both in the laboratory and the field, in order to establish the best ways of representing these complex reactions.[(5)]

16. A number of estuary models have been developed which make assumptions of uniformity of conditions throughout a cross-section when there are marked salinity gradients. The uniformity assumption significantly reduces the number of calculations required in the model but there is inevitably a loss in predictive precision. A non-uniform vertical distribution suggests that the flow conditions may change between the surface and the estuary bed. If a model assumes an average current and effluent distribution over depth, then the effect of differences in flow are implicitly combined into the diffusion coefficient and this factor tends to dominate the diffusion coefficient produced by the turbulent flow. Any changes in the flow conditions such as those which might be caused by an increased freshwater flow, have the effect of altering the diffusion coefficient. Thus, the results from this type of simplified model are likely to be erroneous unless the diffusion coefficients can be obtained for many different states of tide and freshwater run off. At the upstream end of estuaries, particularly those which are narrow and shallow, the cross-sections are likely to be fairly uniform and the predictions of a simplified model may be reasonably dependable.

17. Close to the position at which material is discharged, it is unlikely that conditions will be uniform. A continuous discharge will stream away from the outfall as a plume formed by the ebb or flood tide current. Such local effects cannot be easily taken into account in an estuary model and concentration levels in the plume may be much higher than those predicted. These higher concentrations are a hazard to aquatic life and each discharge has to be examined very carefully to establish the degree of risk. The likely concentration levels and the extent of these plumes may be estimated from the dilution of tracers released at the discharge point[(6)] or by theoretical formulae[(7)].

EFFECTS ON AQUATIC LIFE

18. Once the model is capable of producing data on the concentrations of toxic materials and the variation of these with time, one needs to know how the animals in the estuary will be affected by these changes. The effects of exposure to toxic materials in effluents may be acute and cause deaths (usually within 96 hours), chronic and not necessarily lethal, or the toxic substance may accumulate in the animal tissues (Fig. 1). In an estuary a toxic substance may affect aquatic life through any or all of the above mechanisms, the effect being governed by the concentration of toxic material, the species of animal, the period of exposure and many other variables in the environment such as salinity, oxygen concentration, temperature and pH.

19. Laboratory bioassays are carried out to provide data on how the organisms under study will respond to a range of toxic conditions.

20. The acute toxicity is usually defined as that concentration which kills 50% of the test fish in 96 hours, this LC_{50} (96 hours) has been and still is used extensively to prevent catastrophes such as large fish kills in the water to which the effluents are discharged. Obviously acute tests do not indicate whether the toxic material will have chronic toxic effects, or cause mortalities over periods longer than 96 hours. In the past it has been normal to use the 96 hour LC_{50}'s in conjunction with an application factor of at least 10 as the basis for prevention of long term effects of toxic discharges. Recently, it has been suggested that the application factors that should be applied generally lie between 1% and 40% of the determined 96 hour LC_{50} value. With the development of extended acute toxicity tests, chronic toxicity tests and tests measuring accumulation and loss of toxic material[8,9,10] it is now possible to say with increasing certainty whether a toxic material distributed in an estuary in a particular way is likely to have undesirable long term effects. But the 96 hour acute toxicity test still remains as the primary basis for determining acceptable levels of discharge for common industrial wastes.

21. There are certain substances which it is essential to prevent entering the estuarine environment even at very low concentrations. The Oslo Convention for the Prevention of Marine Pollution[11] lists in Annex 1 a number of groups of substances the dumping of which is prohibited at sea, and it would seem reasonable that these same groups of substances should not be continuously disposed of to the sea via estuaries. The substances which fall into this category are those materials which, themselves or their breakdown products, accumulate into and cause damage to living organisms e.g. mercury, and certain organo-halogen compounds. Of course it is rarely possible to reach nil discharge levels of any material but it is clear that with these substances the maximum possible effort must be applied to ensure the minimum possible discharge.

CONCLUSION

22. A mathematical model can be prepared for any estuary and from this the 'pollution budget' prepared, whatever the degree of pollution in the estuary. However, the criteria which are used as a basis for judgement of acceptable levels are often different, and do depend on the quality of the estuarine waters.

23. For an unpolluted estuary in which there are no clear harmful biological effects, the aim proposed by the Royal Commission is that aquatic life and the amenity interest should be fully protected.

24. On the other hand, for estuaries where detrimental biological effects are apparent, the Royal Commission gives two simple criteria. These criteria are, the ability to support on the mud bottom the fauna essential for sustaining sea fisheries, and the ability to allow the passage of migratory fish at all states of the tide.

25. When consents are being granted for new discharges or plans are being made for the reduction of old ones, the drawing up of a 'pollution budget' provides a basic framework within which decisions can be made. Obviously, it is not possible to predict the precise capacity of an estuary and there is always a need to allow for unforseen developments. Therefore, the practical approach should not be to authorise the use of the full predicted capacity of the estuary initially, but, rather this predicted capacity should be approached gradually over a number of years. During this time both chemical and ecological monitoring should be continued to ensure that the changing load is not producing effects greater than those predicted.

REFERENCES

1. Department of the Environment. Royal Commission on Environmental Pollution. Third Report. H.M.S.O., London, 1972.
2. H. B. Fischer and E. R. Holley, *Water Resources Research* 1971, **7**, 1, 46-51.
3. K. F. Bowden, in Estuaries, G. H. Lauff (Ed.), American Assoc. for the Advancement of Science, Pub. no. 83, 1967.
4. G. D. Hobbs, 5th. Int. Conf. on Wat. Pollut. Res. III-8, 1970.
5. M. J. Barrett, *Proc. Roy. Soc. London,* B, 1972, **180**, 511-520.
6. J. E. Foxworthy, Univ. of Southern California, Allan Hancock Foundation Report 68-1, 72pp.
7. Y. L. Lau, *Water Res.,* 1972, **6**, 7, 749-758.
8. J. B. Sprague, Measurement of Pollutant Toxicity to Fish—I, Bioassay Methods for Acute Toxicity, *Water Res.,* 1969, **3**, 793-821.
9. ——, Measurement of Pollutant Toxicity to Fish—II, Utilising and Applying Bioassay Results, *Water Res.,* 1970, **4**, 3-32.
10. ——, Measurement of Pollutant Toxicity to Fish—III, Sublethal Effects and 'Safe' Concentrations, *Water Res.,* 1971, **5**, 245-266.
11. Convention for the Prevention of Marine Pollution by Dumping from Ships and Aircraft, Oslo, Feb. 1972, H.M.S.O. Misc. No. 2 (1972).

DISCUSSION

LEWIS In order to illustrate how a mathematical model and information on effluent toxicity might be combined to assist with the management of an estuary, a simple flow diagram has been prepared (Fig. 3). This suggests how the consequences to aquatic life of proposed changes in pollutant loads might be assessed in advance. In all instances, monitoring is considered essential for establishing baseline data so that the significance and cause of variations in the condition of the estuary can be properly judged.

HOWELLS (Natural Environment Research Council). Conditions in estuaries vary continuously in time and space, as described in the papers. The standard LC50 toxicity test, on which so many standards are based, is a static test, and does not make any allowance for variability in conditions. It is well known that animals respond differently to a steady state concentration or condition than to conditions which vary. Toxicity tests should take account to the sort of variability which is seen in the field if they are to provide a valid guide to toxicity in the field. It will obviously be a very difficult thing to do, but it must be tackled.

There is also the difficulty that the animal's response is different at different phases in its life history. For example, adult salmon are very strong swimmers. It is known that in the Columbia river in the U.S.A.,

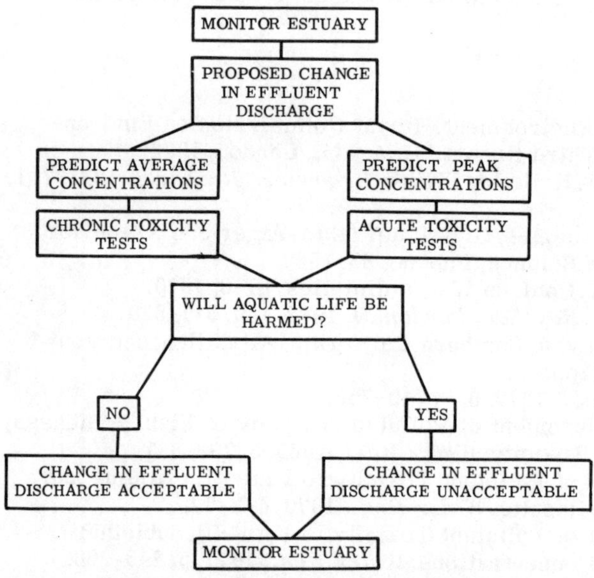

Fig. 3. Flow diagram for the Pollution Budget approach

where there are some fairly old nuclear power stations, there is a very hot but relatively non-dispersed effluent. The adult salmon can swim through this potentially toxic temperature apparently without harm because they pass through quickly and have sufficient means of biological recovery so that there is no permanent damage. On the other hand, the juvenile salmon on their downward migration are not such strong swimmers. They cannot pass the warm region without damage and they are affected by smaller temperature rises. Because of the inherent difficulty of formulating laboratory toxicity tests to take account of all these factors one is driven back to *in situ* biological monitoring with all its difficulties.

STEPHENSON There are immediate practical problems to be solved. Waste must be got rid of the best possible way. Solutions must be found using the best knowledge available. There is no information on the affect of salinity, temperature etc., on toxicity testing. If it were, it would be used. Toxicity tests are carried out on various stages in the life history of the predominant animals in estuaries.

THURSTON (I.C.I.). Could the Authors say how factors of safety are employed to cover the gaps in available information?

STEPHENSON When using toxicity data, an order of magnitude at least is usually allowed when fixing standards for an estuary.

Acute toxicity of individual components and of the mixed discharge would be determined, and then if there is no information as to the effect of life-long exposure an initial dilution would be recommended so as to give a safety factor of at least one order of magnitude of dilution below the acute toxic level. We would then hope that in most situations the subsequent dilution would be very rapid. With materials which have now been clearly defined as having sub-lethal effects, are biocumulative or are non-biodegradable, a level of nil would be suggested. For example, this would apply to mercury and certain organic halide compounds. Of course, 'nil' levels cannot be achieved. One can merely reduce the level to what is economically feasible.

SAUNDERS (National Environment Research Council.) Why is the LC50 rather than the TLM50 (Median Threshold Level) the criterion of toxicity? Secondly, what animals do you select as your guide line to toxicity within the estuary? And thirdly, how do you cope with mixtures?

STEPHENSON In reply to the third question, in the case of effluents we tend to test the whole effluent for toxicity of the mixture rather than the toxicity of the individual components. Where it is a new discharge and if it is not possible to mix up a sample of the projected discharge we would make adaptive assumptions probably down to concentrations of a tenth of the acute toxic level. The number of recorded synergistic effects is not large. It is not true that most of the common industrial effluents have an additive function, and some sort of a safety factor is generally adequate.

LC50 is that concentration which over a given period is toxic to 50% of the animals. In referring to a TLM the speaker is probably suggesting that what should be measured is not merely the acute toxicity median level but that concentration at which toxicity ceases to occur at all. Many of the toxicity tests are continued until an asymptotic level is reached. But in general terms calculations are made in terms of LC50 simply because the majority of the data which is available on toxicity of these compounds is quoted in terms of LC50s.

BAKER (Oil Pollution Research Unit). There is a drawback to many toxicity tests in that death is the criterion of toxicity. An example of misleading results with this type of test occurred with limpets in Milford Haven. Some toxicity tests were performed on limpets using oil spill cleaning detergent to find LC50 for that detergent. However, it was then noticed that a much lower concentration than this was sufficient to cause the limpets to drop off their rocks. When these were transferred to a tank of aerated sea water they recovered eventually. In the natural environment, once they have dropped off, they will be killed very quickly by predators. Death as the criterion of toxicity is not always appropriate.

STEPHENSON This criticism is valid, but data on safe levels are scarce. Long term tests at sub-lethal levels are carried out at Brixham measuring both accumulation and chronic toxicity effects. One can get quite severe toxic effects during laboratory experiments with some materials at concentrations lower than an order of magnitude below their acute toxic level. But as a general principle if the main area of concern with these compounds is with non-biodegradable materials which may have very long term effects, the number of substances is likely to be small. One example is the very high levels of detergent used in treating oil slicks.

HIGHER (Portsmouth City Council). The local authority finds itself in the position of defending against applications by traders to discharge. This is the wrong way round. The onus should be on the trader to show that his waste would not be harmful in the receiving water. The necessary expertise is more likely to be with the trader than with the local authority.

WISDOM (Avon and Dorset River Authority). There is a general tendency to make industry show that their effluents are safe rather than for the River Authority to prove that they are harmful. This is a question of legislation. In practical terms though, it cannot be that every small firm will have its own toxicity testing centre and other facilities.

STEPHENSON The I.C.I. Brixham Laboratory offers a service to other industries and to local authorities, advising on pollution problems. This service is being used.

HAMILTON (Institute for Marine Environmental Research). Current methods employed in acute toxicity testing are often of limited value.

There are new techniques in which it is possible to observe physiological responses in relation to stress, and there are various other methods based upon enzyme reaction rates which are still at the research stage. These are being established as standard methods to be used in the field and should be seriously considered in the future for problems associated with trade effluent.

STEPHENSON This sort of information is now becoming available but it is a recent development. It has not been possible for industry to control its discharges on the basis of this sort of test procedures. The real question is that if one can show a fall of 10% in their enzyme activity, what does this mean to the fish? Is it within its normal response range or not? The background information that one requires as a basis for legisation and good control is not available on this sort of test procedure at the present time.

TAYLOR (Department of Oceanography, University of Southampton). Could we hear a little more about the mathematical modelling techniques used? You seem to imply that you were using a model which involved averaging over a tidal cycle. This is a crude way of modelling. If you were seriously interested in developing a reasonably good numerical model on pollution problems in estuaries, there is no technical reason now why a three-dimensional time dependent model simulating turbulent flow should not be developed using concepts like eddy diffusivity and mixing length to obtain a reasonable picture. This would not be a particularly expensive operation in comparison to the size of the problem.

LEWIS Both sophisticated and simple models are employed. Perhaps the most sophisticated model we have is that for the Tees which is developed by Hobbs [12] which is two-dimensional. Three-dimensional models are still impractical because the storage capacity required is beyond the limits of computers currently available in this country. The Meteorological Office have a computer which can do three-dimensional models but this is very large indeed. Simple models are still of use with minor problems for which a general indication of the effect is adequate.

REFERENCE

12. Hobbs, G.D. and Fawcett, A., 'Two-Dimensional Estuarine Models', Symposium on Mathematical and Hydraulic Modelling of Estuarine Pollution, W.P.R.L., Stevenage, 1972.

EXPERIENCES IN ESTUARY MONITORING

T. L. Shaw B. Sc. (Eng), Ph. D., C. Eng., M.I.C.E., M.I.W.E., M.A.S.C.E.

Department of Civil Engineering
University of Bristol

INTRODUCTION

1. The study of fluid mechanics on a global scale must be the ultimate challenge for this branch of applied science. In the case of the oceans, dynamic fluid behaviour results from temporal and spatial changes of many factors, including atmospheric pressure, temperature, evaporation, wind profile, the tides, ocean geometry, salinity variations, etc. Truly this is a formidable array, and it is of the greatest credit to the 'meteorological oceanographer' that much of the primary behaviour of the oceans is now beginning to unfold.

2. Nevertheless, the study of oceanography lags behind its atmospheric counterpart. Although the latter may be no more straightforward to comprehend, it does have the double advantage that it has had, and still retains, a more immediate relevance to our society, and also that it can be monitored, at close-quarters if necessary, in a variety of now well-proven and comparatively cheap and simple ways. In comparison, the sea is less amenable to observation, essentially because its density is over 800 times that of air at sea level; pressures equivalent to many tens of atmospheres result at even comparatively shallow water depths, whereas the decay with height is slow in air with the limiting reduction, of course, of one atmosphere.

3. This particular property of water is of much reduced consequence in the restricted depths occurring in estuaries. However, the science of oceanography (if this term may be applied to these peripheral and hence apparently insignificant, coastal features) is not simplified as a result. The complications pertaining to estuarine studies result because these 'interfacial' or 'transitional' regions lie between uni-directional and essentially one-dimensional riverine-flows and the comparatively unrestrained circulations of the oceans. All transitional regions in fluid mechanics are fraught with instabilities that render study especially difficult; estuaries are no exception to this pattern.

4. But whereas the fluid mechanic researcher can often afford to ignore transitional flows since these occupy only a very restricted part of most circumstances (i.e., in pipe flow, the defining Reynolds number indicates transition at a value of about 2300 whereas values in excess of 10^6 are common in engineering practice), in the case of estuaries the researcher is attracted by the very real and important function of this particular part of the surface-hydrology/hydrograph phase. Those sympathetic to the topic of this Conference will have been convinced already of the steadily growing economic and social attractions of estuaries, and hence

of their function in the creation of an acceptable society reliant upon concentrated industrial and urban development about their shores. Emphasis on estuarine research must be dictated by this interaction between the user and the focus of his habitual environment, for it is upon such geographical features that many of the world's centres of population at present lie and will continue to be focussed (over 60% of the population in towns in U.K. of over 10,000 people live on estuaries, and seven of the ten largest cities in U.S.A. are similarly sited).

SOCIETY'S ROLE FOR ESTUARIES

5. In order to recognise the needs of estuarine research it is first necessary to attempt to identify the duties that the estuary may be called upon to perform; the capacity of any particular estuary to achieve this function must then be assessed in the light of sufficient knowledge regarding its characteristics. The next two sections of this paper refer respectively to the dynamics of estuaries and to monitoring, from which it may be concluded that, although estuaries may be classified according to certain parameters, no two should be regarded as even approximately similar prior to carrying out detailed and thorough investigations.

6. The pressure imposed on estuaries by society takes four major forms; (a) thermal discharges from power station cooling water outfalls, (b) sewage discharges, (c) industrial wastes, (d) agricultural wastes. To these must be added the natural discharges of fresh water and sediment from catchment erosion, plus waves resulting from wind action, tides, and possibly an input of marine sediments from the sea.

7. Each input to an estuarine system tends to be local in origin, which influences at least its short-term fate within the estuary. Each pollutant will be dispersed in a manner according to the interaction between its own characteristics and those of the estuary, which implies that dispersion behaviour is related to the position of discharge within the estuary. In the case of man-made pollutants, the properties upon which dispersion behaviour is dependent relate essentially to the position, velocity (quantity) and density at which the effluent is discharged, although in the case of some materials it is not so much their immediate behaviour as their long-term fate that may be of primary importance.

ESTUARY DYNAMICS

8. The students' first teaching on the fluid mechanics of estuaries reveals that, since these are comparatively shallow and wide bodies of water exposed to the atmosphere, and experience gradients of surface level according to tidal action, their motion will be controlled primarily by the force of gravity, i.e., their motion will be described in some, perhaps mystical way, by Froude's Law. This is not an inappropriate basis to embark on studies of estuarine motion, yet it yields only part of the story.

9. Although tidal oscillations usually control the macro water movements, density differences resulting from a salinity gradient between inflowing rivers and open sea concentrations can often provoke marked stratification (layering), particularly close to the river mouth. The salinity gradient at any point will not be constant, but will change according to the state (amplitude, phase) of the tide and the fresh water discharge of the river; to a small extent, these factors will also influence the form of the tidal oscillation within the estuary.

10. Wave action can also be responsible for changes to the salinity structure of an estuary, encouraging mixing and possibly magnifying tidal range if accompanied by on-shore winds plus a sympathetic barometric pressure gradient. However, teaching does not run to a quantitative description of the interaction of these three major factors—tides, waves and fresh water flows—in its assessment of estuarine behaviour, from which it should not be presumed that the matter is unimportant but rather that reliable and meaningful data is in short supply, hence he who cares to dwell on the subject must, in large measure, resort to self-instruction (the proliferation of references on the subject makes this alone a substantial occupation; Silberman [1] and U.S. Environmental Protection Agency[2] expose the subject in some detail).

ESTUARY MONITORING

11. Two points must be made immediately, firstly that almost without exception all estuaries have so far received inadequate study, and secondly, as referred to earlier, monitoring in the sea to yield sufficient information must rank as one of the most challenging of field tasks. These two considerations are undoubtedly related, yet motivation to more intensive and meaningful studies is now of such a pitch as to over-rule all other considerations. For instance the Department of the Environment has recently completed in Liverpool Bay one of the most thorough ad hoc British studies, in this case of the capacity of the Bay to absorb sludge dumped from vessels[3,4]. The dispersion and effects of off-shore disposal by pipeline of clay minerals in Cornwall prompted another detailed study[5], and strong brine solution from potash mining in Yorkshire yet another[6]. However these represent but a few of a large number of essentially similar ad hoc coastal surveys that have been carried out, and no doubt there will be many more to follow.

12. But such surveys are not always adequately conclusive, a decision being necessary on information relating to only a limited number of physical circumstances or to a comparatively short chemical or biological exposure of the natural environment to the proposed effluent. Given our present state of knowledge this might be regarded as the only way to proceed, yet it must also be noted that the various estuarine and coastal surveys carried out to date sufficiently demonstrate the presence of common denominators to imply that more, basic studies, if carefully selected, could reveal substantial understanding on the extent to which

similarities in dispersion might be expected. To this end it is encouraging to record that the Natural Environment Research Council has now set up a Working Party on Estuaries; in addition to collating and publishing reviews of research to date on particular British estuaries this NERC body is proposing to rationalise their support of estuarine research in the U.K. The NERC exercise to date has succeeded in demonstrating how incomplete and fragmentary is the evidence for any estuary that has accumulated from the work of individuals over many decades (e.g. 7).

13. As an indicator of what can be achieved by systematic study, the work of the Tay Estuary Research Centre (Dundee University) and the Clyde River Purification Board is to be recommended[8, 9]. Some of the scientific findings of these two bodies are referred to below, but before considering such detail it seems essentially relevant to refer to the fact that each body has only comparatively recently begun to demonstrate a meaningful understanding of the dynamics of their particular estuary after several years of consistent, routine and regularly reassessed studies. The author is in no doubt that the science of coastal oceanography is still sufficiently devoid of comprehensive field information to render comparatively useless, for all but possibly the specific purpose, data collected on an ad hoc basis.

14. The stimulus provided by results collected over a long-period, revealing patterns of estuary motion only after scrupulous attention to scientific detail and with a strong measure of intuition, should act as encouragement for expansion to other estuaries of national importance. That the setting up and operation of such study centres is not straightforward will be referred to later. For the present it is appropriate to confine attention to the parameters that ought to be recorded, their frequency of measurement, and the use of data for analysis and predictive purposes.

15. The recording of data falls into two categories. Reference was made earlier to the various classes of man-made pollutant to which an estuary may be subjected. For field research purposes these suggest monitoring on a macro scale (tidal levels, fresh water flows) as well as on a micro scale (local velocity and salinity gradients, turbulence levels and concentrations of suspended solids). The degree of fine detail actually required is unlikely to be appreciated properly until a substantial amount of coarse information has been pieced together. Since micro-movements are controlled by, and to some extent may themselves influence, movements on a macro-scale, it would seem not only to be dangerous but also possibly to be meaningless to concentrate attention on local density structures and bed forms in isolation from the total environment of which they are a part.

16. Records of tidal levels are now substantial. However, few gauges exist for any purpose other than navigation, and these alone will generally not be sufficient to resolve in adequate detail the instantaneous overall distribution of surface levels. As with each parameter referred to in

this paper, the necessary intensity of surface level monitoring will vary between estuaries and hence it would be dangerous to generalise. Nevertheless between five and ten autographic recorders, if sufficiently sensitive, will demonstrate the need, or otherwise, for additional instruments. The layout needed to give satisfactory information will depend upon current patterns, i.e. upon circulations in the estuary, since levels (surface gradients) and velocities are closely interconnected. One aspect however, which becomes of very considerable importance in many estuaries is the pattern and level of banks off-shore, because the distribution of surface levels over such features at small water depths is largely unknown; hence level gauges on the coast may provide misleading information if not interpreted with caution.

17. Tidal levels coupled with bed profiling should yield a good approximation to at least mean longitudinal velocities and possibly also, with experience, to horizontal and vertical velocity gradients at any cross-section. However some field evidence to qualify such calculations will be advisable, especially when the estuary is unsymmetrical in shape and has extensive shallow areas and large fresh water flows. Admiralty charts provide useful information on velocities, but even this is often inadequate and insufficiently sensitive to the influence of bank movement (an effect scarcely understood, even qualitatively) and fresh water.

18. Salinity measurements taken at selected points throughout an estuary, regularly throughout a range of tidal cycles and fresh water inflows, will be needed, firstly to demonstrate the form and extent of mixing between river flows and the sea, and secondly, to expose any influence that salinity gradients may have on velocity gradients (the latter factor may be very substantial, at least in places within an estuary and also at various states of the tide). Regrettably very little information exists from a sufficient variety of estuaries to demonstrate just how sensitive is the balance of tide, velocity, salinity and estuarine geometry, yet from what information is available it is abundantly clear that an estuarine environment may well be very sensitive to the effects of density, in a manner resembling the control exercised by temperature inversions in the atmosphere.

19. Given a density stratification displayed by a locally marked gradient of salinity, (temperature effects in estuaries are seldom of significance), it is not inappropriate to regard estuary inflows as existing within separate layers[(10)]. However, it is less readily possible, for lack of experimental information, to anticipate the circumstances sufficient to break down stratification even though it is known qualitatively that wave action and tidal currents are capable of causing this. Essentially this lack of understanding concerns the interaction between density gradients, turbulence structure and entrainment. Ellison and Turner[(11)] made a notable contribution to this complex subject, and many further researches, including a continuing effort within the author's Department, are making further progress. The work at Bristol is particularly directed towards

estuarine application, the studies including the effect of suspended solids on interfacial phenomena; laboratory and field applications of laser-based sensors for simultaneous measurement of local velocity, turbulence and suspension concentration should prove a valuable by-product of this programme[12].

PREDICTIVE MODELLING

20. Neither physical nor mathematical modelling should be regarded as a substitute for field studies. As has already been pointed out, detailed surveys of a sufficient variety of circumstances occurring in one estuary will be needed before adequate understanding can be expected. However, this may take too long to complete for many purposes, so that resort must be made to forms of modelling, both to demonstrate estuarine processes and to allow prediction of effluent dispersion.

21. Both physical and mathematical models have their advantages and disadvantages, and there is much to be said for the simultaneous use of each. It is possibly unfortunate that physical models involve substantial capital outlay before results become available, but they retain the advantage of being open for inspection and hence presumably less immediately liable to suspicion. It is encouraging that one of the drawbacks of physical estuary models, that of needing to be of small or preferably zero vertical distortion may in future be less of a restraint as mixing processes, and hence scale effects, become better understood; smaller cheaper and therefore more readily available models should result.

22. Mathematical modelling suffers from the temptation to assume rather more that is compatible with the purpose of the study, in order to attain an answer. Again, there can be no substitute for an astutely chosen field programme, but given such information the modeller's true ability to build and extrapolate soundly can be encouraged. The best type of model for any estuary should only be chosen on the basis of field data and an understanding of the physical processes with which the model will be concerned. Hobbs and Fawcett[13] refer to some of the more advanced types of model and the uses for which these would be appropriate; Pearson and Carter[14] also make many relevant points regarding the choice of models, but their emphasis is more on simplicity.

CONCLUSIONS

23. Sustained industrial pressures for urgent answers to questions involving effluent dispersion will ensure continued emphasis on estuarine studies. For these to be most fruitful and progressive, interested parties would do well to consider the economic return possible from participation in permanent study groups established for appropriate estuaries. Such groups should have the capacity, with special all-party assistance if necessary, to mount sufficiently intensive monitoring programmes at times of special interest (tides, fresh water flows, etc.).

24. The Tay research group is now able to operate in this way[15], and have shown convincingly the reward that is possible from systematic overall surveys. The simultaneous use of many boats for this purpose appears inevitable, although for preliminary data there are good reasons to favour the use of helicopters[16].

25. Tracer injections and subsequent periodic sampling have proved a valuable means of supplementing continuous fixed-point sensing, giving quantitative data on dispersion in conditions applying to particular test situations[17, 18]. By the use of such methods it should become possible to demonstrate the sensitivity of estuarine processes, in particular stratification and vertical and horizontal circulations, to local conditions, hence the capacity and the time-dependence of dispersion for point outfalls. Physical and mathematical models jointly have a most important function to play in this process of understanding, and it is encouraging that much greater emphasis is now being given to the optimum use of these techniques in resolving field problems[19,20].

REFERENCES

1. Silberman, E. and Stefan, H., 'Physical (Hydraulic) Modelling of Heat Dispersion in Large Lakes: A Review of the State of the Art', Argonne Natl. Lab., Rpt. No. ANL/ES-2, 1971.
2. Environmental Pollution Agency (USA), 'Estuary Modelling: an Assessment', Water Pollution Control Research Series 17070, DZV 02/7, 1971.
3. Department of the Environment. Report of the Working Party on the Disposal of Sludge in Liverpool Bay, 'Out of Sight, Out of Mind', H.M.S.O., London, 1972.
4. Best, R., Ainsworth, G., Wood, P. C. and James, J. E., 'Effect of Sewage Sludge on the Marine Environment: A Case Study in Liverpool Bay', *Proc. Inst. Civ. Eng.*, 1973, **55** Pt. 2, 43-66.
5. Green, D. W., 'An Investigation of the Disposal of Micaceous Residue Through a Sea Outfall', *Proc. Inst. Civ. Eng.*, 1972, **53**, 2, 127-145.
6. Warner, Sir F., 'Problems in Establishing a Potash Industry in the North Yorkshire National Park', 9th Brotherton Memorial Lecture, Soc. of Chemical Industry, 1970.
7. Univs. of Bristol and Swansea, and NERC, 'The Severn Estuary and the Bristol Channel: An Assessment of Present Knowledge', NERC Series C, No. 9, 1972.
8. West, J. R., 'Water Movements in the Tay Estuary', *Proc. Roy. Soc. (Edinburgh)*, 1972, **71**, Series B, 115-129.
9. Mackay, D. W., 'A Model for Levels of Dissolved Oxygen in the Clyde Estuary', Symposium on Mathematical and Hydraulic Modelling of Estuarine Pollution, W.P.R.L, Stevenage, 1972.
10. Young, R. M. and Shaw, T. L., 'Some Characteristics of Density Currents Established from Outfalls', 1971, 14th Congress I.A.H.R., Paper A-28.

11. Ellison, T. and Turner, J., 'Turbulent Entrainment in Stratified Flow', *J. Fluid Mechanics*, 1959, **6**, 423.
12. Buttling, S. and Shaw, T. L., 'Predicting the Rate and Pattern of Storage Loss in Reservoirs', 11th Congress Int. Comm. on Large Dams, 1973.
13. Hobbs, G.D. and Fawcett, A., 'Two-Dimensional Estuarine Models', Symposium on Mathematical and Hydraulic Modelling of Estuarine Pollution, W.P.R.L, Stevenage, 1972.
14. Pearson, C.R. and Carter, L., 'The Application of Simple Models for the Prediction of Effluent Dispersal in Estuaries', Symposium on Mathematical and Hydraulic Modelling of Estuarine Pollution, W.P.R.L, Stevenage, 1972.
15. Buller, A. T., McManus, J. and Williams, D. J. A., 'Investigations in the Estuarine Environments of the Tay', Tay Estuary Research Centre, Research Report No. 1, 1971.
16. Winters, A., 'A Desk Study of the Severn Estuary', Symposium on Mathematical and Hydraulic Modelling of Estuarine Pollution, W.P.R.L, Stevenage, 1972.
17. Georgiev, B. and Monev, E., 'Measurements of Large Scale Circulation in Cooling Reservoirs by Means of Radioisotopes', International Symposium on Stratified Flows (Novosibirsk), I.A.H.R., 1972.
18. Talbot, J. W., 'Measurement of Dispersion', Symposium on Mathematical and Hydraulic Modelling of Estuarine Pollution, W.P.R.L., Stevenage, 1972.
19. Elder, R. A. and Wunderlich, W. O., 'Inflow Density Currents in TVA Reservoirs', International Symposium on Stratified Flows (Novosibirsk), I.A.H.R., 1972.
20. Neale, L. C. and Hecker, G. E., 'Model Versus Field Data on Thermal Plumes from Power Stations', International Symposium on Stratified Flows (Novosibirsk), I.A.H.R., 1972.

DISCUSSION

ADEY and BREBBIA (Civil Engineering Department, Southampton University). Mathematical models are at present being developed in the Department of Civil Engineering for the study of coastal water problems. In this contribution a two dimensional mathematical water quality model for vertically well mixed water bodies is presented. The governing equations are discretized using the finite element method.

The two dimensional model was applied to study the dispersion of effluent from a proposed outfall in the eastern part of the Solent. The main item of interest in this study was the coliform distribution from the outfall and its time dependent behaviour under typical (in the present example spring) tide conditions. The eastern part of the Solent was subdivided into 209 fixed node elements with 466 nodes with the smallest element in the area surrounding the outfall (Fig. 1).

The surface velocities during the 13 hours of the spring tide were obtained from a recent extensive study of the Solent. The turbulent diffusivity coefficients depend in general on position and time. In the computer program two coefficients were considered: the longitudinal one, in the direction of flow and the transverse in the normal direction to this vector. From experimental observations with dyes reported in a comprehensive report of J. D. and D. M. Watson the following values were taken,

$$D \text{ (longitudinal)} = 5 \text{ m}^2/\text{s}.$$

$$D \text{ (transverse)} = 0.05 \text{ m}^2/\text{s}.$$

valid for calm conditions.

In order to represent the decay of the coliform population in the mathematical model it was necessary to define a first order rate of decay coefficient. As no experimental data were available for the conditions in the Solent a value of 0.4×10^{-5}/s (0.33/day) was used as this value appeared typical. The discharge from the outfall was defined as 6000 l/s with a coliform density of 1.67×10^6 coli/100 ml remaining constant over the tidal cycle.

The results shown in Fig. 2 are for the 4th spring tidal cycle in the model. They are essentially similar to those obtained for the 7th cycle (not shown here) and indicate that for the type of decay used in this model four cycles are sufficient. The contours plotted by the computer are coliform densities of 50, 10 and 1×10^3 coli/100 ml, concentrated

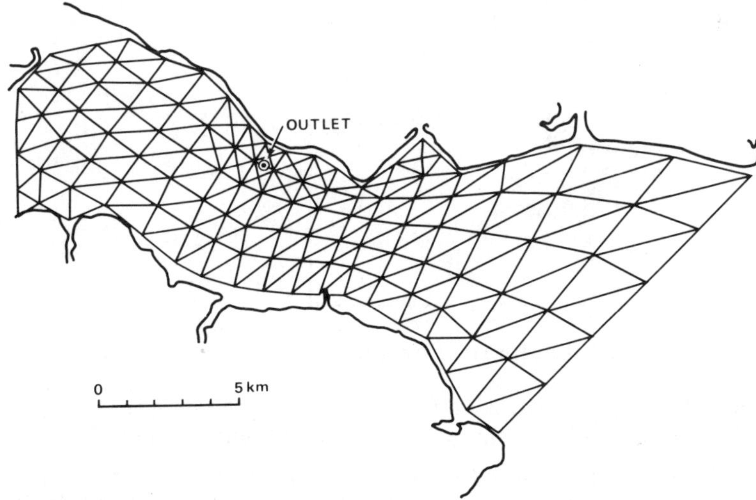

Fig. 1. Finite element grid

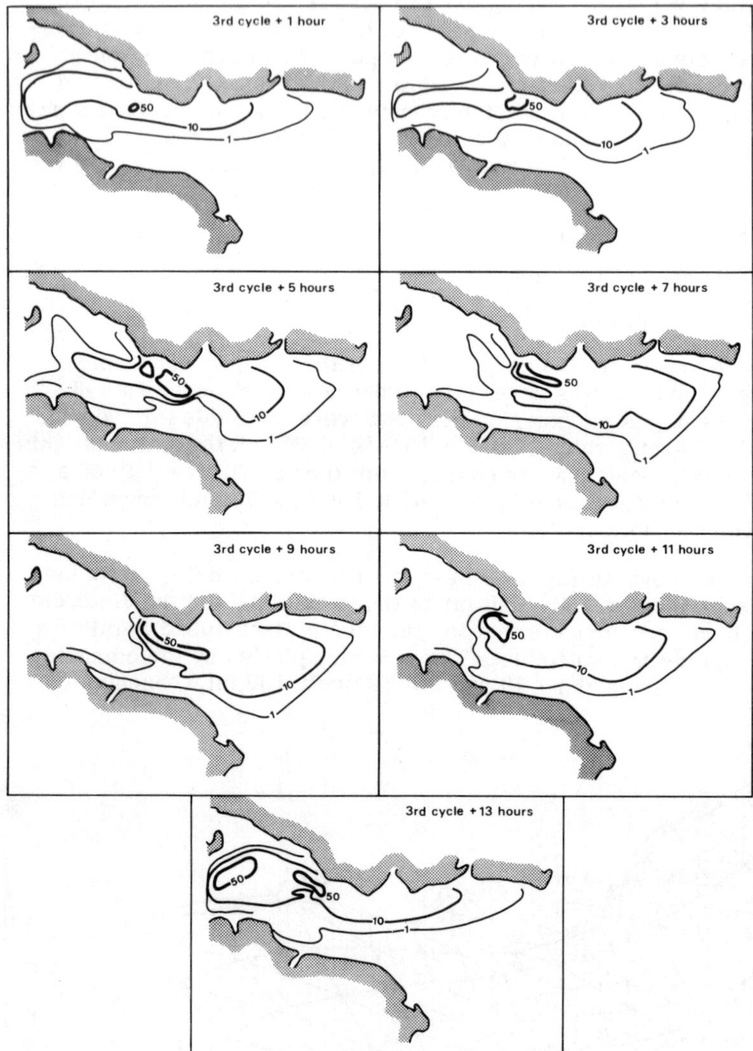

Fig. 2. Spring tide results (4th tidal cycle)

into a depth of 1 m. They still have to be divided by the representative depth to obtain the concentrations which would be found in the field.

The acceptable level to coliform count is still a subject of much discussion but the above results show how relatively simple mathematical models can be used to predict effluent dispersion.

We wish to thank the South Hampshire Plan Advisory Committee and J. D. and D. M. Watson for making available field data on the Solent.

CAMP (J. D. and D. M. Watson). In the work presented by Adey, it is not clear for what criteria the outfall was being designed. Also, what wind conditions were assumed?

ADEY (Southampton University). The results presented were for calm conditions. The wind effect is included in the model through the tidal velocities which are computed allowing for air/water interaction.

CHARLTON (Tay Estuary Research Centre). The Tay tidal model, which is a physical model of the whole estuary, is currently being updated to run in parallel with a mathematical model. The necessity for good prototype data to run a mathematical model has been emphasized, and this cannot be stressed too strongly. Work is therefore proceeding at Dundee to determine the variation of dispersion coefficients with other functions such as turbulence and velocity so that we can include variable coefficients throughout the elements of the mathematical model for pollution studies.

Reference has been made to the Tay Estuary Research Centre. Some descriptive comments may be in order for the benefit of other potential workers. In 1967, after the opening of the Tay road bridge, the University of Dundee purchased the Southern terminal of the Tay ferry to form the headquarters of the Tay Estuary Research Centre. It provides a sloping quay, a quiet anchorage, and the terminal buildings have been converted into laboratories, staff rooms, workshop and boat sheds. Anyone wishing to do any work of this nature must have a facility for launching boats and personnel available to get on to the water quickly when the weather permits.

Our main boat is a catamaran 9 m long and powered by twin diesels. Between the hulls is a large hatchway through which instruments can be lowered. It has a speed of about 8 knots and in a modest sea is quite stable. There are two other boats with cabins. One, of length 5.5 m can do 30 knots, and has been extremely useful. An open boat is not a good idea for this sort of work.

The business of gathering information very often requires combined efforts, and we have some experience of this, as it is impossible to gather information throughout the estuary using our own resources alone. Last year a collaborative effort was mounted, with assistance from two Purification Boards and various other interested parties. Nine boats were used, including the Tay lightship. These were manned throughout the estuary and obtained observations over whole tidal cycles. About 50 people were involved. This is a very good example of collaboration between various authorities.

MACKAY (Clyde River Purification Board). There is a distinction which has to be drawn between research in an estuary and control of pollution. The Tay Estuary Research Group are doing an excellent job.

A whole series of scientific workers are finding new information, writing papers, getting Ph. D's and so on. This may go on for five or ten years, but eventually somebody will be left with the job of doing a fairly dull repetitious thing, scrutinising new industries as they come in with their effluent problems. Somebody has to understand the estuary, know what a proposed new effluent will do, and thus fix standards. To be in a position to do this will require continuous effort, using large well-equipped vessels on investigations lasting weeks or months. This is not extravagance on the part of the pollution control authority, but an essential function.

TUCK (C.E.G.B.). Experience with investigations of heated water discharge problems has shown that relatively simple mathematical formulations are capable of giving predictions which are entirely adequate for many purposes. The detail of a three-dimensional model is not needed. This is the situation as it appears to one who is not a mathematician, but only the user of mathematical services.

LIST OF PARTICIPANTS

Ackroyd, F.	Lincolnshire River Authority, Boston, Lincs.
Addy, J.	IBM UK Laboratories Ltd., Winchester, Hants.
Alabaster, J. S.	Water Pollution Research Laboratory, Stevenage.
Anderson, J.	Dept. of Local Government, Dublin, Ireland.
Arnold, J. L.	Welsh Office, Cardiff, Glam.
Ashman, P. S.	Tyneside Joint Severage Board, Newcastle upon Tyne.
Askew, F. A.	Westminster Dredging Co. Ltd., Southampton
Baker, Dr. J.	Oil Pollution Research Unit, Orielton, Pembrokeshire.
Barrow, J. F.	Oxfordshire County Council, Oxford.
Bene, S. E.	Esso Petroleum Co. Ltd, Fawley, Hants.
Birch, A. F.	Government of Northern Ireland, Ministry of Development, Belfast.
Bosker, C. M.	Willcox, Raikes & Marshall, Birmingham.
Bottomley, Dr. P. E.	Trent River Authority, Fisheries Dept., Nottingham.
Boyle, O.	Department of Local Government, Dublin, Ireland.
Bradfield, R. E. N.	Atkins Research & Development, Epsom, Surrey.
Buckland, P. J.	Kent River Authority, Maidstone, Kent.
Bullwinkle, J. E.	Lemon & Blizard, Southampton.
Burrows, Dr. E. M.	University of Liverpool, Department of Botany.
Burrows, R. S.	Breanoc Field Study Centre, St. Agnes, Cornwall.
Butters, K.	Greater London Council, Dept. of Public Health Engineering.
Cairns, P.	D. Balfour & Sons, Newcastle upon Tyne.
Camp, I. C.	J. D. & D. M. Watson, High Wycombe, Bucks.
Chambers, A.	A. P. Mason, Pittendrigh & Partners, Newcastle upon Tyne.
Chandler, P. M.	Lemon & Blizard, Southampton.
Charlesworth, H.	Pike's Farm House, Wokingham, Berks.
Charlton, Dr. J. A.	Department of Civil Engineering, University of Dundee.
Chubb, C. J.	South West Wales River Authority, Llanelli, Carms.
Cockerill, T. H.	Welland and Nene River Authority, Peterborough.
Collins, A. E.	Greater London Council, Dept. of Public Health Engineering.
Covill, Dr. R. W.	18 Moor Park, Exmouth, Devon.
Crompton, L. A.	Cleansing Service (Southern Counties) Ltd., Southampton.

D'Alton, E. P.	Cork Corporation, Cork, Eire.
Dodd, Dr. E. N.	4 Glebe Park Avenue, Havant, Hants.
Doe, H. W.	Yarmouth (Isle of Wight) Harbour Commissioners, Isle of Wight.
Downing, Dr. A. L.	Binnie and Partners, Westminster, London, S. W. 1.
Doyle, A. G.	Union Carbide U.K. Ltd., London.
Drummond, M.	Manor of Cadland, Fawley, Southampton.
Dunn, J. N.	Southern Water Authority, Eastleigh, Hants.
Eastwood, P. K.	Kent River Authority, Maidstone, Kent.
Edwards, Dr. A. M. C.	Yorkshire River Authority, Leeds.
El-Gindy, A. A. H.	Oceanography Department, Liverpool University.
Ellis, M. J.	Sussex River Authority, Brighton.
Ferguson, W. R.	Edinburgh Corporation, City Engineer's Department.
Fitzgerald, R. D.	Waterhouse & Partners, Rowlands Gill, Co. Durham.
Flynn, B.	Middlesex Polytechnic, Enfield, Middx.
Freeman, A.	Newport County Council, Mon.
Furley, R. J.	Port of London Authority, London.
Gardner, G.	Waterhouse & Partners, Rowlands Gill, Co. Durham.
Gillham, R. M.	University of Liverpool Veterinary Field Station, Wirral, Cheshire.
Gilson, H. C.	Committee for Environmental Conservation, Marlborough, Wilts.
Greenland, A. J.	John Taylor & Sons, London, S. W. 1.
Griffin, D. J.	Wallace, Evans & Partners, Penarth, Glam.
Hamilton, Dr. E. K.	Institute for Marine Environmental Research, Plymouth.
Harney, T. J.	Springville House, Blackrock Road, Cork, Ireland.
Harris, R. B.	Howard Humphreys & Sons, Epsom, Surrey.
Hartley, P.	Willcox, Raikes & Marshall, Birmingham.
Hawes, F. B.	C.E.G.B., Planning Department, London.
Hayman-Joyce, J. G.	Livesey & Henderson, Guildford, Surrey.
Higher, G. E.	Portsmouth C.B.C., City Engineer's Dept., Southsea.
Hopkin, G. M.	Glamorgan River Authority, Bridgend.
Houghton, D. R.	Central Dockyard Laboratory, H.M. Naval Base, Portsmouth.
Howells, Dr. G.	Natural Environment Research Council, London, W. C. 2.
Humby, E.	Department of Civil Engineering, Portsmouth Polytechnic.
Hurley, Dr. F. J.	Youghal Carpets, Carrigtwohill, Co. Cork.
Ingham, G. S. N.	Bristol Avon River Authority, Bath.

James, A. E.	Glaxo Laboratories Limited, Ulverston, Lancs.
James, J. E.	Anglian Water Authority, Huntingdon.
Jarman, R. T.	C.E.G.B., South Western Region, Bristol.
Jennings, D. W.	Scientific Branch, Hogsmill Valley Sewage Treatment Works, Kingston upon Thames.
Jobbins, D.	Greater London Council, Dept. of Public Health Engineering.
Jones, B. D.	International Synthetic Rubber Co. Ltd., Southampton.
Jones, R. G.	South West Wales River Authority, Llanelli, Carms.
Jordan, A. W.	Brighton Marina Co. Ltd., Brighton.
Jury, G. T.	Newport I.W. Borough Council, Isle of Wight.
Knight, D. O.	Bristol Avon River Authority, Bath.
Lewis, Dr. R. E.	Imperial Chemical Industries Ltd., Brixham, Devon.
Little, J. W.	Brighton Marina Co. Ltd., Brighton.
Little, F. J.	Banff, Moray & Nairn River Purification Board, Elgin, Scotland.
Long, P. J.	Department of the Environment, London.
Lovett, M.	John H. Haiste and Partners, Leeds.
Machon, F. J.	Water Research Association, Marlow, Bucks.
Mackay D. W.	Clyde River Purification Board, East Kilbride, Glasgow.
Marstrand, P. K.	University of Sussex, Falmer, Brighton.
Martin, C.	L. G. Mouchel & Partners, Bath, Som.
Meredith, Dr. S. C.	Department of Zoology, University of Liverpool.
Meredith, W. H.	Howard Humphreys & Sons, Epsom, Surrey.
Mills, G. R.	C.E.G.B., Fawley, Southampton.
Mitchell, Dr. R.	Water Resources Board, Reading, Surrey.
Mornement, P. C.	Bertlin & Partners, Redhill, Surrey.
Mott, P. G.	Hunting Surveys Limited, Boreham Wood, Herts.
Murphy, D. F.	Department of Fisheries, Dublin, 1.
McBratney, C. H.	Dept. of the Environment, Water Management Service, Vancouver, Canada.
McCandlish, S. G.	Dept. of Chemical Engineering, University of Exeter.
McDermott, D. P.	Sligo Corporation, Sligo, Ireland.
McGregor, Dr. A.	City of Southampton, Dept. of Community Health.
Nebrensky, Dr. J.	Chemical Engineering Dept., University College of Swansea.
Newton, J. R.	Water Pollution Research Laboratory, Stevenage, Herts.
Nunn, C. M.	Lemon & Blizard, Southampton.
Oakley, H. R.	J. D. & D. M. Watson, London.
Offord, R. S.	Rofe Kennard & Lapworth, Cardiff.

O'Mahony, J.K.	Dublin Port and Docks Board, Dublin, 1.
Osborne, D.	South Hampshire Main Drainage Board, Eastleigh, Hants.
O'Sullivan, J.M.	Cork Harbour Commissioners, Cork, Ireland.
O'Sullivan, M.C.	Consulting Engineer, Cork, Eire.
Owen, A.F.	Dee & Clwyd River Authority, Chester.
Owens, J.M.	IBM UK Laboratories Ltd., Winchester, Hants.
Page, M.J.	J.D. & D.M. Watson, Liverpool, 2.
Parsell, R.J.	Nature Conservancy, Coastal Ecology Research Station, Norwich.
Pavel, J.	Edinburgh Corporation, City Engineer's Department.
Perry, J.M.	Renton Howard Wood Partnership, London.
Pharoah, D.F.H.	Binnie and Partners, London.
Phillips Dr. G.C.	Unilever Limited, Aberdeen, Scotland.
Pratley, D.A.	Esso Chemical Limited, Fawley, Southampton.
Pretorius D.J.	Re-Chem International Ltd., Southampton.
Primrose, J.D.	British Transport Docks Board, Southampton.
Puttock, M.A.	Sir Frederick Snow & Partners, London.
Ramsden, P.B.	Isle of Wight River Authority, Newport, Isle of Wight.
Ravdal, E.	Norwegian Institute for Water Research, Oslo 3, Norway.
Raymer, M.D.	City of Southampton Corporation, Southampton.
Read, P.A.	Napier College of Science & Technology, Edinburgh, Scotland.
Riordan, Dr. J.C.	Pfizer ChemicalCorporation, Ringaskiddy, Co. Cork, Ireland.
Rismondo, R.	Technital, Mestre (Venezia), Italy.
Roberts, A.G.	Whitstable Urban District Council, Engineer & Surveyor's Dept.
Rodda, D.W.	Department of the Environment, London.
Saunders, P.J.	Natural Environment Research Council, London.
Savage, P.D.V.	C.E.G.B., Marchwood Laboratories, Southampton.
Shaw, Dr. T.L.	Department of Civil Engineering, University of Bristol.
Shilston, M.T.	Sir William Halcrow & Partners, London.
Smith, H.J.	Cumberland River Authority, Carlisle.
Smith, P.R.	Department of Micro Biology, Galway, Ireland.
Soulsby, P.G.	Hampshire River Authority, Eastleigh, Hants.
Speight, H.	Hampshire River Authority, Eastleigh, Hants.
Staples, I.	University of Stirling, Scotland.
Stark, G.T.C.	Great Ouse River Authority, Cambridge.
Senders, S.	Danish Isotope Centre, Copenhagen V, Denmark.
Stephenson, Dr. R.R.	Imperial Chemical Industries Ltd., Overgang, Brixham, Devon.

Stone, J. T.	L. G. Mouchel & Partners, Bath, Somerset.
Stoner, J. H.	Somerset River Authority, Bridgwater, Somerset.
Sutton, P.	Esso Petroleum Co. Ltd., Fawley, Hants.
Taylor, R. J.	Ward, Ashcroft & Parkman, Liverpool.
Tetlow, J. A.	Essex River Authority, Chelmsford, Essex.
Thurston, E. F.	I.C.I. Limited, Mond Division, Winnington, Northwich, Cheshire.
Tuck, G. E.	C.E.G.B., Planning Department, London.
Van Den Broek, W.	City of London Polytechnic, Dept. of Biological Sciences.
Vollbrecht, Dr. K.	Deutsches Hydrographisches Institut, Hamburg, Germany.
Wakefield, J. A.	Coastal Anti-Pollution League Ltd., Bath, Somerset.
Walkey, J. M.	Harvey, McGill & Hayes, Exeter, Devon.
Watts, L.	Fareham Urban District Council, Fareham, Hants.
Webber, N. B.	Department of Civil Engineering, University of Southampton.
Welsh, W. T.	Solway River Purification Board, Dumfries, Scotland.
Wharfe, J. R.	City of London Polytechnic, Dept. of Biological Sciences.
Whelan, J.	M. C. O'Sullivan, Consulting Engineer, Cork, Eire.
White, N.	International Synthetic Rubber Co. Ltd., Southampton.
Whitehead, C.	Havant & Waterloo U.D.C., Technical Services Department.
Williamson, T.	Lothians River Purification Board, Colinton, Edinburgh.
Wisdom, A. S.	Athene, Boundary Drive, Colehill, Wimborne, Dorset.
Woodcock, D. H. A.	Devon River Authority, Exeter.
Woods, D.	International Synthetic Rubber Co. Ltd., Southampton.
Wright, S. L.	Hampshire River Authority, Eastleigh, Hants.

INDEX

Acts of Parliament
  Clean Rivers (Estuaries and Tidal Waters) 1960, 2. 5
  Oil in Navigable Waters, 2. 9
  Prevention of Oil Pollution 1971, 2. 9
  Public Health (Drainage of Trade Premises) 1937, 2. 8
  Rivers Pollution Prevention 1876, 2. 2
  Rivers (Prevention of Pollution) 1951-61, 2. 3, 2. 5, 2. 12
  Rivers (Prevention of Pollution) (Scotland) 1951, 9. 1
  Sea Fisheries Regulation 1966
  Water 1973, 1. 5, 2. 3
Aesthetics, 1. 21, 1. 25, 1. 27, 12. 1, 12. 10
Air photography, 10. 3
Air survey, 6. 17, 10. 14
Airborne sensors, 10. 1
Algae, 6. 5
Amenity, 12. 1
Ammonia, 5. 20
Anti-foulants, 8. 3, 8. 7
Arsenic, 8. 3

Bacteria, pollution by, 3. 1
  mortality of, 3. 17, 14. 9
Barnacles, 4. 3
Bathing, 3. 5
Beach pollution, 11. 7, 11. 15
Bembridge, 1. 1
Biological survey, 4. 1, 9. 50, 11. 11
Biological diversity, 4. 12
Boil, 1. 20, 9. 28, 12. 2
Brent geese, 1. 15
Bridport, 12. 1

Chemical pollution, 3. 3, 9. 51
Chichester Harbour, 1. 15
Cholera, 3. 2, 11. 4, 11. 7
Clams, 3. 13, 4. 3
Clostridium perfringens, 3. 15, 9. 13
Clyde Estuary, 8. 5, 11. 1
Coliform, 3. 6, 3. 9, 7. 19, 9. 4, 9. 14, 9. 58, 11. 13, 14. 9
Colour photography, 10. 7
Cooling water, 4. 2, 4. 7, 9. 59, 10. 16
Copper, 8. 3, 8. 7, 8. 8
Cord grass, 1. 15
Crepidula fornicata, 4. 4

Diffusion coefficient, 13. 4, 14. 9
Dispersion, 5. 19
Dissolved oxygen, 9. 37, 11. 13
Double tides, 5. 6
Dredging, 4. 1, 5. 12, 6. 21
Dumping, 2. 10, 11. 1
Dyer, 5. 16

E. coli, 3. 5, 3. 14, 9. 4, 9. 61, 11. 17
Ecology, 6. 7, 7. 7
English Channel, 5. 4, 5. 22
Esso, 1. 19

Fat, 12. 8
Flocculation, 6. 8, 6. 17, 6. 20
Forth Estuary, 9. 1
Fouling organisms, 4. 6
Fresh water inflow, 5. 10, 6. 2 14. 4

Geomorphology, 5. 1
Gosport boil, 1. 20
Grease, 12. 7

Heated water discharge, 3. 5, 3. 18, 4. 7, 4. 10, 5. 26, 14. 12
Hobbs, 14. 6
Hurst Point, 1. 1

Infra-red photography, 10. 10
Initial dilution, 12. 2
Isle of Wight, 1. 19, 5. 1
Itchen River, 3. 5, 5. 3

Japanese seaweed, 4. 6
Jeger Committee, 1. 16, 1. 18

Land reclamation, 4. 2
Langstone Harbour, 1. 15, 6. 4, 6. 13, 7. 9
Lead, 3. 3, 3. 14, 8. 4, 9. 40, 11. 3
Leptospirosis, 3. 13
Limnoria, 4. 6
Litter, 1. 24, 11. 8
Liverpool Bay, 6. 21, 10. 14, 14. 3

Medical Research Council, 1. 26, 3. 5, 9. 59, 11. 17
Mercenaria, 4. 3, 4. 9, 4. 10, 4. 13, 5. 22
Mercury, 3. 3, 3. 13, 8. 3, 8. 7, 13. 8
Minor infections, 3. 15, 11. 17
Mitchell, 4. 5
Models, 14. 6
  hydraulic, 5. 16, 13. 4
  mathematical, 5. 16, 5. 21, 5. 26, 13. 4, 13. 8, 13. 11, 14. 8
Monitoring, 10. 14, 14. 3
Mud, 6. 6, 6. 20, 6. 21
Multiple use, 1. 10
Multi-spectral photography, 10. 8
Mya, 4. 10, 4. 13

Natural Environment Research Council, 14. 4
Nitrogen, 9. 41, 11. 3
Needles, 1. 1
Nutrients, 4. 10, 4. 11, 11. 17

Odour, 12. 2
Oil, 1. 19, 1. 24, 2. 9, 3. 16, 6. 20, 8. 2, 10. 11, 10. 17, 11. 7, 12. 13
Organisation of surveys, 11. 10, 11. 16
Organochlorine compounds, 3. 3, 11. 1, 11. 5, 13. 9
Oslo Convention, 2. 10, 13. 6
Ostrea edulis, 4. 4
Oxygen demand, 13. 1
Oysters, 4. 4, 5. 22

Phosphate, 9. 41, 9. 58, 11. 3
Plankton, 4. 3
Planning, 7. 1, 11. 11, 13. 1
Poliomyelitis, 3. 5, 11. 17
Pollution, 4. 11, 4. 12
  bacterial, 3. 1
  budget, 13. 1
  by heat, 3. 5, 10. 6
  by noise, 3. 5
  by waste, 3. 4
  causes of, 2. 1, 9. 8
  chemical, 3. 3, 9. 51
  definition of, 8. 1
  viral, 3. 1
Pollution monitoring from the air, 10. 4
Portsmouth, 1. 1, 1. 19, 5. 1
Portsmouth Harbour, 4. 6, 5. 4, 5. 12, 7. 8

Radar, 10. 11
Raymont, 4. 2, 4. 3
Recreation, 1. 1, 1. 16, 7. 5
Ripples, 6. 7
Royal Commission on Environmental Pollution, 13. 1

Salinity, 5. 10, 5. 20, 11. 3, 14. 4
  gradient, 6. 2
Salmonellae, 3. 2, 3. 11, 9. 10, 9. 47
Sampling methods, 9. 10
Sampling programmes, 11. 3
Sand waves, 6. 7, 6. 14, 6. 18
Sargassum muticum, 4. 10
Satellites, 10. 3, 10. 17
Sea fisheries, 2. 9
Sediment size, 6. 15, 6. 16
Sedimentary processes, 6. 1
  effect of biological activity on, 6. 5
Selsey Bill, 1. 1
Serratia indica, 12. 13
Sewage effluent, 1. 2, 4. 2, 12. 1
Sewage outfall, 1. 6, 1. 27, 4. 10, 7. 9, 7. 21, 8. 9, 9. 1, 9. 61, 11. 1, 12. 14, 14. 8
Sewage slick, 9. 9, 9. 47, 12. 1, 12. 9, 12. 14
Sewage treatment, 1. 29, 7. 19, 9. 60, 12. 16
Sewerage of South Hampshire, 7. 18

Shell fish, 3. 5, 3. 13, 11. 4
sampling of 3. 10, 9. 49
Shipping, 4. 2
Shipboard wastes, 2. 13
Shoreline surveys, 9. 7, 9. 10, 11. 7
Slick, 12. 1-12. 12, 12. 13
Slipper limpet, 4. 4
Sludges, 8. 6, 11. 15
Soap, 12. 8
Solent, 1. 1, 1. 26, 5. 1, 5. 5, 5. 19, 5. 25, 7. 9
Solent Sailing Conference, 1. 2, 1. 17
South Hampshire Plan, 1. 1, 1. 25, 7. 3
Southampton Water, 1. 10, 3. 4, 3. 12, 4. 5, 5. 3, 5. 9, 7. 9
Spartina, 1. 15
Spithead, 5. 7
Standards, 1. 24, 2. 11, 9. 59, 9. 60, 11. 18, 11. 19
for bathing waters, 9. 46
Storm water overflows, 7. 23
Structure plans, 7. 1
Surface film, 11. 6, 12. 7
Surveys, 9. 58, 11. 15, 11. 19
Suspended solids, 6. 6, 10. 16, 11. 3

Tay Estuary, 5. 24, 6. 4, 14. 4, 14. 11
Temperature, 4. 7, 4. 11, 11. 3
Test River, 3. 9, 5. 3, 7. 9
Thermal imagery, 10. 10, 10. 16
Tidal excursion, 5. 10, 5. 24
Tidal flow, 5. 19, 5. 24
Tidal model, 5. 16
Tidal volume, 5. 9, 6. 2
Tide recording, 14. 4
Tides, 5. 4, 9. 52
Tin, 8. 3, 8. 9
Toxic metals, 11. 9
Toxicity, 13. 1, 13. 6, 13. 8-13. 10
Toxins, 8. 2
Tracers, 11. 4
Tubeworms, 6. 5
Typhoid, 3. 2, 11. 14, 11. 17

Virus
pollution by, 3. 1, 11. 3, 11. 17
Vibrio parahaemolyticus, 3. 13, 3. 15

Water analyses, 3. 6
Water supply, 7. 7
Watson, J. D. and D. M., 5. 16
Weed, 9. 59
Weston Shore, 3. 5
Wood borers, 4. 6

Yacht basins, 1. 11
Yarmouth, 1. 15

Zinc, 8. 3